MECHANICS OF MATERIAL FORCES

Advances in Mechanics and Mathematics

VOLUME 11

MECHANICS OF MATERIAL FORCES

Edited by

PAUL STEINMANN
University of Kaiserslautern, Germany

GÉRARD A. MAUGIN
Université Pierre et Marie Curie, Paris, France

Springer

Library of Congress Cataloging-in-Publication Data

Mechanics of material forces / edited by Paul Steinmann, Gérard A. Maugin.
p. cm. — (Advances in mechanics and mathematics ; 11)
Includes bibliographical references.

1. Strength of materials. 2. Strains and stresses. 3. Mechanics, Applied. I. Steinmann, Paul, Dr.-Ing. II. Maugin, G. A. (Gérard A.), 1944– III. Series.

TA405.M512 2005
620.1´123—dc22

2005049040

AMS Subject Classifications: 74-06, 74Axx, 74R99, 74N20, 65Z05

ISBN 978-1-4419-3879-4 e-ISBN 978-0-387-26261-1

Printed in the United States of America.

9 8 7 6 5 4 3 2 1

springeronline.com

Contents

Preface

The notion dealt with in this volume of proceedings is often traced back to the late 19th-century writings of a rather obscure scientist, C.V. Burton. A probable reason for this is that the painstaking deciphering of this author's paper in the *Philosophical Magazine* (Vol.33, pp.191-204, 1891) seems to reveal a notion that was introduced in mathematical form much later, that of local structural rearrangement. This notion obviously takes place on the *material manifold* of modern continuum mechanics. It is more or less clear that seemingly different phenomena – phase transition, local destruction of matter in the form of the loss of local ordering (such as in the appearance of structural defects or of the loss of cohesion by the appearance of damage or the extension of cracks), plasticity, material growth in the bulk or at the surface by accretion, wear, and the production of debris – should enter a common framework where, by pure logic, the material manifold has to play a prominent role. Finding the mathematical formulation for this was one of the great achievements of J.D. Eshelby. He was led to consider the apparent but true motion or displacement of embedded material inhomogeneities, and thus he began to investigate the "driving force" causing this motion or displacement, something any good mechanician would naturally introduce through the duality inherent in mechanics since J.L. d'Alembert. He rapidly remarked that what he had obtained for this "force" – clearly not a force in the classical sense of Newton or Lorentz, but more a force in the sense of those "forces" that appear in chemical physics (thermodynamical forces), was related to the divergence of a quantity known since David Hilbert in mathematical physics, namely, the so-called energy-momentum tensor of field theory, or to be more accurate, the purely spatial part of this essentially four-dimensional object. Similar expressions were to appear in the study of the "force" that causes the motion of a dislocation (Peach and Koehler) and the driving force acting on the tip of a progressing macroscopic crack (Cherepanov, Rice). Simultaneously, this tensor, now rightfully called an Eshelby stress tensor, but identified by others as a tensor of *chemical potentials* (a tensorial generalization of the Gibbs energy), was appearing in the theory of fluid mixtures (Ray Bowen, M. Grinfeld) and with its jump at discontinuity surfaces. It would be dangerous to pursue this historical disquisition further. If we did, we would enter a period where most active contributors to the field are still active at the time of this writing and are in fact contributors to the present book. Still, a few general trends can be emphasized. The late 1980s and the early 1990s witnessed a more comprehensive approach to the notion of local structural rearrangement (the local change of material configuration resulting from some physico-chemical process) and the recognition of the tensor introduced by Eshelby in the treatment of interface phenomena. With these works, a whole industry started to develop along different paths, sustained by different viewpoints, and dealing with various applications, while identifying a general background. Although certainly not an entirely new field *per se*, it started to organize itself with the introduction of a nomenclature. This is often the result of personal idiosyncrasies. But two essential expressions have come to be accepted, practically on an equal footing: that of *material forces*, and that of *configurational forces*, both of which expressions denote the thermodynamical forces that drive the development of the above-mentioned structural rearrangements and the defect motions. While we are not sure that these two terms are fully equivalent, it became obvious to the co-editors of this volume in the early 2000s that time was ripe for a symposium fully pledged to this subject matter. The format of the EUROMECH Colloquia was thought to be well adapted because of its specialized focus and the relatively small organizational effort required for its set up. Midway between Paris and central Germany was appropriate for the convenience of transport and some neutrality. All active actors, independently of their peculiar viewpoints, were invited, and most came, on the whole resulting in vivid discussions between gentlemen.

The present book of proceedings well reflects the breadth of the field at the time of the colloquium in 2003, and it captures the contributions in the order in which they were presented, all orally, at the University of Kaiserslautern, Germany, May 21–24, 2003, on the occasion of the EUROMECH Colloquium 445 on the "Mechanics of Material Forces". The notion of material force acting on a variety of defects such as cracks, dislocations, inclusions, precipitates, phase boundaries, interfaces and the like was extensively discussed. Accordingly, typical topics of interest in continuum physics are kinetics of defects, morphology changes, path integrals, energy-release rates, duality between direct and inverse motions, four-dimensional formalism, etc. Especially remarkable are the initially unexpected recent developments in the field of numerics, which are generously illustrated in these proceedings. Further developments have already taken place since the colloquium, and many will come, no doubt. Students and researchers in the field will therefore be happy to find in the present volume useful basic material accompanied by fruitful hints at further lines of research, since the book offers, with a marked open-mindedness, various interpretations and applications of a field in full momentum.

Last but not least our sincere thanks go to everybody who helped to make this EUROMECH Colloquium a success, in particular to the co-workers and students at the Lehrstuhl für Technische Mechanik of the University of Kaiserslautern who supported the Colloquium during the time of the meeting. Very special thanks go to Ralf Denzer who took the responsibility, and the hard and at times painful work, to assemble and organize this book.

Paris, June 2004. Gérard A. Maugin
Kaiserslautern, June 2004. Paul Steinmann

Contributing Authors

Franz-Joseph Barthold University of Dortmund, Germany

Arkadi Berezovski Tallinn Technical University, Estonia

Manfred Braun Universität Duisburg-Essen, Germany

Michael Brünig Universität Dortmund, Germany

Sanda Cleja-Tigoiu University of Bucharest, Romania

Serguei Dachkovski University of Bremen, Germany

Alexandre Danescu Ecole Centrale de Lyon, France

Cristian Dascalu Université Joseph Fourier, Grenoble, France

Ralf Denzer University of Kaiserslautern, Germany

Antonio DiCarlo Università degli Studi "Roma Tre", Italy

Marcelo Epstein The University of Calgary, Canada

Jean-Francois Ganghoffer École Nationale Supérieure en Électricité et Mécanique, Nancy, France

Krishna Garikipati University of Michigan, Ann Arbor, USA

Morton E. Gurtin Carnegie Mellon University, Pittsburgh, USA

George Herrmann Stanford University, USA

Vassilios K. Kalpakides University of Ioannina, Greece

Reinhold Kienzler University of Bremen, Germany

Ellen Kuhl University of Kaiserslautern, Germany

Ragnar Larsson Chalmers University of Technology, Göteborg, Sweden

Markus Lazar Université Pierre et Marie Curie, Paris, France

Gérard A. Maugin Université Pierre et Marie Curie, Paris, France

Ralf Müller Technische Universität Darmstadt, Germany

Paolo Podio-Guidugli Università di Roma "Tor Vergata", Italy

Lalaonirina Rakotomanana Université de Rennes, France

Pascal Sansen Ecole Supérieure d'Ingénieurs en Electrotechnique et Electronique, Amiens, France

Miroslav Šilhavý Mathematical Institute of the AV ČR, Prague, Czech Republic

Erwin Stein University of Hannover, Germany

Claude Stolz Ecole Polytechnique, Palaiseau, France

Helmut Stumpf Ruhr-Universität Bochum, Germany

Bob Svendsen University of Dortmund, Germany

Carmine Trimarco University of Pisa, Italy

Anna Vainchtein University of Pittsburgh, USA

Chien H. Wu University of Illinois at Chicago, USA

I

4D FORMALISM

Chapter 1

ON ESTABLISHING BALANCE AND CONSERVATION LAWS IN ELASTODYNAMICS

George Herrmann
Division of Mechanics and Computation, Stanford University
g.herrmann@dplanet.ch

Reinhold Kienzler
Department of Production Engineering, University of Bremen
rkienzler@uni-bremen.de

Abstract By placing time on the same level as the space coordinates, governing balance and conservation laws are derived for elastodynamics. Both Lagrangian and Eulerian descriptions are used and the laws mentioned are derived by subjecting the Lagrangian (or its product with the coordinate four-vector) to operations of the gradient, divergence and curl. The 4 x 4 formalism employed leads to balance and conservation laws which are partly well known and partly seemingly novel.

Keywords: elastodynamics, 4 x 4 formalism, balance and conservation laws

Introduction

The most widely used procedure to establish balance and conservation laws in the theory of fields, provided a Lagrangian function exists, is based on the first theorem of Noether [1], including an extension of Bessel–Hagen [2]. If such a function does not exist, e. g., for dissipative systems, then the so–called Neutral–Action method (cf. [3], [4]) can be employed. For systems which do possess a Lagrangian function, the Neutral–Action method leads to the same results as the Noether procedure including the Bessel–Hagen extension.

In addition to the above two methods, a third procedure consists in submitting the Lagrangian L to the differential operators of *grad*, *div* and *curl*, respectively. The latter two operations are applied to a vector $\mathbf{x}L$ where $\mathbf{x}$ is the position vector. This procedure has been successfully used in elastostatics in material space (cf. [4]), leading to the path–independent integral commonly known as $\mathbf{J}$, M and $\mathbf{L}$.

Thus it appeared intriguing to investigate the results of the application of the three differential operators to the elastodynamic field, considering the time not as a parameter, as is usual, but as an independent variable on the same level as the three space coordinates. To make all four coordinates of the same dimension, the time is multiplied by some characteristic velocity, as is done in the theory of relativity. Results are presented both in terms of a Lagrangian and an Eulerian formulation.

1. ELASTODYNAMICS IN LAGRANGIAN DESCRIPTION

1.1. LAGRANGIAN FUNCTION

We consider an elastic body in motion with mass density $\rho_o = \rho_o(X^J)$ in a reference configuration. We identify the independent variables ξ^μ ($\mu = 0, 1, 2, 3$) as

$$\begin{aligned} \xi^0 &= \hat{t} = c_o t \,, \\ \xi^J &= X^J, \qquad\qquad J = 1, 2, 3 \,. \end{aligned} \tag{1}$$

The independent variable $\hat{t}$ and not t is used in order for all independent variables to have the same dimensions, where c_o is some velocity. In the special theory of relativity, $c_o = c =$ *velocity of light* is used, to obtain a Lorentz–ivariant formulation in Minkowski space. In the non–relativistic theory, c_o may be chosen arbitrarily, e. g., as some characteristic wave speed or it may be normalized to one. The independent variables $\xi^J = X^J$ ($J = 1, 2, 3$) are the space–coordinates in the reference configuration of the body.

Although $\xi^0 = \hat{t}$ and $\xi^J = X^J$ have the same dimensions, the time t is an independent variable and is not related to the space–coordinates by a proper time τ as in the theory of relativity. Therefore, we will deal with Galilean–invariant objects, for "vectorsänd matrices that will be called (cf. [5]) "tensors", although the formulation is not covariant. Special care has to be used when dealing with *div* and *curl*, where time– and space–coordinates are coupled.

The current coordinates are designated by x^i and play the role of dependent variables or fields. They are defined by the mapping or motion

$$x^i = \chi^i(\hat{t}, X^J), \qquad i = 1,2,3 \;. \tag{2}$$

The derivatives are

$$\frac{\partial x^i}{\partial \xi^0} = \frac{\partial x^i}{c_o \partial t}\Big|_{X^j \ fixed} = \frac{1}{c_o} v^i \;, \tag{3}$$

$$\frac{\partial x^i}{\partial X^J} = F^i{}_J \tag{4}$$

with the physical velocities v^i and the deformation gradient $F^i{}_J$ as the Jacobian of the mapping (2). For later use, the determinant of the Jacobian is introduced

$$J_F = det\left[F^i{}_J\right] > 0 \;. \tag{5}$$

The Lagrangian function that will be treated further is thus identified as

$$L = L(\hat{t}, X^J; x^i, v^i, F^i{}_J) \;. \tag{6}$$

In more specific terms, we postulate the Lagrangian to be the kinetic potential

$$L = T - \rho_0(W + V) \tag{7}$$

with the densities of the

- kinetic energy $\quad T = \rho_o(X^J)v^i v_i \;,$
- strain energy $\quad \rho_o W = \rho_o W(\hat{t}, X^J; F^i{}_J) \;,$ (8)
- force potential $\quad \rho_o V = \rho_o V(\hat{t}, X^J; x^i) \;.$

The various derivatives of L allow their identifications in our specific case as indicated

$$\underbrace{\frac{\partial L}{\partial x^i} = \rho_o f_i}_{Volume\ force,} \tag{9}$$

$$\underbrace{\frac{\partial L}{\partial v^j} = c_o \rho_o v_j}_{Momentum,} \tag{10}$$

$$\underbrace{\frac{\partial L}{\partial (F^i{}_J)} = -P^J{}_i}_{First\ Piola-Kirchhoff\ stress,} \tag{11}$$

$$\underbrace{\left.\frac{\partial L}{\partial \hat{t}}\right|_{expl.} = j_o \frac{1}{c_o}}_{Energy\ source,} \tag{12}$$

$$\underbrace{\left.\frac{\partial L}{\partial X^J}\right|_{expl.} = j_J}_{Inhomogeneity\ force.} \tag{13}$$

The Euler–Lagrange equations

$$E_j(L) = \frac{\partial}{\partial t}(\frac{\partial L}{\partial v^j}) + \frac{\partial}{\partial X^I}(\frac{\partial L}{\partial {F^j}_I}) - \frac{\partial L}{\partial x^j} = 0 \tag{14}$$

are the equations of motion

$$\frac{\partial}{\partial t}(\rho_o v_j) - \frac{\partial}{\partial X^I}({P^I}_j) - \rho_o f_j = 0 \, . \tag{15}$$

1.2. APPLICATION OF *GRAD*

We first evaluate the gradient of L, i.e.,

$$\frac{\partial L}{\partial \xi^\mu} = \left(\frac{\partial L}{\partial \xi^\mu}\right)\Big|_{expl.} + \frac{\partial L}{\partial x^i}\frac{\partial x^i}{\partial \xi^\mu} + \frac{\partial L}{\partial \left(\frac{\partial x^i}{\partial \xi^\nu}\right)}\frac{\partial}{\partial \xi^\mu}\left(\frac{\partial x^i}{\partial \xi^\nu}\right) . \tag{16}$$

After rearrangement (details are given in [6]) we find

$$\frac{\partial}{\partial \xi^\nu}\left[L\delta^\nu_\mu - \frac{\partial L}{\partial \left(\frac{\partial x^j}{\partial \xi^\nu}\right)}\frac{\partial x^j}{\partial \xi^\mu}\right] = j_\mu + E_j(L)\frac{\partial x^j}{\partial \xi^\mu} \, . \tag{17}$$

Along solutions of the field equations, the last term of (17) vanishes due to (14). The expression in square brackets is abbreviated by ${T^\nu}_\mu$ as

$${T^\nu}_\mu = L\delta^\nu_\mu - \frac{\partial L}{\partial (\frac{\partial x^j}{\partial \xi^\nu})}\frac{\partial x^j}{\partial \xi^\mu} \, , \tag{18}$$

which is identical to the energy–momentum "tensor" discussed by Morse & Feshbach [5]. The various components of ${T^\nu}_\mu$ may be identified as follows.

$$\underbrace{-T^0_0 = H = T + \rho_o(W + V)}_{Hamiltonian\ (total\ energy),} \tag{19}$$

$$\underbrace{-T^0{}_J \;=\; c_o\rho v_j F^j{}_J \;=\; c_o R_J}_{\textit{Field– (or wave–) momentum density,}} \tag{20}$$

$$\underbrace{-T^I{}_0 \;=\; -\frac{1}{c_o}\, P^I{}_j v^j \;=\; -\,\frac{1}{c_o} S^I}_{\textit{Field intensity, or energy flux,}} \tag{21}$$

$$\underbrace{-T^I{}_J \;=\; (\rho_o(W \,+\, V) \,-\, T)\delta^I_J \,-\, P^I{}_j F^j{}_J \;=\; b^I{}_J}_{\textit{Material momentum or Eshelby tensor.}} \tag{22}$$

It is noted that if $j_o = 0$, then $\partial T^\nu{}_0/\partial\xi^\nu = 0$, stating that energy is conserved. Further, if $j_J = 0$, i. e., the body is homogeneous, then $\partial T^\nu{}_J/\partial\xi^\nu = 0$, stating that the material momentum, involving the Eshelby tensor, in conserved. This is stated in explicit formulae

$$\frac{\partial T^\nu{}_0}{\partial\xi^\nu} \;=\; 0 \;=\; -\frac{1}{c_o}\left[\frac{\partial H}{\partial t} - \frac{\partial}{\partial X^J}(P^J{}_j v^j)\right] , \tag{23}$$

$$\frac{\partial T^\nu{}_I}{\partial\xi^\nu} \;=\; 0 \;=\; -\left[\frac{\partial}{\partial t}(\rho_o v_j F^j{}_I) + \frac{\partial}{\partial X^J} b^J{}_I\right] . \tag{24}$$

1.3. APPLICATION OF *DIV*

We next evaluate the divergence of the four–"vector" $\xi^\mu L$, i. e.,

$$\frac{\partial(\xi^\mu L)}{\partial X^\mu} \;=\; \delta^\mu_\mu L \,+\, \xi^\mu \frac{\partial L}{\partial \xi^\mu} . \tag{25}$$

In space–time $\delta^\mu_\mu = 4$. Using (14) - (18) we find

$$\frac{\partial}{\partial\xi^\nu}\left[\xi^\mu T^\nu{}_\mu \,+\, x^i \frac{\partial L}{\partial\left(\frac{\partial x^i}{\partial\xi^\nu}\right)}\right] \;=\; 4L \,+\, \frac{\partial L}{\partial x^i} x^i \,+\, j_\mu \xi^\mu . \tag{26}$$

If the Lagrangian function is homogeneous of degree 2, i.e., for a linearly elastic body, then the relation

$$L \;=\; \frac{1}{2}\,\frac{\partial L}{\partial\left(\frac{\partial x^i}{\partial\xi^\nu}\right)}\frac{\partial x^i}{\partial\xi^\nu} \tag{27}$$

holds and (26) can be further modified to (again, for details see [6])

$$\frac{\partial}{\partial\xi^\nu}\left[\xi^\mu T^\nu{}_\mu \,-\, x^i \frac{\partial L}{\partial\left(\frac{\partial x^i}{\partial\xi^\nu}\right)}\right] \;=\; -\rho_o f_i x^i \,+\, j_\mu \xi^\mu . \tag{28}$$

It is seen that a conservation law exists only, if L does not depend either on ξ^μ explicitly or on x^i. The expression in square brackets is abbreviated by V^ν and might be called virial of the system. After introducing the notations (9) - (13) and (19) - (22) its components are given by

$$\begin{aligned} V^0 &= (-tH - X^J \rho v_j F^j{}_J - x^j \rho_o v_j) c_o \,, \\ V^I &= t P^I{}_j v^j - X^J b^I{}_J + x^j P^I{}_j \,, \end{aligned} \tag{29}$$

and, under the restrictions mentioned above, the conservation law holds

$$\frac{\partial V^0}{\partial \hat{t}} + \frac{\partial V^I}{\partial X^I} = 0 \,. \tag{30}$$

Eq. (30) may be rearranged to

$$\hat{t}\, \frac{\partial T^\nu{}_0}{\partial \xi^\nu} + X^l\, \frac{\partial T^\nu{}_I}{\partial \xi^\nu} = 0 \,, \tag{31}$$

which states in words that

$\hat{t}$ times conservation of energy plus X^l times conservation of material momentum equals a divergence–free expression.

This is a statement similar to that given by Maugin [7].

1.4. APPLICATION OF *CURL*

The rotation of a vector $\xi_\mu L$ in space–time is represented by

$$\frac{\partial}{\partial \xi^\nu} (\epsilon^{\gamma\delta\mu\nu} \xi_\mu L) \tag{32}$$

with the completely skew–symmetric, fourth–rank permutation tensor $\epsilon^{\gamma\delta\mu\nu}$. Proceeding similarly as in the previous sections we find

$$\frac{\partial}{\partial \xi^\lambda} \left[\epsilon^{\gamma\delta\mu\nu} \xi_\mu T^\lambda{}_\nu \right] = \epsilon^{\gamma\delta\mu\nu} (T_{\mu\nu} + \xi_\mu j_\nu) \,. \tag{33}$$

It is seen that a conservation law exists, provided L does not depend explicitly on ξ^ν, and, in addition $T_{\mu\nu}$ be symmetric, i. e., $T_{\mu\nu} = T_{\nu\mu}$. From (20) - (22) we see that this condition is not satisfied here. Thus, the application of the *curl* operator does not lead to a conservation law. Considering $\mu = 0$ and $\nu = J$ in equation (33), in the absence of inhomogeniety forces $j_\nu = 0$ and volume forces $f_i = 0 (V = 0)$, we find

$$\frac{\partial}{\partial \xi^\lambda}(c_o t T^\lambda_J) - \frac{\partial}{\partial \xi^\lambda}(X_J T^\lambda_0)$$

$$= T^0{}_J + c_o t \left[\frac{\partial T^0{}_J}{\partial \hat{t}} + \frac{\partial T^I{}_J}{\partial X^I} \right] - X_J \left(\frac{\partial T^0_0}{\partial \hat{t}} + \frac{\partial T^I_0}{\partial X^I} \right) - T_J{}^0$$

$$= -c_o \rho v_j F^j{}_J - \frac{1}{c_o} v_j P^j{}_J \qquad (34)$$

$$= -(\textit{field intensity} + \textit{field momentum density}).$$

Thus rotation in space and time results in a balance rather than a conservation law.

2. ELASTODYNAMICS IN EULERIAN DESCRIPTION

2.1. LAGRANGIAN FUNCTION AND THE ENERGY–MOMENTUM TENSOR

In this description, the role of the dependent and independent variables is reversed as compared to the Lagrangian description above, i. e.,

$$\mathcal{L} = \mathcal{L}\left(c_o t, x^j; X^J, X^J{}_{,\nu}\right). \qquad (35)$$

We introduce a different symbol $\mathcal{L}$ for the Lagrangian function in Eulerian description to indicate, firstly that the functional dependence is different from L and, secondly, that the Lagrangian $\mathcal{L}$ is the action per unit of volume of the actual configuration whereas L is the action per unit of volume of the reference configuration. Thus, with (5), the following relation is valid

$$\mathcal{L} = J_F L. \qquad (36)$$

By contrast to the Lagrangian description, the "tensor" $\mathcal{T}^\mu{}_\nu$ (corresponding to $T^\mu{}_\nu$ in the Lagrangian description) can not be written down in a straight forward manner, but is rather the result of a certain power–law expansion of relevant quantities. Since a detailed derivation may be found in [8] it will not be repeated here. We merely recall that the Euler–Lagrange equations now represent conservation (or balance) of material momentum (Eshelby) and the "tensor" $\mathcal{T}^\mu{}_\nu$ represents the mass–stress "tensor", which may be written out as

$$\mathcal{T}^{\mu\nu} = \begin{bmatrix} \mathcal{T}^{00} & \mathcal{T}^{0j} \\ \mathcal{T}^{i0} & \mathcal{T}^{ij} \end{bmatrix} = \begin{bmatrix} \rho c_o^2 & \rho v^j c_o \\ \rho v^i c_o & \rho v^i v^j - \sigma^{ij} \end{bmatrix} . \tag{37}$$

Instead of the mass density in the reference configuration ρ_o, the mass density of the current configuration, ρ , is used with the interrelation

$$\rho_o = J_F \rho . \tag{38}$$

The connection between the first Piola–Kirchhoff stress $P^I_{\ j}$ of the Lagrangian description and the Cauchy stress $\sigma^i_{\ j}$ is

$$\sigma^i_{\ j} = J_F^{-1} F^i_{\ I} P^I_{\ j} , \tag{39}$$

where J_F^{-1} is the Jacobian of the inverse transformation $X^K = \chi^{K^{-1}}(c_o t, x^k)$, cf.[7].
Differentiation of $\mathcal{T}^{\mu\nu}$ leads now first to

$$\frac{\partial \mathcal{T}^{\nu 0}}{\partial \xi^\nu} = 0 = c_o \left(\frac{\partial \rho}{\partial t} + \frac{\partial}{\partial x_j}(\rho v^j) \right) , \tag{40}$$

which represents conservation of mass, and next to

$$\frac{\partial \mathcal{T}^{\nu j}}{\partial \xi^\nu} = \frac{\partial}{\partial t}(\rho v^j) + \frac{\partial}{\partial x^i}(\rho v^i v^j - \sigma^{ij}) = 0 , \tag{41}$$

which represents conservation of physical momentum, i. e., the equations of motions, which can be written as

$$\rho \frac{D v^j}{D t} - \frac{\partial \sigma^{ij}}{\partial x^i} = 0 \tag{42}$$

with

$$\frac{D()}{Dt} = \frac{\partial()}{\partial t} + v^i \frac{\partial()}{\partial x^i} \tag{43}$$

provided mass is conserved.

2.2. APPLICATIONS OF *DIV* AND *CURL*

Since the associated Lagrangian function is not homogeneous of any degree, the operation of *div* does not lead to a conservation law, but rather to the balance law

$$\frac{\partial}{\partial t}(x_j \rho v^j + c_o^2 t \rho) + \frac{\partial}{\partial x^i}\left[x_j(\rho v^i v^j - \sigma^{ij}) + c_o^2 t \rho v^i \right]$$

$$= \rho c_o^2 + (\rho v^i v_i - \sigma^i_{\ i}) = \mathcal{T}^\nu_{\ \nu} . \tag{44}$$

The source term is equal to the trace of the mass–stress "tensor".

The "tensor" $\mathcal{T}^{\mu\nu}$ is, however, symmetric which indicates that the operation of *curl* will lead to a conservation law of the form

$$\frac{\partial}{\partial t}\left[t\rho v^j - x^j\rho\right] + \frac{\partial}{\partial x^i}\left[t(\rho v^i v^j - \sigma^{ij}) - x^j\rho v^i\right] = 0\,, \qquad (45)$$

or expressed in words

t times equation of motion minus x^j times conservation of mass equals a divergence–free expression.

Concluding Remarks

The principal results obtained are summarized in Table 1. It is seen that the 4 x 4 formalism has revealed a remarkable structure of the basic balance and conservation laws of elastodynamics, which contains both well known as a well as seemingly novel relations valid in either physical or material space. A more detailed derivation, including examples, is to be found in a forthcoming publication [6].

References

[1] Noether, E. (1918). *Invariante Variationsprobleme.* Nachrichten der Königlichen Gesellschaft der Wissenschaften, Mathematisch–Physikalische Klasse 2, 235-257.

[2] Bessel-Hagen, E. (1921). *Über die Erhaltungssätze der Elektrodynamik.* Mathematische Annalen 84, 258-276.

[3] Honein, T., Chien, N. and Herrmann, G. (1991). *On conservation laws for dissipative systems.* Physical Letters A 155, 223-224.

[4] Kienzler, R. and Herrmann, G. (2000). *Mechanics in Material Space.* Berlin: Springer.

[5] Morse, P. M. and Feshbach, H. (1953). *Methods of Theoretical Physics.* New York: McGraw-Hill.

[6] Kienzler, R. and Herrmann, G. (2003). *On conservation laws in elastodynamics.* Submitted for publication.

[7] Maugin, G. A. (1993). *Material Inhomogeneities in Elasticity.* London: Chapman & Hall.

[8] Kienzler, R. and Herrmann, G. (2003). *On the four-dimensional formalism in continuum mechanics.* Acta Mechanica 161, 103-125.

Elastodynamics in

Lagrangian	description	Eulerian	description
Euler–Lagrange eqs.	Balance involving *phys.* momentum (equations of motion)	Euler–Lagrange eqs.	Balance involving *mat.* momentum (Eshelby relations)
$E_j(L) = 0$		$E_J(\mathcal{L}) = 0$	
Gradient $\dfrac{\partial T^{\nu 0}}{\partial \xi^\nu} = 0$	Conservation of energy	*Gradient* $\dfrac{\partial \mathcal{T}^{\nu}{}_{0}}{\partial \xi^\nu} = 0$	Conservation of mass
Gradient $\dfrac{\partial T^{\nu}{}_{J}}{\partial \xi^\nu} = 0$	Balance involving *mat.* momentum (Eshelby relations)	*Gradient* $\dfrac{\partial \mathcal{T}^{\nu}{}_{J}}{\partial \xi^\nu} = 0$	Balance involving *phys.* momentum (equations of motion)
Divergence:	leads to a *conservation* law	Divergence:	leads to a *balance* law
Curl:	leads to a *balance* law	Curl:	leads to a *conservation* law

Table 1: Juxtaposition of the results in Lagrangean and Eulerian descriptions

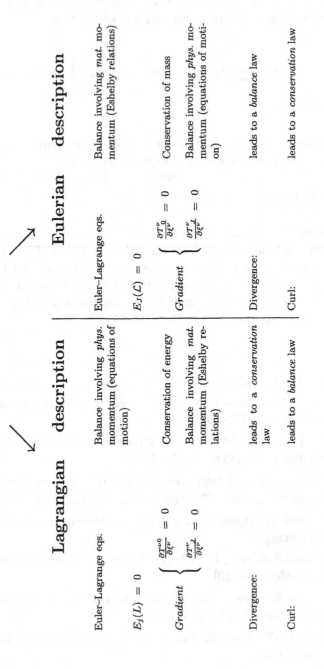

Chapter 2

FROM MATHEMATICAL PHYSICS TO ENGINEERING SCIENCE

Gérard A. Maugin

Université Pierre et Marie Curie, Laboratoire de Modélisation en Mécanique, UMR 7607, 4 place Jussieu, case 162, 75252 Paris Cedex 05, France

Abstract The theory of configurational - or material - forces smells good of its mathematical-physical origins. This contribution outlines this characteristic trait with the help of a four-dimensional formalism in which energy and canonical momentum go along, with sources that prove to be jointly consistent in a dissipation inequality. This is what makes the formulation so powerful while automatically paving the way for an exploitation of irreversible thermodynamics, e.g., in the irreversible progress of a defect throughout matter

Keywords: configurational forces, invariance, irreversible thermodynamics, conservation laws

1. INTRODUCTION

Although it now finds applications in typical engineering settings such as the application of finite-element and finite-volume computational techniques, the theory of configurational - or material - forces smells good of its **mathematical-physical origin**. This was transparent in Eshelby's original works but even more so in the author's works. The newly entertained relationship with computational techniques based on mathematical formulations akin to **general conservation laws** (weak form such as the principle of virtual power, classical volume balance laws) is not so surprising.

In fact, both the deep physical meaning and the practical usefulness of the critical expressions (e.g., of thermodynamical driving forces) obtained within this framework stem from the intimate relationship built from the start between these expressions and a **field-theoretical invariance** or lack of invariance. Symmetries and their eventual breaking are the physically most profound and intellectually puzzling tenets in modern physics. That engineering applications fit into this general picture is a comforting view of the **mutual enrichment** of pure science and modern engineering. This contribution outlines these features with the help of a four-(3+1) dimensional formalism in which energy and canonical momentum go along, with sources that prove to be jointly consistent in a dissipation inequality. This is what makes the formulation so powerful while automatically paving the way for an exploitation of **irreversible thermodynamics**, e.g., in the irreversible progress of a defect throughout matter.

The wealth and wide range of applications opened by the configurational vision of mechanics or mathematical physics is extremely impressive. Among the now traditional applications we find:

i.) The study of the progress of "**defects**" or objects considered as such, or even macroscopic physical manifestations of the like. Among these we have: dislocations, disclinations ; cracks, growth, phase transformations, plasticity (pseudo-material inhomogeneities). This is exemplified and documented in several synthetic works, e. g., [1]-[5].

ii.) The applications to **dynamical systems** rewritten in the appropriate form : Canonical balance laws applied to the perturbed motion of soliton complexes (solitonic systems governed by systems of partial differential equations - works by Maugin-Pouget-Sayadi-Kosevich et al in both continuous and discrete frameworks - *etc*) representing localized solutions, cf. [6]-[8];

and more recently

iii.) Applications to **various numerical schemes**, e.g., in *Finite-differences* (e.g., Christov and Maugin [6]; in *Finite-elements* (e.g., Braun [9], Mueller, Gross, Maugin [10]-[11], Steinmann *et al* [12], *etc*, and in *Finite-volumes /continuous cellular automata* (e.g., works by: Berezovski and Maugin [13].

Points (i) and (iii) above are richly illustrated by other contributions in this volume. We shall focus attention on the original field-theorretical framework (already exposed in lecture notes, Maugin and Trimarco [14]), and the thermomechanical formulation that contains the essence of the subject matter.

In a different style but of great interest are also the presentation and applications developed by Kienzler and Herrmann [15] - also these authors' contributions in the lecture notes mentioned in ref. 14).

2. THE FIELD-THEORETICAL FRAMEWORK

Basic fields are noted $\phi^\alpha, \alpha = 1,2,3,...$ They depend on a space-time parametrization (X^K, t) where in the classical continuum physics of deformable solids X^K, K=1,2,3 are material coordinates and t is Newton's absolute time. Examples of such fields are the classical *direct* motion $x^i = \bar{x}^i(X^K, t)$, the micromotion $\chi^i = \bar{\chi}^i(X^K, t)$ of micromorphic media, and electromagnetic potentials $(\phi, \mathrm{A})(X^K, t)$. Note in the latter case that the fields are what are called "potentials" by physicists. It is essential here to clearly distinguish between the fields *per se* and the space-time parametrization. In standard continuum mechanics the placement **x** in the actual configuration at time t is the three-dimensional field. In the absence of dissipation, relying on a Hamiltonian variational formulation, the basic Euler-Lagrange equations of "motion", (one for each field or for each component of multidimensional field) read formally

$$\delta \phi^\alpha : E_\alpha = 0 \quad , \quad \alpha = 1,2,3,.. \tag{1}$$

Noether's identity - that we prefer to refer to as Noether-Ericksen identity [16] - is written as (note the summation over α; $\nabla_R = \partial / \partial \mathbf{X}$)

$$\sum_\alpha E_\alpha . (\nabla_R \phi^\alpha)^T = 0 \tag{2}$$

This is a co-vectorial equation on the material manifold M^3. It is referred to as the equation of *canonical* or *material momentum* because, according to d'Alembert's principle, it is clearly generated by a variation of the material coordinates X^K. Equation (2), unlike (1), governs the whole set of fields α. That is, it has the same ontological status as the following "theorem of kinetic energy" that is obtained from the set (1) by an equivalent time-like operation:

$$\sum_\alpha E_\alpha . \left(\frac{\partial \phi^\alpha}{\partial t} \right) = 0 \tag{3}$$

Although this clearly is the time-like equation associated with the space-like equation (2), this is not exactly the *energy theorem*, for it should be combined with a statement of the first law of thermodynamics in order to introduce the notions of internal energy and heat. Ultimately, this manipulation will yield a *balance of entropy*, perhaps for an energy quantity such as entropy multiplied by temperature: θS, where θ is the absolute temperature and S is the entropy per unit reference volume (see below). Both this equation and eqn. (2) are canonical (of a general ever realized form) and, in the presence of dissipative processes, do not refer to necessarily strictly conserved quantities. This will be seen hereafter. For the time being, in order to emphasize the role of this type of equation as governing the *whole* physical system under consideration and not only the classical degrees of translation of a standard deformable medium, we note a few examples of expressions of the canonical (or material) momentum:

- general analytical-mechanics formula:

$$P = -\rho_0 \sum_{\alpha} \left(\frac{\partial \phi^{\alpha}}{\partial t} \right) . (\nabla_R \phi_{\alpha}) \tag{4}$$

- classical deformable solid:

$$P = -\rho_0 \mathbf{F}^T . \mathbf{v} \ ; \ \mathbf{v} = \frac{\partial}{\partial t} \bar{\mathbf{x}} \ , \ \mathbf{F} = \nabla_R \bar{\mathbf{x}} \ . \tag{5}$$

where $\rho_0(\mathbf{X})$ is the matter density at the reference configuration, $\mathbf{v}$ is the physical velocity, and $\mathbf{F}$ is the direct motion gradient. Here the canonical or material momentum simply is the pullback, changed of sign, of the physical (mechanical) momentum $\mathbf{p} = \rho_0 \mathbf{v}$.

- deformable solid with a rigid microstructure (micropolar medium) [17]

$$P = -\rho_0 \left(\mathbf{F}^T . \mathbf{v} + \left(\nabla_R \bar{\chi} \right) . \mathbf{I} . \frac{\partial}{\partial t} \dot{\bar{\chi}} \right) \tag{6}$$

where $\mathbf{I}$ is some kind of rotational inertia tensor.

- * *deformable electromagnetic solid* (full electrodynamics) [1],[14]

$$P = -\left(\rho_0 \mathbf{F}^T . \mathbf{v} - \frac{1}{c} \Pi \times \mathbf{B} \right) = \rho_0 \mathbf{C} . \mathbf{V} + \frac{1}{c} \Pi \times \mathbf{B} \tag{7}$$

where Π and **B** are *material* electric polarization and magnetic induction. In this case the "physical" electromagnetic electromagnetic momentum is $\mathbf{p}^{em} = \frac{1}{c}\mathbf{E}\times\mathbf{B}$, where **E** and **B** are "physical" electric field and magnetic induction).

This series of examples emphasizes the global-system nature of the canonical balance laws of continuum physics insofar as the balance of canonical momentum is concerned.

3. CANONICAL BALANCE LAWS

We now consider a standard continuous solid (no microstructure, no electromagnetic fields) but in the presence of heat and dissipative processes. The canonical balance laws still are the fundamental balance laws of thermomechanics (momentum and energy) expressed **intrinsically in terms of a good space-time parametrization**. In a relativistic background this would be the conservation - or lack of conservation - of the canonical energy-momentum tensor first spelled out in 1915 and 1918 by David Hilbert and Emmy Noether on a variational basis. Here we consider free energies per unit reference volume such as

$$W = \overline{W}(\mathbf{F},\theta,\alpha;\mathbf{X}) \tag{8}$$

for an anisotropic, *possibly anelastically inhomogeneous* material in finite strains, whose basic behavior is elastic, but it may present combined anelasticity. The arguments of the supposedly sufficiently smooth function W are: **F:** deformation gradient, θ: thermodynamical temperature, α: set of internal variable of state representative of a macroscopically manifested irreversible behavior; **X:** material coordinates. The so-called *laws of* thermodynamical *state* are given by

$$\mathbf{T} = \frac{\partial \overline{W}}{\partial \mathbf{F}}, S = -\frac{\partial \overline{W}}{\partial \theta}, A = -\frac{\partial \overline{W}}{\partial \alpha} \tag{9}$$

Then at any *regular* material point **X** in the body B, we have the following ***local balance equations for* mass, linear momentum, and energy**:

$$\left.\frac{\partial \rho_0}{\partial t}\right|_{\mathbf{X}} = 0, \tag{10}$$

$$\left.\frac{\partial \mathbf{p}}{\partial t}\right|_{\mathbf{X}} - div_R \mathbf{T} = \mathbf{0}\,, \tag{11}$$

$$\left.\frac{\partial (K+E)}{\partial t}\right|_{\mathbf{X}} - \nabla_R.(\mathbf{T}.\mathbf{v} - \mathbf{Q}) = 0\ , \tag{12}$$

These equations are presented here in the so-called *Piola-Kirchhoff formulation,* with an $(\mathbf{X},t)$ space-time parametrization, but the components of eqn. (11) are still in **physical space**, so that it is not an intrinsic formulation. We remind the reader of the following definitions and relations:

$$\mathbf{x} = \chi(\mathbf{X},t), \tag{13}$$

$$\mathbf{F} = \left.\frac{\partial \chi}{\partial \mathbf{X}}\right|_{t} = \nabla_R \chi\,, \mathbf{v} = \left.\frac{\partial \chi}{\partial t}\right|_{\mathbf{X}}\,, \tag{14}$$

$$\mathbf{p} = \rho_0 \mathbf{v}\,, K = \frac{1}{2}\rho_0 \mathbf{v}^2\ , \tag{15}$$

$$E = W + S\theta\,. \tag{16}$$

Equation (10)-(12) are **strict conservation laws** (no source terms), because we assume, for the sake of simplicity, that there are neither external body force acting nor energy input per unit volume. In these conditions the entropy equation and the dissipation inequality read [18]

$$\theta \left.\frac{\partial S}{\partial t}\right|_{\mathbf{X}} + \nabla_R.\mathbf{Q} = \Phi^{intr}\ \ ,\ \ \Phi^{intr} := A\dot{\alpha}\ \ ,\ \ \dot{\alpha} \equiv \left.\frac{\partial \alpha}{\partial t}\right|_{\mathbf{X}}\,, \tag{17}$$

and

$$\sigma_B = \theta^{-1}\left(\Phi^{Intr} - \mathbf{S}.\nabla_R\,\theta\right) \geq 0\ ;\ \mathbf{S} \equiv \mathbf{Q}/\theta, \tag{18}$$

with the continuity condition

$$\mathbf{Q}(\mathbf{F},\theta,\alpha;\nabla_R\theta;\mathbf{X})\to\mathbf{0} \quad as \quad \nabla_R\theta\to\mathbf{0}. \tag{19}$$

Canonical equation of linear momentum

This is obtained by *projecting canonically* eqn. (11) onto the material manifold M^3 of points **X** constituting the body. The now classical - but rather trivial - result (first obtained by the author and co-workers) is

$$\left.\frac{\partial \mathbf{P}}{\partial t}\right|_{\mathbf{X}} - \left(div_R\mathbf{b} + \mathbf{f}^{inh}\right) = \mathbf{f}^{th} + \mathbf{f}^{intr}, \tag{20}$$

where the ***canonical momentum*** (here of purely mechanical and translational nature- cf. eqns. (4)-(7)) is given by

$$\mathbf{P} = -\mathbf{p}.\mathbf{F} = \rho_0\mathbf{C}.\mathbf{V} \tag{21}$$

A *Lagrangian function* density is formally introduced by:

$$L = L^{th} = K - W, \tag{22}$$

and there have been defined the ***Eshelby material stress:***

$$\mathbf{b} = -\left(L^{th}\mathbf{1}_R + \mathbf{T}.\mathbf{F}\right), \tag{23}$$

and various ***"material forces"*** (i.e., quantities having the physical dimension of forces but all co-vectors on the material manifild M^3):

$$\mathbf{f}^{inh} = \left.\frac{\partial L^{th}}{\partial \mathbf{X}}\right|_{expl} \; ; \; \mathbf{f}^{th} = S\nabla_R\theta \; , \; \mathbf{f}^{intr} = A\left(\nabla_R\alpha\right)^T. \tag{24}$$

while the *Cauchy-Green finite strain* and the *material* velocity are defined by:

$$\mathbf{C} = \mathbf{F}^T.\mathbf{F} \quad , \quad \mathbf{V} = -\mathbf{F}^{-1}.\mathbf{v} = \left.\frac{\partial \chi^{-1}}{\partial t}\right|_x. \tag{25}$$

In particular, the first of eqn. (24) is explicitly given by

$$\mathbf{f}^{inh} = (\nabla_R \rho)\left(\frac{1}{2}\mathbf{v}^2\right) - \left.\frac{\partial \overline{W}}{\partial \mathbf{X}}\right|_{\mathbf{F},\theta,\alpha\ fixed}. \qquad (26)$$

Thus the "force" $\mathbf{f}^{inh}$ captures indeed the explicit **X**-dependency and deserves its naming as *material force of inhomogeneity*, for short *inhomogeneity force*. This is the first cause for the momentum equation (20) to be *inhomogneous* (i.e., to have a source term) while the original - in physical space - momentum equation (11) was a true conservation law.

An inhomogeneity force is a *directional indicator* of the changes of material properties (this holds also at the sharp interface between components in a composite body). What is more surprising is that a *spatially nonuniform state of temperature* $(\nabla_R \theta \neq \mathbf{0})$ causes a similar effect, i.e., the *material* thermal force $\mathbf{f}^{th}$ acts just like a true material inhomogeneity in so far as the balance of canonical (material) momentum is concerned, cf. Epstein and Maugin [19]. It seems that Bui [20] was the first to uncover such a thermal term while studying fracture although in the small-strain framework and not in the material setting. Finally, any internal variable of state α that has not reached a spatially uniform state at point $\mathbf{X}$, $\nabla_R \alpha \neq \mathbf{0}$, has a similar effect in the equation of canonical momentum through the *intrinsic* material force $\mathbf{f}^{intr}$ [17]. We call such material forces, material forces of quasi- or *pseudo-inhomogeneity* [5]. Note that any additional variable put in the functional dependency of the free energy W will cause a similar effect. It is only in the pure materially homogeneous elastic case (W depending only on **F**) that the balance of canonical momentum is also a strict conservation law.

Note also that the material stress tensor **b** is not symmetric in a traditional sense. If the Cauchy stress is symmetric (that is the case in the present example), **b** is only symmetric with respect to **C** considered as the deformed metric on M^3, i.e.,

$$\mathbf{Cb} = \mathbf{b}^T\mathbf{C}. \qquad (27)$$

Transformation of the energy equation (canonical form for dissipative processes)

Pushing the temperature under the time differentiation in the first of eqns. (17) we readily obtain the following enlightening form:

$$\left.\frac{\partial(S\theta)}{\partial t}\right|_{\mathbf{X}} + \nabla_R.\mathbf{Q} = \Phi^{th} + \Phi^{int},\ \Phi^{\text{th}} := S\frac{\partial\theta}{\partial t}. \quad (28)$$

which is to be compared to eqn. (20), i.e.,

Comparing the source terms in the right-hand sides of eqns. $(28)_1$ and (29), we acknowledge that these two equations are none other than the time-like and space-like components of a unique four-dimensional Cartesian formulation in which **P** and θS are the space and time components of a unique four-vector [21]. This more generally hints at a true 4d analytical mechanics of dissipative continua.

4. CONCLUSION

This contribution had for purpose to emphasize the parallel and complementary roles played by the balance of canonical momentum and energy (the latter written in a specific way) in the mechanics of materials; It is this essentially four-(3+1) dimensional formalism, clearly inspired by mathematical physics, which shows that all reasonings made on the thermodynamics of driving forces in all types of applications (progress of defects, dynamics of nonlinear waves, improvement of numerical schemes) must be simultaneously carried out on the material forces of interest and the accompanying expended power. This, in particular, will guarantee that the corresponding dissipation, if any, is indeed the product of a material force and a material velocity, cf. [21].

5. ACKNOWLEDGMENT

The author benefits from a Max Planck Award for international cooperation (2002-2005).

6. REFERENCES

1. Maugin, G.A., 1993. Material Inhomogeneities in Elasticity. Chapman and Hall, London.
2. Gurtin, ME., 2000. Configurational Forces as Basic Concepts of Continuum Physics. Springer, New York.
3. Epstein, M., Maugin , G.A., 2000. Thermodynamics of Volumetric Growth in Uniform Bodies. Intern. J. Plasticity 16, 951-978.

4. Epstein, M., Maugin, G.A., 1995. On the geometrical material structure of anelasticity. Acta Mechanica 115, 119-131.
5. Maugin, G.A., 2003. Pseudo-plasticity and Pesudo-inhomogeneitiy Effects in Materials Mechanics (Truesdell Memorial Volume). J. of Elasticity, Vol. 71, 81-103.
6. Maugin, G.A., Christov, C.I., 2001. Nonlinear waves and conservation laws (Nonlinear duality between elastic waves and quasi-particles). In: Selected Topics in Nonlinear Wave Mechanics Phenomena, Christov, C.I., Guran, A. (eds) (Birkhauser, Boston, 2001) 100-142.
7. Maugin, G.A., 1999. Nonlinear Waves in Elastic Crystals, Oxford Texts in Applied Mathematics, (Oxford University Press, U.K, 1999).
8. Bogdan, M.M., Maugin, G.A., 2003. Exact discrete breather solutions and conservation laws of lattice equations. Proc. Est. Acad. Sci. Math. Phys. 52, 76-84.
9. Braun, M., 1997. Configurational forces induced by finite-element discretization. Proc. Est. Acad. Sci. Math. Phys. 46, 24-31.
10. Mueller, R., Kolling, S., Gross, D., 2002. On configurational forces in the context of the finite-element method. Int. J. Num. Meth. Engng. 53, 95-104.
11. Mueller, R. Maugin. G.A., 2002. On material forces and finite element discretizations. Computational Mechanics 29, 52-60.
12. Steinmann, P. *et al* 2003 (in this volume).
13. Maugin, G.A., Berezovski A., 2000. Thermoelasticity of inhomogeneous solids and finite-volume dynamic computations. In: Contributions to Continuum Theories (K.Wilmanski's Festschrift). Albers, B. (ed) (W.I.A.S., Berlin, 2000), 166-173.
14. Maugin, G.A., Trimarco, C., 2001. Elements of field theory in inhomogeneous and defective materials, in: Configurational Mechanics of Materials, Kienzler, R., Maugin, G.A. (Eds), (Springer, Wien 2001), 55-171.
15. Kienzler, R., Herrmann, G., 2000. Mechanics in Material Space (with applications to defect and fracture mechanics). Springer, Berlin.
16. Maugin, G.A., 1996. On Ericksen-Noether identity and material balance laws in thermoelasticity and akin phenomena. In: Contemporary Research in the Mechanics and Mathematics of Materials (J. L. Ericksen's 70th Anniversary Volume). Batra, R.C., Beatty, M.F. (eds), (C.I.M.N.E., Barcelone, 1996) 397-407.
17. Maugin, G.A., 1998. On the structure of the theory of polar elasticity. Phil. Trans. Royal. Soc. Lond. A356, 1367-1395.
18. Maugin, G.A., 1997. Thermomechanics of inhomogeneous-heterogeneous systems: applicationd to the irreversible progress of two and three-dimensional defects. ARI 50, 41-86.
19. Epstein, M., Maugin, G.A., 1995. Thermoelastic material forces: definition and geometric aspects. C.R. Acad. Sci. Paris, II-320, 63-68.
20. Bui, H.D., 1978. Mécanique de la Rupture Fragile (Masson Editeurs, Paris, 1978).
21. Maugin, G.A., 2000. On the universality of the thermomechanics of forces driving singular sets. Arch. Appl. Mech. (Jubilee volume) 70, 31-45.

II

EVOLVING INTERFACES

Chapter 3

THE UNIFYING NATURE OF THE CONFIGURATIONAL FORCE BALANCE

Eliot Fried
Department of Mechanical Engineering
Washington University in St. Louis
St. Louis, MO 63130-4899
efried@me.wustl.edu

Morton E. Gurtin
Department of Mathematical Sciences
Carnegie Mellon University
Pittsburgh, PA 15213-3890
mg0c@andrew.cmu.edu

1. INTRODUCTION: SOME CONDITIONS FOR NONMATERIAL INTERFACES

The past half-century has seen much activity among materials scientists and mechanicians concerning nonmaterial interfaces, a central outcome being the realization that such problems generally result in an interface equation over and above those that follow from the classical balances for forces, moments, and mass. In *two space-dimensions* with the interface a *curve* $\mathcal{S}$, this extra interface condition takes a variety of forms, the most important examples being:

Mullins's equation. This is a geometric equation,

$$bV = \psi K \tag{1}$$

for the respective motions of an isotropic grain-boundary and an isotropic grain-vapor interface, neglecting evaporation-condensation. Here V and K are the (scalar) normal velocity and curvature of the grain-boundary (or interface) $\mathcal{S}$, while ψ, b, ρ, and L are strictly positive constants, with ψ the interfacial free-energy (density), b a kinetic modulus (or, recipro-

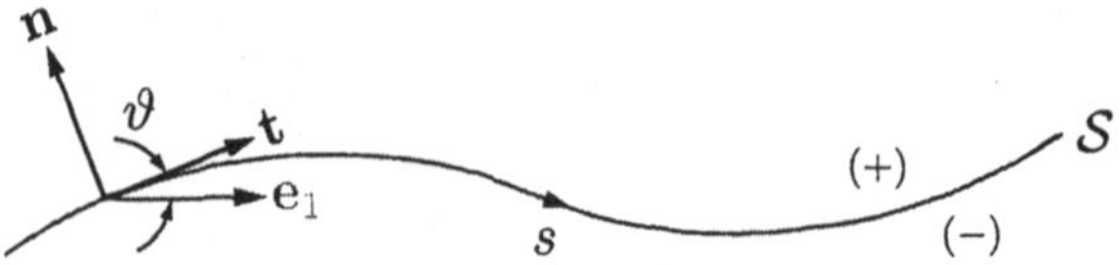

Figure 3.1 The interface $\mathcal{S}$. Our convention is that K be positive on concave upward portions of $\mathcal{S}$. The symbols $(\pm)$ label the phases on the two sides of the interfaces.

cal mobility), ρ the atomic density of the solid, and L the mobility for Fickean diffusion within $\mathcal{S}$. The argument of Mullins (1956) in support of (1) is physical in nature and based on work of Smoluchowski (1951), Turnbull (1951), and Beck (1952).

Variationally derived chemical potential. Working within a framework that neglects deformation and mass transport and invoking an assumption of local equilibrium, Herring (1951) defines the chemical potential U of a solid-vapor interface as the variational derivative of the total free energy with respect to variations in the configuration of the interface. Following Herring, Wu (1996), Norris (1998), and Freund (1998) generalize earlier work of Asaro and Tiller (1972) and Rice and Chuang (1981) to compute the chemical potential U of a solid-vapor interface in the presence of deformation, allowing for interfacial stress. Their result is given by

$$U = \Psi - \mathbf{Sn}\cdot\mathbf{Fn} - (\psi - \bar{\sigma}\lambda)K - \frac{\partial\tau}{\partial s}. \qquad (2)$$

Here Ψ is the bulk free-energy (density); $\mathbf{S}$ is the bulk Piola stress; $\mathbf{F} = \nabla\mathbf{y}$ is the deformation gradient (with $\mathbf{y}$ the deformation); $\mathbf{n}$ and $\mathbf{t}$ are the interface normal and tangent (Figure 3.1); K is the curvature; s denotes arc length; ψ is the interfacial free-energy (density); and $\lambda = |\mathbf{Ft}|$ is the interfacial stretch. The result (2) is derived variationally. It hinges on the assumption that the vapor pressure and vapor free-energy vanish, and is based on standard bulk constitutive relations in the solid and a constitutive equation $\psi = \hat{\psi}(\lambda, \vartheta)$ for the interfacial free-energy, and $\bar{\sigma}$ and τ defined by

$$\bar{\sigma} = \frac{\partial\hat{\psi}}{\partial\lambda}, \qquad \tau = \frac{\partial\hat{\psi}}{\partial\vartheta}, \qquad (3)$$

with ϑ the counterclockwise angle to the interface tangent $\mathbf{t}$.

Kinetic Maxwell equation. This is a condition

$$\left[\!\left[\Psi - \sum_\alpha \rho^\alpha\mu^\alpha - \mathbf{Sn}\cdot\mathbf{Fn}\right]\!\right] = b(\vartheta)V. \qquad (4)$$

for a propagating coherent phase interface between two phases composed of atomic species $\alpha = 1, 2, \ldots, N$. Here, μ^α and ρ^α are the chemical potentials and atomic densities in bulk, $[\![\varphi]\!] = \varphi^+ - \varphi^-$ is the jump of a bulk field φ across the interface, and, as in $(1)_1$, $b(\vartheta)$ is a constitutively determined kinetic modulus. The kinetic Maxwell condition was first obtained by Heidug and Lehner (1985), Truskinovsky (1987), and Abeyaratne and Knowles (1990), who ignored atomic diffusion but allowed for inertia. Their derivations are based on determining the local energy dissipation associated with the propagation of the interface and then appealing to the second law. When $b = 0$, (4) reduces to the classical Maxwell equation

$$\left[\!\left[\Psi - \sum_\alpha \rho^\alpha \mu^\alpha - \mathbf{Sn} \cdot \mathbf{Fn} \right]\!\right] = 0 \tag{5}$$

first derived variationally by Larché and Cahn (1978).[1]

Leo–Sekerka relation. This is a condition for an interface in *equilibrium.* Relying on a variational framework set forth by Larché and Cahn (1978) (cf., also, Alexander and Johnson, 1985; Johnson and Alexander, 1986), Leo and Sekerka (1989) consider coherent and incoherent solid-solid interfaces as well as solid-fluid interfaces. For an interface between a vapor and an alloy composed of N atomic species, neglecting vapor pressure and thermal influences, the Leo–Sekerka relation takes the form

$$\sum_\alpha (\rho^\alpha - \delta^\alpha K)\mu^\alpha = \Psi - \mathbf{Sn} \cdot \mathbf{Fn} - (\psi - \bar{\sigma}\lambda)K - \frac{\partial \tau}{\partial s}. \tag{6}$$

The relation (6) is based on a constitutive equation $\psi = \hat{\psi}(\lambda, \vartheta, \vec{\delta}\,)$, with $\vec{\delta}$ the list of interfacial atomic densities δ^α, $\alpha = 1, 2, \ldots, N$, supplemented by the definitions

$$\bar{\sigma} = \frac{\partial \hat{\psi}}{\partial \lambda}, \qquad \tau = \frac{\partial \hat{\psi}}{\partial \vartheta}, \qquad \mu^\alpha = \frac{\partial \hat{\psi}}{\partial \delta^\alpha}. \tag{7}$$

2. THE NEED FOR A CONFIGURATIONAL FORCE BALANCE

One cannot deny the applicability of the interface conditions discussed above; nor can one deny the great physical insight underlying their derivation. But in studying these derivations one is left trying to ascertain the status of the resulting equations (1), (2), (4), (5), and (6): are they balances, or constitutive equations, or neither?[2] This and the

[1] Cf. also Eshelby (1970), Robin (1974), Grinfeld (1981), James (1981), and Gurtin (1983), who neglect compositional effects.

[2] Successful theories of continuum mechanics are typically based on a clear separation of balance laws and constitutive equations, the former describing large classes of materials, the latter describing particular materials.

disparity between the physical bases underlying their derivations would seem to at least indicate *the absence of a basic unifying principle.*

That additional configurational forces may be needed to describe phenomena associated with the material itself is clear from the seminal work of Eshelby (1951, 1956, 1970, 1975), Peach and Koehler (1950), and Herring (1951) on lattice defects. But these studies are based on variational arguments, arguments that, by their very nature, cannot characterize dissipation. Moreover, the introduction of configurational forces through such formalisms is, in each case, based an underlying constitutive framework and hence restricted to a particular class of materials.[3]

A completely different point of view is taken by Gurtin and Struthers (1990),[4] who — using an argument based on invariance under observer changes — conclude that a configurational force balance should join the standard (Newtonian) force balance as a *basic* law of continuum physics. Here the operative word is "basic". Basic laws are by their very nature independent of constitutive assumptions; when placed within a thermodynamic framework such laws allow one to use the now standard procedures of continuum thermodynamics to develop suitable constitutive theories.

3. A FRAMEWORK FOR THE STUDY OF EVOLVING NONMATERIAL INTERFACES

A complete theory of evolving interfaces in the presence of deformation and atomic transport may be developed using a framework based on: (a) *standard (Newtonian) balance laws for forces and moments* that account for standard stresses in bulk and within the interface; (b) an *independent balance law for configurational forces* that accounts for configurational stresses in bulk and within the interface;[5] (c) *atomic balances*, one for each atomic species, that account for bulk and surface diffusion; a mechanical (isothermal) version of the first two laws of thermodynamics in the form of a *free-energy imbalance* that accounts for temporal changes in free-energy, energy flows due to atomic transport,

---

[3]A vehicle for the discussion of configurational forces within a dynamical, dissipative framework derives configurational force balances by manipulating the standard momentum balance, supplemented by hypereslastic constitutive relations (e.g., Maugin, 1993). But such derived balances, while interesting, are satisfied automatically whenever the momentum balance is satisfied and are hence superfluous.

[4]This work is rather obtuse; better references for the underlying ideas are Gurtin (1995, 2000).

[5]As extended by Davì and Gurtin (1990), Gurtin (1991), Gurtin and Voorhees (1995), and Fried and Gurtin (1999, 2003) to account for atomic transport.

and *power expended by both standard and configurational forces*; (d) *thermodynamically consistent constitutive relations* for the interface and for the interaction between the interface and its environment.

One of the more interesting outcomes of this approach is an explicit relation for the configurational surface tension σ in terms of other interfacial fields; viz.,

$$\sigma = \psi - \sum_{\alpha} \delta^{\alpha} \mu^{\alpha} - \bar{\sigma}\lambda. \tag{8}$$

This relation, a direct consequence of the free-energy imbalance applied to the interface, is a basic relation valid for all isothermal interfaces, independent of constitutive assumptions and hence of material; it places in perspective the basic difference between the configurational surface tension σ and standard surface tension $\bar{\sigma}$. There is much confusion in the literature concerning surface tension σ and its relation to surface free-energy ψ. By (8), we see that these two notions coincide if and only if standard interfacial stress as well as interfacial atomic densities are negligible.

4. THE NORMAL CONFIGURATIONAL FORCE BALANCE AND THE DISSIPATION INEQUALITY

The configurational force balance for the interface takes the simple form

$$\frac{\partial \mathbf{c}}{\partial s} + \mathbf{g} + [\![\mathbf{C}]\!]\mathbf{n} = \mathbf{0}. \tag{9}$$

Here $\mathbf{c}$ is the configurational surface stress, $\mathbf{g}$ is a dissipative internal force associated with the rearrangement of atoms at the interface, and $\mathbf{C}$ is the configurational stress in the solid. The tangential and normal components $\sigma = \mathbf{c} \cdot \mathbf{t}$ and $\tau = \mathbf{c} \cdot \mathbf{n}$ of $\mathbf{c}$ are the configurational *surface tension* and the configurational shear; thus, in contrast to more classical discussions, the surface tension actually represents a *force* tangent to the interface, with no *a priori* relationship to surface energy. The theory in bulk shows $\mathbf{C}$ to be the Eshelby tensor

$$\mathbf{C} = \Big(\Psi - \sum_{\alpha} \rho^{\alpha} \mu^{\alpha}\Big)\mathbf{1} - \mathbf{F}^{\mathsf{T}}\mathbf{S}, \tag{10}$$

a relation that bears comparison with (8). Of most importance is the component

$$\sigma K + \frac{\partial \tau}{\partial s} + \mathbf{n} \cdot [\![\mathbf{C}]\!]\mathbf{n} + g = 0, \qquad g = \mathbf{g} \cdot \mathbf{n}, \tag{11}$$

of (9) normal to the interface, as this is the component relevant to the motion of the interface.

For an interface in the presence of deformation and atomic transport, the normal configurational force balance (11), when combined with (8), (10), and the standard force, moment, and atomic balances, yields the *normal configurational force balance* (Fried and Gurtin, 2003)

$$\sum_{\alpha}([\![\rho^{\alpha}]\!] + \delta^{\alpha}K)\mu^{\alpha} = [\![\Psi - \mathbf{Sn}\cdot\mathbf{Fn}]\!] + (\psi - \bar{\sigma}\lambda)K + \frac{\partial\tau}{\partial s} + g, \qquad (12)$$

with $\bar{\sigma}$ the standard surface-tension. *This balance is basic, as its derivation utilizes only basic laws*; as such it is independent of material.

The free-energy imbalance localized to the interface using the basic balances yields the *interfacial dissipation inequality*

$$\overset{\square}{\psi} - \bar{\sigma}\overset{\square}{\lambda} - \tau\overset{\square}{\vartheta} - \sum_{\alpha}\mu^{\alpha}\overset{\square}{\delta^{\alpha}} + \sum_{\alpha}h^{\alpha}\frac{\partial\mu^{\alpha}}{\partial s} + gV \leq 0, \qquad (13)$$

with a superposed box denoting the time derivative following the normal trajectories of $\mathcal{S}$; this inequality, which is also basic, is used as a starting point for the discussion of constitutive relations.

5. RELATION OF THE NORMAL CONFIGURATIONAL FORCE BALANCE TO THE CLASSICAL EQUATIONS

Each of the interface conditions in §1 may be derived — without assumptions of local equilibrium — within the framework set out in §3.

Leo–Sekerka relation. If we take $g \equiv 0$, then the normal configurational force balance (12) reduces to the Leo–Sekerka relation (6). The relation (6) follows rigorously as an Euler–Lagrange equation associated with the variational problem of minimizing the total free-energy of a solid particle surrounded by a vapor. Thus, for solid-vapor interfaces in equilibrium, the format adopted here is completely consistent with results derived variationally. The Leo–Sekerka relation (6) (or similar relations for other types of phase interfaces) is typically applied, as is, to dynamical problems, often with an accompanying appeal to an hypothesis of "local equilibrium", although the precise meaning of this assumption is never spelled out. The more general framework leading to the normal configurational force balance (12) would allow for a nonequilibrium term $-g$, with $g = -b(\vartheta)V$, on the right side of (6). The question as to when the Leo–Sekerka relation is applicable in dynamical situations is equivalent to the question as to when the internal force g is negligible.

Our more general framework provides an answer to this question: *for sufficiently small length scales g cannot be neglected*, because the term emanating from g in the evolution equations for the interface is of the same order of magnitude as the other kinetic term in these equations, which results from accretion (Fried and Gurtin, 2004, §26.3).

Variationally derived chemical potential. We restrict attention to a single atomic species, take $g = 0$, and neglect the adatom density. Further, consistent with the variational treatment we assume that the vapor pressure and vapor free-energy (and hence the standard and configurational forces) vanish in the vapor. This allows us to replace the jumps by negative interfacial limits from the solid (so that $[\![\Psi]\!]$ becomes $-\Psi^- = -\Psi$, and so forth). Then the normal configurational force balance reduces to (2) with $U = \rho\mu$. The chemical potential U is, by its very definition, a potential associated with the addition of material at the solid-vapor interface, without regard to the specific composition of that material. As such, U cannot be used to discuss alloys. As with the Leo–Sekerka relation, the more general framework discussed here allows for a kinetic term $b(\vartheta)V$ on the right side, and hence for a nonequilibrium chemical potential.

Mullins's equation and the kinetic Maxwell equation. If in (12) we neglect deformation, adatoms, and all fields related to the bulk material, and consider constitutive relations of the form $\psi = \hat{\psi}(\vartheta)$ and

$$\tau = \frac{\partial\hat{\psi}}{\partial\vartheta}, \qquad g = -b(\vartheta)V, \tag{14}$$

with $b(\vartheta) \geq 0$ a kinetic modulus, then the dissipation inequality (13) is satisfied and the normal configurational force balance (11) reduces to the *curvature-flow equation*

$$b(\vartheta)V = \big[\psi(\vartheta) + \psi''(\vartheta)\big]K, \tag{15}$$

proposed by Uwaha (1987) and independently, using configurational forces, by Gurtin (1988); for an isotropic material, (15) reduces to Mullins's equation (1).

Similarly, the kinetic Maxwell equation (4) follows from (11) upon neglecting atomic transport as well as interfacial structure and taking $g = -b(\vartheta)V$.

6. A FINAL REMARK

Interface conditions that in other theories play the role of the normal configurational force balance are typically based on an assumption of local equilibrium or on a chemical potential derived as a variational derivative of the total free-energy with respect to variations in the configuration of the interface. By their very nature, such variational paradigms *cannot involve the normal velocity V*. To the contrary, a framework based on a configurational force balance allows for nonequilibrium terms of this form.

References

Abeyaratne, R., Knowles, J.K., 1990. *J. Mech. Phys. Solids* **38**, 345–360.

Alexander, J.I.D., Johnson, W.C., 1985. *J. Appl. Physics* **58**, 816–824.

Asaro, R.J., Tiller, W.A., 1972. *Metall. Trans.* **3** 1789–1796.

Eshelby, J.D., 1951. *Phil. Trans. Royal Soc. Lond. A* **244**, 87–112.

Eshelby, J.D., 1956. In *Progress in Solid State Physics 3* (eds. F. Seitz, D. Turnbull), Academic Press, New York.

Eshelby, J.D., 1970. In *Inelastic Behavior of Solids* (eds. M. F. Kanninen, W. F. Alder, A. R. Rosenfield, R. I. Jaffe), McGraw-Hill, New York.

Eshelby, J.D., 1975. *J. Elast.* **5**, 321–335.

Fried, E., Gurtin M.E., 1999. *J. Stat. Phys.* **95**, 1361–1427.

Fried, E., Gurtin M.E., 2003. *J. Mech. Phys. Solids* **51**, 487–517.

Fried, E., Gurtin M.E., 2004. *Adv. Appl. Mech.*, in press.

Freund, L.B., 1998. *J. Mech. Phys. Solids* **46**, 1835–1844.

Grinfeld, M., 1981. *Lett. Appl. Engin. Sci.* **19**, 1031–1039.

Gurtin, M.E., 1983. *Arch. Rat. Mech. Anal.* **84**, 1–29.

Gurtin, M.E., 1988. *Arch. Rat. Mech. Anal.* **104**, 185–221.

Gurtin, M.E., 1991. *Zeit. angewandte Math. Phys.* **42**, 370–388.

Gurtin, M.E., 1995. *Arch. Rat. Mech. Anal.* **131**, 67–100.

Gurtin, M.E., 2000. *Configurational Forces as Basic Concepts of Continuum Physics.* Springer, New York.

Gurtin, M.E., Struthers, A., 1990. *Arch. Rat. Mech. Anal.* **112**, 97–160.

Gurtin, M.E., Voorhees, P. W., 1993. *Proc. Roy. Soc. Lond. A* **440**, 323–343.

Heidug, W., Lehner, F.K., 1985. *Pure Appl. Geophys.* **123**, 91–98.

Herring, C., 1951. In *The Physics of Powder Metallurgy* (ed. W. E. Kingston), McGraw-Hill, New York.

James, R.D., 1981. *Arch. Rat. Mech. Anal.* **77**, 143–176.

Johnson, W.C., Alexander, J.I.D., 1985. *J. Appl. Phys.* **59**, 2735–2746.

Larché, F.C., Cahn, J.W., 1978. *Acta Metall.* **26**, 1579–1589.

Leo, P., Sekerka, R.F., 1989. *Acta Metall.* **37**, 5237–5252.

Maugin, G.A., 1993. *Material Inhomogeneities in Elasticity.* Chapman and Hall, London.

Mullins, W.W. 1956. *J. Appl. Physics* **27**, 900–904.

Norris, A.N., 1998. *Int. J. Solids Struct.* **35**, 5237–5253.

Peach, M.O., Koehler, J.S., 1950. *Phys. Rev.* **80**, 436–439.

Rice, J.R., Chuang, T.J., 1981. *J. American Cer. Soc.* **64**, 46–53.

Robin, P.-Y.F., 1974. *Ann. Miner.* **59**, 1286–1298.

Smoluchowski, R., 1951. *Phys. Rev.* **83**, 69–70.

Truskinovsky, L.M., 1987. *J. Appl. Math. Mech.* (*PMM*) **51**, 777–784.

Turnbull, D., 1952. *J. Metals.* **3**, 661–665.

Uhuwa, M., 1987. *J. Cryst. Growth* **80**, 84–90.

Wu, C.H., 1996. *J. Mech. Phys. Solids* **44**, 2059–2077.

Chapter 4

GENERALIZED STEFAN MODELS

Alexandre Danescu
Départment Mécanique des Solides
Ecole Centrale de Lyon, France
danescu@ec-lyon.fr

Abstract In this paper we discuss models able to account for discontinuities of the temperature field across a immaterial interface. Our theory is based on a scalar interfacial field and cover two previously proposed theories of Fried and Shen [6] and Dascalu and Danescu [4]. We briefly present an application to the problem of solidification of an under-cooled liquid.

Keywords: Stefan models, phase transitions, interfaces, free-boundaries.

Introduction

Recently, a theory accounting for discontinuous temperature and velocity across a immaterial moving interface was proposed in Fried and Shen [6]. The starting point is the interfacial version of the balance of energy combined with the interfacial version of the imbalance of entropy. The resulting interfacial dissipation inequality suggests supplemental interfacial relation associated with the mass exchange, velocity slip and jump in the temperature field at the interface.

The aim of this paper is to present a generalized framework that account for discontinuous temperature field. We show that using an additional superficial field we are able to construct a general theory accounting for discontinuous temperature field that contains as particular cases both results of [6] and [4]. We illustrate the use of the supplemental interfacial relation in the problem of the solidification of a super-cooled liquid, in a classic constitutive context.

1. INTERFACIAL BALANCE AND IMBALANCE LAWS

In this section following [5], [6], we recall briefly the interfacial versions of balance and imbalance laws. We denote by $\mathcal{D}(t)$ the region occupied by the material body at time t, and by $\mathcal{S}(t)$ the nonmaterial surface across which the velocity $\boldsymbol{v}$ and the temperature θ may experience finite jump discontinuities. We use $\boldsymbol{n}(\boldsymbol{x},t)$ for the unit normal to $\mathcal{S}(t)$ in $\boldsymbol{x} \in \mathcal{S}(t)$ and $V(\boldsymbol{x},t)$ for the scalar normal velocity $\boldsymbol{\nu} \cdot \boldsymbol{n}$ of $\mathcal{S}(t)$ at $\boldsymbol{x} \in \mathcal{S}(t)$. If a is a field on $\mathcal{D}(t)$, discontinuous across $\mathcal{S}(t)$, for $\boldsymbol{x} \in \mathcal{S}(t)$ we use $a^{\pm}$ for the limits

$$a^{\pm}(\boldsymbol{x},t) = \lim_{h\to 0} a(\boldsymbol{x} \pm h\boldsymbol{n}(\boldsymbol{x},t),t)) \tag{1}$$

and we denote $[\![a]\!]$ and $\langle a \rangle$ the jump and the mean value of a across $\mathcal{S}(t)$. We use frequently the following algebraic identity

$$[\![ab]\!] = [\![a]\!]\langle b \rangle + \langle a \rangle [\![b]\!]. \tag{2}$$

Mass balance. If ρ and $\boldsymbol{v}$ denote respectively the mass density and the motion velocity the interfacial mass balance is

$$[\![\rho(V - \boldsymbol{v} \cdot \boldsymbol{n})]\!] = 0 \tag{3}$$

and following [6] we define the *mass flux* across $\mathcal{S}$ through

$$\mathsf{m} = \langle \rho(V - \mathbf{v} \cdot \mathbf{n}) \rangle. \tag{4}$$

Momentum balance. If $\boldsymbol{T}$ denotes the Cauchy stress, in the absence of body forces the momentum balance across $\mathcal{S}$ is

$$[\![\rho(V - \boldsymbol{v} \cdot \boldsymbol{n})\boldsymbol{v}]\!] + [\![\boldsymbol{T}]\!]\boldsymbol{n} = \boldsymbol{0}, \tag{5}$$

or, using (3)

$$\mathsf{m}[\![\boldsymbol{v}]\!] + [\![\boldsymbol{T}]\!]\boldsymbol{n} = \boldsymbol{0} \tag{6}$$

Energy balance. Neglecting supplies, if e and $\boldsymbol{q}$ denotes the internal energy density and the heat flux the energy balance across $\mathcal{S}$ can be written as

$$[\![\rho(e + \frac{1}{2}\boldsymbol{v} \cdot \boldsymbol{v})(V - \boldsymbol{v} \cdot \boldsymbol{n})]\!] + [\![\boldsymbol{T}\boldsymbol{v}]\!] \cdot \boldsymbol{n} - [\![\boldsymbol{q}]\!] \cdot \boldsymbol{n} = 0. \tag{7}$$

Once again using (3) and (6), we rewrite (7) as

$$\mathsf{m}[\![e]\!] + \langle \boldsymbol{T}\boldsymbol{n} \rangle \cdot [\![\boldsymbol{v}]\!] - [\![\boldsymbol{q}]\!] \cdot \boldsymbol{n} = 0. \tag{8}$$

If $\mathsf{P} = \boldsymbol{I} - \boldsymbol{n} \otimes \boldsymbol{n}$ denotes the interfacial projection, following [6], we call $\mathsf{s} = \mathsf{P}[\![\boldsymbol{v}]\!]$ and $\mathbf{t} = \mathsf{P}\langle \boldsymbol{T}\boldsymbol{n}\rangle$ respectively the *interfacial velocity slip* and *interfacial friction.* After rearrangement, the balance of energy can be written as

$$\mathsf{m}[\![e - \frac{1}{\rho}\boldsymbol{T}\boldsymbol{n}\cdot\boldsymbol{n} + \frac{1}{2}(V - \boldsymbol{v}\cdot\boldsymbol{n})^2]\!] + \mathbf{t}\cdot\mathsf{s} - [\![\boldsymbol{q}]\!]\cdot\boldsymbol{n} = 0. \tag{9}$$

Let θ, η denote respectively the absolute temperature and the entropy density; using the free energy density defined as $\Psi = e - \theta\eta$, we obtain

$$\mathsf{m}[\![\Psi - \frac{1}{\rho}\boldsymbol{T}\boldsymbol{n}\cdot\boldsymbol{n} + \frac{1}{2}(V - \boldsymbol{v}\cdot\boldsymbol{n})^2]\!] + \mathbf{t}\cdot\mathsf{s} + \mathsf{m}[\![\theta\eta]\!] - [\![\boldsymbol{q}]\!]\cdot\boldsymbol{n} = 0. \tag{10}$$

Imbalance of entropy Neglecting specific external entropy supply the interfacial version of the entropy imbalance is

$$[\![\frac{1}{\theta}\boldsymbol{q}\cdot\boldsymbol{n}]\!] - \mathsf{m}[\![\eta]\!] \geq 0. \tag{11}$$

The interfacial dissipation inequality is then obtained using (11) in (10). We present briefly the classical case in the following subsection.

1.1. CONTINUOUS TEMPERATURE FIELD

When the temperature field is continuous across $\mathcal{S}$ the use of (11) in (10) is obvious; we rewrite (10) as

$$\mathsf{m}[\![\Psi - \frac{1}{\rho}\boldsymbol{T}\boldsymbol{n}\cdot\boldsymbol{n} + \frac{1}{2}(V - \boldsymbol{v}\cdot\boldsymbol{n})^2]\!] + \mathbf{t}\cdot\mathsf{s} = \theta([\![\frac{\boldsymbol{q}}{\theta}]\!]\cdot\boldsymbol{n} - \mathsf{m}[\![\eta]\!]). \tag{12}$$

and taking into account (11) we are lead to the classical interfacial dissipation inequality (see also [1], [7], [8], [10], [11])

$$\mathsf{m}[\![\Psi - \frac{1}{\rho}\boldsymbol{T}\boldsymbol{n}\cdot\boldsymbol{n} + \frac{1}{2}(V - \boldsymbol{v}\cdot\boldsymbol{n})^2]\!] + \mathbf{t}\cdot\mathsf{s} \geq 0. \tag{13}$$

2. DISCONTINUOUS TEMPERATURE MODELS

2.1. THE MODEL OF FRIED AND SHEN

Fried and Shen introduced in [6] the *scaled temperature jump*

$$\mathsf{j} = \frac{[\![\theta]\!]}{\langle\theta\rangle} = -\frac{[\![1/\theta]\!]}{\langle 1/\theta\rangle} \tag{14}$$

and rewrite (10) as

$$\mathsf{m}[\![\Psi - \frac{1}{\rho}\boldsymbol{n}\cdot\boldsymbol{T}\boldsymbol{n} + \frac{1}{2}(V - \boldsymbol{v}\cdot\boldsymbol{n})^2]\!] + \mathsf{s}\cdot\mathsf{t} + \mathsf{j}\,[\mathsf{m}\langle\theta\eta\rangle - \langle\boldsymbol{q}\cdot\boldsymbol{n}\rangle] = \langle\frac{1}{\theta}\rangle^{-1}([\![\frac{\boldsymbol{q}}{\theta}\cdot\boldsymbol{n}]\!] - \mathsf{m}[\![\eta]\!]). \quad (15)$$

which leads to an interfacial dissipation inequality in the form

$$\mathsf{em} + \mathsf{t}\cdot\mathsf{s} + \mathsf{jh} \geq 0, \quad (16)$$

where $\mathsf{e} = [\![\Psi - \frac{1}{\rho}\boldsymbol{n}\cdot\boldsymbol{T}\boldsymbol{n} + \frac{1}{2}(V - \boldsymbol{v}\cdot\boldsymbol{n})^2]\!]$ denotes the *interfacial energy release* and $\mathsf{h} = \mathsf{m}\langle\theta\eta\rangle - \langle\boldsymbol{q}\cdot\boldsymbol{n}\rangle$ the *interfacial heating.* Fried and Shen in [6] provide a rigorously treatment of supplemental constitutive relations on the interface based on (16), and invariance under superposed rigid changes of observer, lead to (see also [2], [10])

$$\mathsf{e} = \mathsf{e}(\mathsf{m}, |\mathsf{s}|, \mathsf{j}), \quad \mathsf{t} = \mathsf{t}(\mathsf{m}, |\mathsf{s}|, \mathsf{j})\frac{\mathsf{s}}{|\mathsf{s}|}, \quad \mathsf{h} = \mathsf{h}(\mathsf{m}, |\mathsf{s}|, \mathsf{j}). \quad (17)$$

Two questions arise with respect to the above theory; the first one concerns the interfacial heating. More precisely, the first term of the interfacial heating contains the mass flux m and can be equally regarded as a part of the interfacial energy release. The second question concerns the first factor in the right-hand-side of (15). Alternative choices, leading to different theories, were proposed in Danescu and Dascalu [4] and[1] Šilhavý [9] and we shall develop on this issue in the following.

2.2. THEORY BASED ON THE TEMPERATURE JUMP

A straightforward computation shows that an equivalent form for (10) is [9], [4]

$$\mathsf{m}[\![\Psi - \frac{1}{\rho}\boldsymbol{n}\cdot\boldsymbol{T}\boldsymbol{n} + \frac{1}{2}(V - \boldsymbol{v}\cdot\boldsymbol{n})^2]\!] + \mathsf{s}\cdot\mathsf{t} + [\![\theta]\!]\left[\mathsf{m}\langle\theta\rangle - \langle\frac{\boldsymbol{q}}{\theta}\rangle\cdot\boldsymbol{n}\right] = \langle\theta\rangle([\![\frac{\boldsymbol{q}}{\theta}\cdot\boldsymbol{n}]\!] - \mathsf{m}[\![\eta]\!]) \quad (18)$$

The right-hand-side of (18) is still positive and following the line developed by Fried and Shen we rerwrite the interfacial dissipation inequality as

$$\mathsf{em} + \mathsf{t}\cdot\mathsf{s} + \hat{\jmath}\hat{\mathsf{h}} \geq 0, \quad (19)$$

where $\hat{\mathsf{j}}$ is the temperature jump and $\hat{\mathsf{h}}$ is the interfacial heating. In this case we are lead to constitutive relations similar to (17) in the form

$$\mathsf{e} = \mathsf{e}(\mathsf{m}, |\mathsf{s}|, \hat{\mathsf{j}}), \quad \mathsf{t} = \mathsf{t}(\mathsf{m}, |\mathsf{s}|, \hat{\mathsf{j}})\mathsf{s}/|\mathsf{s}|, \quad \hat{\mathsf{h}} = \mathsf{h}(\mathsf{m}, |\mathsf{s}|, \hat{\mathsf{j}}). \quad (20)$$

3. EQUIVALENT TEMPERATURE OF THE INTERFACE

A straightforward computation shows that (12) and (18) are special cases of a more general form of the interfacial balance of energy which reads

$$\mathsf{m}[\![\Psi - \frac{1}{\rho}\boldsymbol{n}\cdot\boldsymbol{T}\boldsymbol{n} + \frac{1}{2}(V - \boldsymbol{v}\cdot\boldsymbol{n})^2 + \eta(\theta - \overline{\theta})]\!] + \mathbf{s}\cdot\mathbf{t} -$$

$$-[\![\frac{\boldsymbol{q}}{\theta}(\theta - \overline{\theta})\cdot\boldsymbol{n}]\!] = \overline{\theta}([\![\frac{\boldsymbol{q}}{\theta}\cdot\boldsymbol{n}]\!] - \mathsf{m}[\![\eta]\!]), \tag{21}$$

which introduces a **positive scalar interfacial field** $\overline{\theta}$. We call $\overline{\theta}$ **the equivalent temperature of the interface** and we regard this concept as to be prescribed by a *constitutive function.* The particular choices

$$\overline{\theta} = \langle\theta\rangle, \quad \overline{\theta} = \langle 1/\theta\rangle, \tag{22}$$

lead to (18) and (15) respectively, but other special constitutive choices, like for example[2]

$$\overline{\theta} = \langle\theta^r\rangle\langle\frac{1}{\theta^{1-r}}\rangle^{-1} \tag{23}$$

are possible, leading to different theories. We discussed in [3] the physical meaning of the superficial field $\overline{\theta}$ and we showed that $\overline{\theta}$ is the temperature of the interface which *a priori* may be different from both limits $\theta^{\pm}$.

4. INTERFACIAL DISSIPATION INEQUALITY

Taking into account (11) and (21) we conclude that the interfacial dissipation inequality holds in the form

$$\underbrace{\mathsf{m}\,[\![\Psi - \frac{1}{\rho}\boldsymbol{n}\cdot\boldsymbol{T}\boldsymbol{n} + \frac{1}{2}(V - \boldsymbol{v}\cdot\boldsymbol{n})^2 + \eta(\theta - \overline{\theta})]\!]}_{\text{interfacial energy release}} + \mathbf{s}\cdot\mathbf{t} - [\![\frac{\boldsymbol{q}}{\theta}(\theta - \overline{\theta})\cdot\boldsymbol{n}]\!] \geq 0 \tag{24}$$

where the interfacial energy release includes also the term $\mathsf{m}[\![\eta(\theta - \overline{\theta})]\!]$.

We shall focus in the following on a pure thermal problem in a motionless body so that the dissipation inequality is reduced to

$$\mathsf{m}[\![\Psi - \eta(\theta - \overline{\theta})]\!] - [\![\frac{\boldsymbol{q}}{\theta}(\theta - \overline{\theta})\cdot\boldsymbol{n}]\!] \geq 0 \tag{25}$$

In the case of a continuous temperature field the above relation is just $\mathsf{m}[\![\Psi]\!] \geq 0$, and holds if and only if a supplemental *kinetic relation* in the form $[\![\Psi]\!] = b\mathsf{m}$, with $b \geq 0$ is given. When the temperature is discontinuous, we still postulate a supplemental constitutive relation for the interfacial energy release but we interpret the inequality

$$-[\![\frac{\boldsymbol{q}}{\theta}(\theta-\overline{\theta})\cdot\boldsymbol{n}]\!]\geq 0 \tag{26}$$

as a thermodynamic restriction on the constitutive function $\overline{\theta}$. The analog of (26) in the bulk is the classical residual inequality $\boldsymbol{q}\cdot\mathrm{grad}\theta\leq 0$.

In special situations when the constitutive function $\overline{\theta}$ is prescribed (26) provides an additional interfacial relation. For example, if (23) holds, a straightforward computation in [3] shows that (26) is equivalent to

$$\langle\frac{\boldsymbol{q}}{\theta^r}\rangle\cdot\boldsymbol{n}=-\gamma\left[\frac{[\![\theta^r]\!]}{\langle\theta^r\rangle}+\frac{[\![\theta^{1-r}]\!]}{\langle\theta^{1-r}\rangle},\right] \tag{27}$$

for some positive function γ.

This shows that compatibility of a given interfacial structure, i.e. a prescribed $\overline{\theta}$, with the interfacial dissipation inequality (26), imposes a supplemental constitutive relation like (27).

5. SOLIDIFICATION OF AN UNDER-COOLED LIQUID

The next section is devoted to the solidification of an under-cooling liquid. To illustrate the role of the additional relation at the interface, we shall provide a sufficient condition for the existence of travelling wave solutions in a particular constitutive context.

We start recalling the field equations and the interfacial conditions; for a pure thermal problem the field equation is the balance of energy,

$$\rho\dot{e}=-\mathrm{div}\boldsymbol{q}, \tag{28}$$

and the associated interface balance

$$\rho V[\![e]\!]=[\![\boldsymbol{q}\cdot\boldsymbol{n}]\!], \tag{29}$$

supplemented, for some positive α, by

$$[\![\Psi-\eta(\theta-\overline{\theta})]\!]=\alpha\rho V,\qquad [\![\frac{\boldsymbol{q}}{\theta}]\!]\cdot\boldsymbol{n}\geq\frac{1}{\overline{\theta}}[\![\boldsymbol{q}]\!]\cdot\boldsymbol{n}. \tag{30}$$

Constitutive assuptions: We shall assume following [6] constitutive relations in the form

$$\Psi(\theta)=\begin{cases} c_s\theta(1-\ln(\theta/\theta_s)) & \text{in the solid phase,}\\ c_l\theta(1-\ln(\theta/\theta_l))+l & \text{in the liquid phase,}\end{cases} \tag{31}$$

where c_s, c_l, θ_s, θ_l and l are positive constants. The imbalance of entropy in the bulk gives

$$\eta(\theta) = -\Psi'(\theta) = \begin{cases} c_s \ln(\theta/\theta_s) & \text{in the solid phase,} \\ c_l \ln(\theta/\theta_l) & \text{in the liquid phase,} \end{cases} \tag{32}$$

and

$$e(\theta) = \Psi(\theta) - \theta\eta(\theta) = \begin{cases} c_s\theta & \text{in the solid phase,} \\ c_l\theta + l & \text{in the liquid phase.} \end{cases} \tag{33}$$

Constitutive relations for the heat flux are assumed in the form

$$\boldsymbol{q} = \begin{cases} -k_s \mathrm{grad}\theta & \text{in the solid phase,} \\ -k_l \mathrm{grad}\theta & \text{in the liquid phase,} \end{cases} \tag{34}$$

for k_s and k_l positive and we assume that there exists a unique *transition temperature* such that

$$\theta_l < \theta_\star < \theta_s, \qquad \Psi_s(\theta_\star) = \Psi_l(\theta_\star). \tag{35}$$

For $\overline{\theta}$ we shall focus here only on the case $r = 1/2$ in (23), but the general case $0 < r < 1$ can be treated in a very similar manner. Relation (27) provides

$$\langle \frac{\mathbf{q}}{\theta^{1/2}} \rangle \cdot \mathbf{n} = -2\gamma \frac{[\![\theta^{1/2}]\!]}{\langle \theta^{1/2} \rangle}. \tag{36}$$

We investigate the existence of travelling waves solutions for the system (28), (29), (30.1) and (36) in the constitutive framework described by (31) and (34). We restrict to cases when the interface is a plane that propagates with the scalar velocity V and we look for solutions $\theta(\boldsymbol{x}, t)$ depends only on $\xi = \boldsymbol{x} \cdot \boldsymbol{n} - Vt$ in a case when the under-cooled liquid in located at $\xi > 0$ while the solid at $\xi < 0$. We assume a far field condition θ_∞ in the liquid phase such that

$$\theta_\infty < \theta^+ < \theta^- < \theta^\star. \tag{37}$$

The temperature field

$$\theta(\boldsymbol{x}, t) = \begin{cases} \theta^- & \text{if } \xi < 0, \\ \theta_\infty + (\theta^+ - \theta_\infty)\exp\left(-c_l V \rho \xi / k_l\right)) & \text{if } \xi > 0, \end{cases} \tag{38}$$

with $\theta^- = (c_l\theta_\infty + l)/c_s$ satisfies (28) and (29), while relations (30.1) and (36) can be rewritten as

$$\rho V = -\frac{4\gamma}{c_l} \frac{(\theta^+)^{1/2}}{\theta^+ - \theta_\infty} \frac{(\theta^+)^{1/2} - (\theta^-)^{1/2}}{(\theta^+)^{1/2} + (\theta^-)^{1/2}}, \tag{39}$$

$$c_l(\theta^+ - \theta_\infty) - \sqrt{\theta^+\theta^-}\left[c_l \ln(\theta^+/\theta_l) - c_s \ln(\theta^-/\theta_s)\right] = \alpha\rho V. \qquad (40)$$

Equations (39) and (40) form a system for the unknowns θ^+ and V. We note that (39) shows that V is positive when θ^+ satisfies (37). We shall provide in what follows a simple sufficient condition for the existence of a travelling wave solution. Substitution of ρV from (39) in (40) gives, after rearrangement, a single equation for θ^+; we obtain

$$\ln\theta^+ - C = \frac{1}{\sqrt{\theta^+\theta^-}}\left[\theta^+ - \theta_\infty + \frac{2\alpha\gamma}{c_l^2}\frac{(\theta^+)^{1/2}}{\theta^+ - \theta_\infty}\frac{(\theta^+)^{1/2} - (\theta^-)^{1/2}}{(\theta^+)^{1/2} + (\theta^-)^{1/2}}\right]. \qquad (41)$$

for $C = \ln\theta_l + \frac{c_l}{c_s}\ln(\theta^-/\theta_s)$. We denote the right-hand-side above $F(\theta^+)$ and note that both sides in (41) are continuous functions on $(\theta_\infty, \theta^-)$. Moreover, we have $\lim_{\theta^+\to\theta_\infty} F(\theta^+) = -\infty$, and $\lim_{\theta^+\to\theta^-} F(\theta^+) = (\theta^- - \theta_\infty)/\theta^-$. It follows directly that if $\ln\theta^- - C < (\theta^- - \theta_\infty)/\theta^-$, equation (41) has at least one solution in $(\theta_\infty, \theta^-)$.

Notes

1. We acknowledge an interesting discussion with Prof. M. Šilhavý who kindly point out to us his approach on this problem; previous to our formula (18) (and before our result in [4]) he used the same approach in [9] to obtain the interfacial dissipation inequality (19).

2. This assumption generalizes both previous cases; indeed, for $r = 0$ and $r = 1$ in (23), we obtain $\overline{\theta} = \langle\frac{1}{\theta}\rangle^{-1}$ and respectively, $\overline{\theta} = \langle\theta\rangle$.

References

[1] Abeyaratne, R., Knowles, J.K., Kinetic relations and the propagation of phase boundaries in elastic solids, *Arch. Rational Mech. Anal.*, **114** (119-154), 1991.

[2] Coleman, B.D., Noll, W., The thermodynamics of elastic materials with heat conduction and viscosity, *Arch. Rational Mech. Anal.*, **13** (167-178), 1963.

[3] Danescu, A, Generalized Stefan models acounting for discontinuous temperature field, to appear in *Continuum Mechanics and Thermodynamics.*

[4] Dascalu, C., Danescu, A., Thermoelastic driving forces on singular surfaces, *Mech Res. Comm.*, **29** (507-512), 2002.

[5] Fried, E., Energy release, friction and supplemental relations at phase interface, *Continuum Mech. Thermodyn.*, **7** (111-121), 1995.

[6] Freid, E., Shen, A.Q., Generalization of the Stefan model to allow for both velocity and temperature jumps, *Continuum Mech. Thermodyn.*, **11** (277-296), 1999.

[7] Gurtin, M.E., *Configurational Forces as Basic Concepts of Continuum Physics*, Springer Verlag, 2000.

[8] Gurtin, M.E., On the nature of configurational forces, *Arch. Rational Mech. Anal.*, **131** (67-100), 1995.

[9] Šilhavý, M, *The Mechanics ond Thermodynamics of Continuous Media*, Springer Verlag, 1997.

[10] Truskinovsky, L., Dynamics of nonequilibrium phase boundaries in a heat conducting nonlinear elastic medium, *J. Appl. Math. Mech.*, **51** (777-784), 1987.

[11] Truskinovsky, L., Kinks versus shocks, in *Shock Induced Transitions and Phase Structures in General Media* (ed. R. Fosdick, E. Dunn and M. Slemrod), Springer Verlag, Berlin, 1991.

Chapter 5

EXPLICIT KINETIC RELATION FROM "FIRST PRINCIPLES"

Lev Truskinovsky
Laboratoire de Mécanique des Solides, CNRS-UMR 7649
Ecole Polytechnique, 91128, Palaiseau, France
trusk@lms.polytechnique.fr

Anna Vainchtein
Department of Mathematics
University of Pittsburgh, Pittsburgh, PA, 15260, USA
aav4@pitt.edu

Abstract We study a fully inertial discrete model of a martensitic phase transition which takes into account interactions of first and second nearest neighbors. Although the model is Hamiltonian at the *microscale*, it generates a nontrivial *macroscopic* relation between the velocity of the martensitic phase boundary and the conjugate configurational force. The apparent dissipation is due to the induced radiation of lattice waves carrying energy away from the front.

Keywords: Kinetic relations, lattice waves, radiative damping

Introduction

The fact that a nonzero configurational force is required to sustain a martensitic phase transition reflects inability of the classical continuum elasticity to describe dissipative phenomena inside the transition front where discreteness of the underlying crystal structure cannot be neglected. We recall that in continuum theory a phase boundary can move without friction. At the same time its motion in a lattice can be compared to that of a particle placed in a wiggly (Peierls-Nabarro) landscape: the oscillations of the velocity then lead to the energy transfer

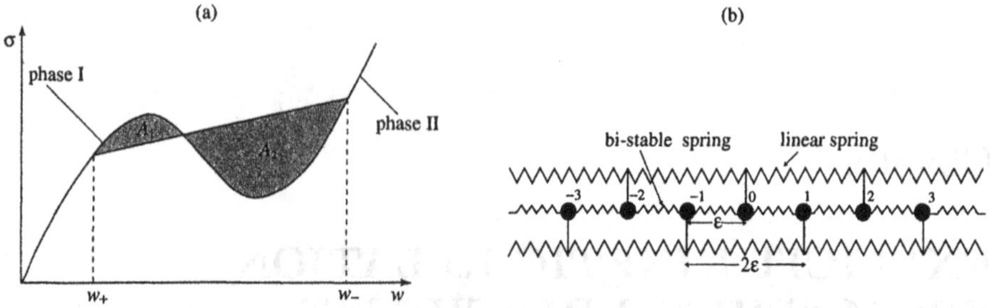

Figure 5.1. (a) The macroscopic stress-strain law and the Rayleigh line for a subsonic phase boundary. The difference $A_2 - A_1$ between the shaded areas is the configurational force G. (b) The discrete microstructure with nearest and next-to-nearest neighbor interactions.

from macro to microscale [14]. At continuum level the emitted short-length lattice waves are invisible, and the radiation is perceived as energy dissipation. Since the rate of energy release at the macroscale remains unspecified, in order to close the system of equations at the macrolevel, one needs to supplement the conservative continuum equations with the dissipative kinetic relation on the moving discontinuity [10, 11]. In this paper we consider the simplest nonlocal discrete model of a martensitic phase boundary allowing one to find the unknown energy release rate explicitly. Following some previous work in fracture [7, 8] and plasticity [1, 3, 4] we use a biparabolic ansatz for the free energy and construct an explicit solution of the discrete problem. We emphasize that our only input information concerns the elasticities of the constitutive elements, and hence the resulting kinetic relation can be considered of the "first principles" type.

1. CONTINUUM MODEL

Consider an isothermal motion of an infinite homogeneous bar with a unit cross-section. Let $u(x,t)$ be the displacement of a reference point x at time t. Then strain and velocity fields are given by $w = u_x(x,t)$ and $v = u_t(x,t)$, respectively. The balances of mass and linear momentum are $v_x = w_t$ and $\rho v_t = (\sigma(w))_x$, where the function $\sigma(w)$ specifies the stress-strain relation. To model martensitic phase transitions, we follow [2] and assume that $\sigma(w)$ is non-monotone as shown in Figure 5.1a. The two monotonicity regions where $\sigma'(w) > 0$ will be associated with material phases I and II. Suppose now that an isolated strain discontinuity propagates along the bar with constant velocity V. On the discontinuity the balance laws reduce to the Rankine-Hugoniot jump conditions

$$\rho V^2 [\![w]\!] = [\![\sigma]\!], \quad \rho V [\![v]\!] = -[\![\sigma]\!], \tag{1}$$

where $[\![f]\!] \equiv f_+ - f_-$ denotes the jump. Conditions (1) must be supplemented by the entropy inequality $\mathcal{R} = GV \geq 0$, where

$$G = [\![\phi]\!] - \{\sigma\}[\![w]\!] \tag{2}$$

is the associated configurational force. Here $\{\sigma\} = (\sigma_+ + \sigma_-)/2$. Given V and the state (v_+, w_+) in front of the moving discontinuity, one can use (1) to determine the state (v_-, w_-) behind. In particular, $(1)_1$ implies that $w_\pm$ lie on the intersection of the curve $\sigma(w)$ and the Rayleigh line with the slope ρV^2, as shown in Figure 5.1a. To satisfy the entropy inequality it is sufficient to require that the difference between the areas A_2 and A_1 shown in Figure 5.1a is nonnegative. It is not hard to see that the macroscopic jump conditions do not provide enough information to specify the velocity of the phase boundary V uniquely.

Although the difficulty with finding V does not arise in the case of supersonic shock waves, it is essential in the case of subsonic phase boundaries, where additional jump condition controlling the rate of dissipation must be provided to ensure that the continuum problem is well posed [6, 12]. The corresponding closing kinetic relation in the form $G = G(V)$ can be either postulated as a phenomenological constitutive relation (e.g. [10, 11]) or derived from a regularized continuum model which usually includes dissipative as well as dispersive terms (e.g. [5, 10]). Below we take a different approach and derive the closing relation from a discrete lattice model represented by an infinite system of coupled ordinary differential equations.

2. DISCRETE MODEL

Consider a chain of particles connected to their nearest neighbors (NN) and next-to-nearest neighbors (NNN) by elastic springs, as shown in Figure 5.1b. In the undeformed configuration the NN and NNN springs have length ε and 2ε, respectively. Let $u_n(t)$, $-\infty < n < \infty$, denote the displacement of nth particle at time t with respect to the reference configuration. In terms of the strain variables $w_n = (u_n - u_{n-1})/\varepsilon$ the dynamic equations take the form

$$m\ddot{w}_n = \phi'_{\mathrm{NN}}(w_{n+1}) - 2\phi'_{\mathrm{NN}}(w_n) + \phi'_{\mathrm{NN}}(w_{n-1}) + \gamma(w_{n+2} - 2w_n + w_{n-2}). \tag{3}$$

Here ϕ_{NN} is the nonlinear and nonconvex NN potential, while the NNN interactions are assumed to be linear: $\phi'_{\mathrm{NNN}}(w) = 2\gamma w$. We seek solutions of (3) in the form of a traveling wave moving with the velocity

V and connecting two states in different phases. Let $x = n\varepsilon - Vt$ and assume that

$$u_n(t) = u(x), \qquad w_n(t) = w(x) = [u(x) - u(x-\varepsilon)]/\varepsilon \tag{4}$$

The system (3) can now be replaced by a single nonlinear advance-delay differential equation for $w(x)$:

$$\begin{aligned} mV^2 w'' = {} & \phi'_{\mathrm{NN}}(w(x+\varepsilon)) - 2\phi'_{\mathrm{NN}}(w(x)) + \phi'_{\mathrm{NN}}(w(x-\varepsilon)) \\ & + \gamma(w(x+2\varepsilon) - 2w(x) + w(x-2\varepsilon)). \end{aligned} \tag{5}$$

The states at $x = \pm\infty$ must correspond to their macroscopic limits

$$w(x) \to w_{\pm} \quad \text{as } x \to \pm\infty. \tag{6}$$

Since we expect emission of elastic waves, the limits in (6) must be understood in the weak sense only.

In order to obtain analytical solution of the discrete problem, we choose NN potential to be biparabolic and symmetric so that $\phi'_{\mathrm{NN}}(w) = K(w - a\theta(w - w_c))$, where $\theta(x)$ is a unit step function. Assume that all springs in the region $x > 0$ are in phase I, while all springs with $x < 0$ are in phase II. Then in nondimensional variables equation (5) can be written as

$$\begin{aligned} & V^2 w'' - w(x+1) + 2w(x) - w(x-1) - \frac{\beta}{4}(w(x+2) - 2w(x) \\ & + w(x-2)) = -\theta(-x-1) + 2\theta(-x) - \theta(1-x), \end{aligned} \tag{7}$$

where $\beta = 4\gamma/K$ is the main nondimensional parameter of the problem. Observe that Eq. (7) is linear in $x < 0$ and $x > 0$ so that the nonlinearity is hidden in the switching condition

$$w(0) = w_c \tag{8}$$

and in the constraints

$$w(x) < w_c \quad \text{for } x > 0, \qquad w(x) > w_c \quad \text{for } x < 0 \tag{9}$$

ensuring that the springs are in proper phases. The problem now reduces to solving (7) subject to (6), (8) and (9). We remark that a related discrete problem with $\beta = 0$ (no NNN interactions) was previously considered in [8, 9].

3. SOLUTION OF THE DISCRETE PROBLEM

Equation (7) can be solved by standard Fourier transform (see [13] for details) yielding

$$w(x) = \begin{cases} w_- + \sum\limits_{k \in M^-} \frac{4\sin^2(k/2)e^{ikx}}{kL'(k)} & x < 0 \\ w_- - \frac{1}{1+\beta-V^2} - \sum\limits_{k \in M^+} \frac{4\sin^2(k/2)e^{ikx}}{kL'(k)} & x > 0, \end{cases} \tag{10}$$

where $L(k) = 4\sin^2(k/2) + \beta \sin^2 k - V^2k^2$ and $M^\pm = \{k : L(k) = 0, \{Im k \gtrless 0\} \bigcup \{Im k = 0, kL'(k) \gtrless 0\}$. The solution can be viewed as a homogeneous state superimposed with the combination of plane waves with phase velocity V and wave numbers given by the zeroes of $L(k)$. In particular, there is a finite number of real roots of $L(k) = 0$ corresponding to radiative modes. To obtain (10), we applied the radiation conditions [1, 7] requiring that all radiative modes with group velocities V_g higher than the interface velocity V appear in front of the moving phase boundary ($x > 0$), while all radiative modes with $V_g < V$ appear behind the front. Since $V_g = V + \frac{L'(k)}{2Vk}$, the relevant radiative modes ahead (behind) the interface must satisfy $kL'(k) > 0$ (< 0).

Equations (10) imply that the limiting states are related by $w_+ = w_- - 1/(1+\beta-V^2)$, which is exactly the macroscopic Rankine-Hugoniot condition $(1)_1$. The switching condition (8) implies that

$$w_\pm = w_c \mp \frac{1}{2(1+\beta-V^2)} + \sum_{k \in N^\pm_{\text{pos}}} \frac{4\sin^2(k/2)}{|kL'(k)|}, \tag{11}$$

where $N^\pm = \{k : L(k) = 0,\ Im k = 0, kL'(k) \gtrless 0\}$. Since both $L(k)$ and $N^\pm$ depend explicitly on V, Eq. (11) provides two additional relations between the velocity of the moving interface and the strains at infinity; one of them is equivalent to $(1)_1$ while the other one generates a nontrivial kinetic relation. To recover the second Rankine-Hugoniot condition $(1)_2$, we recall that given the self-similar ansatz (4), the strain and velocity fields are related through $v(x) - v(x-1) = w'(x)$. Since the right hand side of the latter equation is known explicitly (see (10)), we can again use the Fourier transform to show that the difference between the average velocities at infinity satisfies $v_+ - v_- = V/(1+\beta-V^2)$, This is exactly our macroscopic jump condition $(1)_2$.

4. KINETIC RELATION

Consider the global energy balance in the discrete model

$$\frac{d}{dt}\left\{\sum_{n=-\infty}^{\infty}\left[\frac{v_n^2}{2}+\phi_{\rm NN}(w_n)+\frac{\beta}{2}\left(\frac{w_n+w_{n+1}}{2}\right)^2\right]\right\}=F_n v_n|_{n=-\infty}^{n=\infty}, \quad (12)$$

where $F_n = \phi'_{\rm NN}(w_n) + \frac{\beta}{4}(w_{n-1} + 2w_n + w_{n+1})$ is the total force acting on the nth particle from the left. Since at infinity our solution tends to the homogeneous state plus linear oscillations we use asymptotic orthogonality of the modes and write: $\langle F_n v_n|_{n=-\infty}^{n=\infty}\rangle = \mathcal{P} + \mathcal{P}_0$, where $\langle\cdot\rangle$ denotes the averaging over sufficiently large period, $\mathcal{P} = \sigma_+ v_+ - \sigma_- v_-$ is the macroscopic power supply at $\pm\infty$ and $\mathcal{P}_0$ is the energy carried away by the microscopic lattice waves:

$$-\mathcal{P}_0 = \sum_{k\in N^+} \langle \mathcal{G}_k\rangle_+ (V_g - V) + \sum_{k\in N^-} \langle \mathcal{G}_k\rangle_- (V - V_g). \quad (13)$$

Here $\mathcal{G}_k$ is the sum of kinetic and potential energies per particle carried by the mode k. The average energy density carried by the mode k can be computed from $\langle \mathcal{G}_k\rangle = \langle \mathcal{G} - \mathcal{G}_0\rangle_k$, where $\mathcal{G}(x)$ is the total energy per particle and $\mathcal{G}_0$ is the energy of the limiting homogeneous states. The calculation yields [13] $\langle \mathcal{G}_k\rangle_\pm = \frac{8V^2\sin^2(k/2)}{(L'(k))^2}$ for the the average energy density carried by the radiative wave with $k \in N^\pm_{\rm pos} = \{k \in N^\pm : k > 0\}$. Substituting these explicit relations into (13) and observing that $\mathcal{R} = GV = -\mathcal{P}_0$, we obtain the desired expression for the driving force:

$$G = 4 \sum_{k\in N^\pm_{\rm pos}} \frac{\sin^2(k/2)}{|kL'(k)|}. \quad (14)$$

Since both $L(k)$ and $N^\pm$ are known functions of V, Eq. (14) yields an explicit kinetic relation (see also 11).

Alternatively, we could compute the driving force G by using Eq. (2) for the continuum macromodel. Observe that the macroscopic energy density $\phi(w)$ is related to its microscopic counterparts via $\phi(w) = \frac{1}{2}(1+\beta)w^2 - \theta(w - w_c)(w - w_c)$. By substituting this relation into (2) and using (11), we obtain $G = \frac{1}{2}(w_- + w_+) - w_c = 4\sum_{k\in N^\pm_{\rm pos}} \frac{\sin^2(k/2)}{|kL'(k)|}$, which coincides with (14). This confirms that the macroscopic energy release rate is consistent with the microscopic account of dissipation.

To compute the resulting kinetic relation we need to find at each V all positive real zeroes of $L(k)$. The typical function $V(k)$ is plotted in Figure 5.2a. It possesses an infinite number of local maxima V_i^r

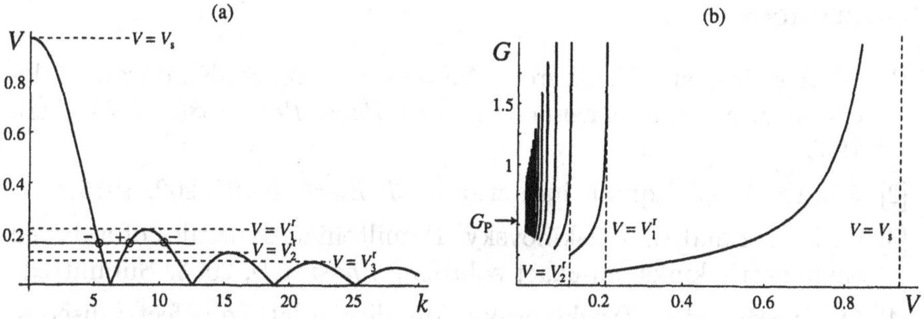

Figure 5.2. (a) Real wave numbers k corresponding to a given interface velocity V with "+" and "−" denoting the sign of $kL'(k)$. Also marked are the resonance velocities V_i^r. (b) Mobility curves $G(V)$. The entire region around the resonances should be excluded. In both graphs $\beta = -1/8$.

(resonance velocities) where $L'(k) = 0$ and the sums in (10) and (14) diverge. These resonances are symmetry-related and disappear when the curvatures of the energy wells are different [3]. Two limiting cases, $V \to 0$ and $V \to V_s$, where $V_s = \sqrt{1+\beta}$ is the macroscopic sound velocity, deserve particular attention. In the zero-velocity limit we obtain [13]

$$G(0) = \frac{1}{2\sqrt{1+\beta}} = G_{\mathrm{P}},$$

which coincides with the Peierls force computed in [14]. The limit $V \to V_s$ depends on β. Assume for determinacy that $-1/4 < \beta \leq 0$. Then one can show that for $V \to V_s$ the only relevant positive real root $k \in N_{\mathrm{pos}}^-(V)$ tends to zero and since in this limit $G = \frac{6}{(1+4\beta)k^2} + \frac{44\beta-1}{10(1+4\beta)^2} + O(k^2)$, we obtain that $G(V) \to \infty$.

In the intermediate range $0 < V < V_s$ the kinetic relation can be obtained numerically by computing the sets $N_{\mathrm{pos}}^{\pm}(V)$. Figure 5.2b shows the typical mobility curves $G(V)$ at $\beta = -1/8$. As expected, in the small-velocity range $0 < V \leq V_1^r$ there is an accumulation of resonances. It can be shown [13] that the corresponding traveling wave solution are not admissible because they violate the condition (9). In this range of average velocities the interface motion may be of a more complex nature, for instance, stick-slip.

Acknowledgments

This work was supported by the NSF grants DMS-0102841 (L.T.) and DMS-0137634 (A.V.).

References

[1] W. Atkinson and N. Cabrera. Motion of a Frenkel-Kontorova dislocation in a one-dimensional crystal. *Phys. Rev.*, 138(3A):763–766, 1965.

[2] J.L. Ericksen. Equilibrium of bars. *J. Elast.*, 5:191–202, 1975.

[3] O. Kresse and L. Truskinovsky. Hamiltonian dynamics of a driven asymmetric kinks: an exact solution. *Physica D*, 2003. Submitted.

[4] O. Kresse and L. Truskinovsky. Mobility of lattice defects: discrete and continuum approaches. *J. Mech. Phys. Solids*, 51(7):1305–1332, 2003.

[5] S.-C. Ngan and L. Truskinovsky. Thermal trapping and kinetics of martensitic phase boundaries. *J. Mech. Phys. Solids*, 47:141–172, 1999.

[6] S.-C. Ngan and L. Truskinovsky. Thermoelastic aspects of nucleation in solids. *J. Mech. Phys. Solids*, 50:1193–1229, 2002.

[7] L. I. Slepyan. Dynamics of a crack in a lattice. *Sov. Phys. Dokl.*, 26(5):538–540, 1981.

[8] L. I. Slepyan. Feeding and dissipative waves in fracture and phase transition II. Phase-transition waves. *J. Mech. Phys. Solids*, 49:513–550, 2001.

[9] L. I. Slepyan and L. V. Troyankina. Fracture wave in a chain structure. *J. Appl. Mech. Techn. Phys.*, 25(6):921–927, 1984.

[10] L. Truskinovsky. Equilibrium interphase boundaries. *Sov. Phys. Dokl.*, 27:306–331, 1982.

[11] L. Truskinovsky. Dynamics of nonequilibrium phase boundaries in a heat conducting elastic medium. *J. Appl. Math. Mech.*, 51:777–784, 1987.

[12] L. Truskinovsky. Kinks versus shocks. In *Shock Induced Transitions and Phase Structures in General Media*, pages 185–229. Springer-Verlag, New York, 1993.

[13] L. Truskinovsky and A. Vainchtein. 2003. In preparation.

[14] L. Truskinovsky and A. Vainchtein. Peierls-Nabarro landscape for martensitic phase transitions. *Phys. Rev. B*, 67:172103, 2003.

III

GROWTH & BIOMECHANICS

Chapter 6

SURFACE AND BULK GROWTH UNIFIED

Antonio DiCarlo
Università degli Studi "Roma Tre"
Mathematical Structures of Materials Physics at DiS
Via Vito Volterra, 62 I-00146 Roma, Italy
adicarlo@mac.com

Solid surfaces can have their physical area changed in two ways, either by creating or destroying surface without changing surface structure and properties per unit area, or by an elastic strain ... along the surface keeping the number of surface lattice sites constant ... .

—J.W. Cahn, 1980

Abstract I have been puzzled for a long time by the unnatural divide between the theory of bulk growth—strikingly underdeveloped—and that for surface growth—much better developed, along apparently independent lines.

Recent advances in growth mechanics (DiCarlo and Quiligotti, 2002) make it now possible to subsume growth phenomena of both kinds under one and the same format, where surface growth is obtained as an infinitely intense bulk growth confined in a layer of vanishingly small thickness.

This has allowed me to recover the results collected in Gurtin, 2000 from the standpoint of DiCarlo and Quiligotti, 2002. In particular, I am able to construe Gurtin's technique of referential control volumes that evolve in time as a special application of the principle of virtual power.

1. INTRODUCTION

Any continuum theory of growth—be it modelled as spread in bulk or concentrated on surfaces—hinges on two key issues: in kinematics, extra degrees of freedom have to be introduced, in order to distinguish growth from deformation; in dynamics, new balance laws have to be provided, apt to govern the evolution of such degrees of freedom.

Both issues are much subtler for bulk than for surface growth, however, demanding a sharper veering from the customary route in continuum mechanics. In my mind, here lies the basic difficulty which hindered surface and bulk growth theories from progressing on a par, within a common frame.

The theory of bulk growth which I (still) prefer is the one set forth in DiCarlo and Quiligotti, 2002.[1] It is the only one I am aware of where the evolution law for bulk growth is obtained as a constitutively augmented balance. In order to make this paper reasonably self-contained, Sect. 2 offers a résumé of that theory, adapted to the present discussion.

As a paradigm for surface growth theories, I take the tract by Gurtin (Gurtin, 2000): it is a convenient reference book, where the role of an independent configurational balance is duly stressed—from a point of view, I may add, which in several respects differs from my own.

The quote from Cahn, 1980 I put in the prologue is a paraphrase of Gibb's discussion of multiphase equilibria. I took it from the introductory chapter of Gurtin, 2000, where it is completed by the following comment from the author: "The creation of surface involves configurational forces, while stretching the surface involves standard forces." Despite the fact that Gibbs, Cahn and Gurtin had only surface phenomena in mind (namely, the evolution of phase interfaces), it is perfectly legitimate—and much to the point—to rephrase the whole statement in terms of bulk phenomena, as follows (italicized words are my own):

> Solid *bodies* can have their physical *volume* changed in two ways, either by creating or destroying *bulk* without changing *bulk* structure and properties per unit *volume*, or by an elastic strain *within the body* keeping the number of *bulk* lattice sites constant. The creation of *bulk* involves configurational *couples*, while stretching the *body* involves standard forces.

By *couple* I mean a tensor quantity—not necessarily skew—having the physical dimensions of length×force. Why configurational couples are germane to bulk growth is explained in Sect. 2.2; explaining how they match with configurational forces, proper to surface growth, is one of the main aims of this paper.

2. A CONTINUUM THEORY OF BULK GROWTH

The growing bodies considered here are standard Cauchy continua: the only kinematic descriptor ascribed to their points is place in ordinary physical space. In order to distinguish growth from deformation, *two* evolving configurations are associated with each body element: its *current* configuration, describing how it is actually placed in space, and its *relaxed* configuration, describing how it "would like" to be placed. The field of relaxed configurations need not be (and usually is not) compatible, not even locally.

This is a good old kinematic idea, primarily introduced to distinguish between elastic and viscoplastic strains by Kröner, 1960 and Lee, 1969, and much later imported into growth modelling by Rodriguez et al., 1994 (see also Taber, 1995). The original contribution by DiCarlo and Quiligotti, 2002 is in dynamics. As summarized in the following, we obtain the evolution law for bulk growth as a constitutively augmented *new* balance, the balance of configurational (or remodelling) couples, *independent* of the standard force balance.

2.1. KINEMATICS

We regard a body as a smooth manifold $\mathcal{B}$ (with boundary $\partial\mathcal{B}$), and call **placement** any smooth embedding

$$p : \mathcal{B} \to \mathcal{E} \tag{1}$$

of the body into the Euclidean place manifold $\mathcal{E}$, whose translation space will be denoted by $\mathrm{V}\mathcal{E}$. Tangent vectors on the body manifold itself are called **line elements**. The set of all line elements attached to a single body-point $b \in \mathcal{B}$ is called the **body element** at b, and denoted $\mathrm{T}_b\mathcal{B}$ (the *tangent space* to $\mathcal{B}$ at b). The union of all body elements is denoted $\mathrm{T}\mathcal{B}$ (the *tangent bundle* of $\mathcal{B}$).

The **body gradient** ∇p of a placement p is a tensor field on $\mathcal{B}$, whose value at any given point b, denoted by $\nabla p|_b$, maps linearly the body element $\mathrm{T}_b\mathcal{B}$ onto $\mathrm{V}\mathcal{E}$. We call **stance** any tensor field of this kind, *be it a gradient or not*. Therefore, a stance is any smooth mapping

$$\mathrm{P} : \mathrm{T}\mathcal{B} \to \mathrm{V}\mathcal{E}\,, \tag{2}$$

such that the restriction $\mathrm{P}|\mathrm{T}_b\mathcal{B}$ is a linear embedding, for all $b \in \mathcal{B}$. If a stance happens to be the gradient of a placement, we say that it is **induced** by that placement: all placement induces a stance, but a general stance is not induced by any placement, not even locally. We describe growth by the time evolution of the **relaxed stance** $\mathbb{P}$, while motion is described as the time evolution of the **actual placement** p.

The **complete motion** of a growing body is a family of pairs $(p, \mathbb{P})$ smoothly parametrized by the *time line* $\mathcal{T}$ (identified with the real line), and the velocity **realized** along that motion at the time $\tau \in \mathcal{T}$ is the pair of fields (a superposed dot denoting time differentiation):

$$(\dot{p}(\tau), \dot{\mathbb{P}}(\tau)\mathbb{P}(\tau)^{-1}) : \mathcal{B} \to \mathcal{V}\mathcal{E} \times (\mathcal{V}\mathcal{E} \otimes \mathcal{V}\mathcal{E}) \,. \tag{3}$$

The linear space of **test velocities** $\mathfrak{T}$, comprising all smooth fields

$$(\mathrm{v}, \mathbb{V}) : \mathcal{B} \to \mathcal{V}\mathcal{E} \times (\mathcal{V}\mathcal{E} \otimes \mathcal{V}\mathcal{E}) \,, \tag{4}$$

will play a central role in the next subsection. The **visible velocity** of body-points (with physical dimensions length/time) is given by the vector field v, while the *tensor* field $\mathbb{V}$ gives the **growth velocity** of the corresponding body elements (with physical dimensions 1/time).

2.2. DYNAMICS

To us, a **force** is primarily a continuous linear real-valued *functional* on the space of test velocities, whose *value* we call the **working** expended by that force.[2] We assume that the total working expended on any test velocity $(\mathrm{v}, \mathbb{V}) \in \mathfrak{T}$ admits the following integral representation:

$$\int_{\mathcal{B}} -(\mathrm{s} \cdot \mathrm{v} + \mathbb{C} \cdot \mathbb{V} + \mathbb{S} \cdot \mathrm{D}\mathrm{v}) + \int_{\mathcal{B}} (\mathrm{b} \cdot \mathrm{v} + \mathbb{B} \cdot \mathbb{V}) + \int_{\partial\mathcal{B}} \mathrm{t}_{\partial\mathcal{B}} \cdot \mathrm{v} \,, \tag{5}$$

where the integrals are taken with respect to the *relaxed* volume and surface area of body elements, and D denotes the *relaxed* gradient:

$$\mathrm{D}\mathrm{v} := (\nabla \mathrm{v})\mathbb{P}^{-1} \,. \tag{6}$$

Because of the compound structure of test velocities (4), the force functional splits additively into a **brute force**, dual to v, and a **remodelling force**, dual to $\mathbb{V}$. Another important splitting is between the **inner** working, given by the first bulk integral in (5), and the **outer** working, given by the remaining sum. The **brute self-force** per unit volume s, the **outer brute bulk-force** per unit volume b, and the **brute boundary-force** per unit area $\mathrm{t}_{\partial\mathcal{B}}$ take values in $\mathcal{V}\mathcal{E}$; the **remodelling self-couple** per unit volume $\mathbb{C}$, the **brute Piola stress** $\mathbb{S}$ (also a specific couple!), and the **outer remodelling couple** per unit volume $\mathbb{B}$ take values in $\mathcal{V}\mathcal{E} \otimes \mathcal{V}\mathcal{E}$.

All balance laws are systematically provided by the **principle of null working**: the total working expended on any test velocity should be zero, i.e., the total force should be the null functional. Skipping the balance of brute forces, which is standard, I present here only the **balance of remodelling couples**:

$$-\mathbb{C} + \mathbb{B} = 0 \quad \text{on } \mathcal{B}. \tag{7}$$

2.3. CONSTITUTIVE THEORY

Our treatment of constitutive issues rests on two pillars (altogether independent of balance): the *principle of material indifference* to change in observer, and the *dissipation principle.* Both of them deliver strict selection rules on the constitutive prescription for the *inner* force. Such *a priori* restrictions do not apply to the outer force, which is regarded as an adjustable control on the process. This inner/outer dichotomy does not pertain to the physics of interactions, but to the limitations of the model: what appears as an outer interaction within a given theory may always be accounted for—in principle—as an inner interaction within a broader and more cumbersome theory. In an all-embracing model there would be no outer interactions at all. In our model of growth mechanics, the *outer* remodelling couple $\mathbb{B}$ has a determinative role whenever growth—be it surface or bulk—is powerfully controlled by non-mechanical phenomena, such as biochemical reactions in living tissues.

I will only flash the outcome of the first principle, as applied in DiCarlo and Quiligotti, 2002, while summarizing with some more detail the machinery of the second one. Material indifference rules out non-trivial values of the brute self-force s and non-symmetric values of the *Cauchy stress* $\mathrm{T} := |\det \mathrm{F}|^{-1}\, \mathbb{S}\, \mathrm{F}^{\mathsf{T}}$, where the **warp**

$$\mathrm{F} := \mathrm{D}p = (\nabla p)\, \mathbb{P}^{-1}. \tag{8}$$

measures how the actual stance, i.e., the body gradient of the actual placement, differs from the relaxed stance. If we further assume that the response of the body element at b filters off from $(p, \mathbb{P})$ all information other than $p|_b$, $\nabla p|_b$, and $\mathbb{P}|_b$, we obtain from the same principle that

$$\mathbb{S}(b,\tau) = \mathrm{R}(b,\tau)\, \check{\mathbb{S}}_b(\mathrm{U}|_b, \mathbb{P}|_b, \tau)\,, \qquad \mathbb{C}(b,\tau) = \check{\mathbb{C}}_b(\mathrm{U}|_b, \mathbb{P}|_b, \tau)\,, \tag{9}$$

the *rotation* R and the *stretch* U being, respectively, the orthogonal and the right positive-symmetric factor of the warp (8): $\mathrm{F} = \mathrm{R}\,\mathrm{U}$.

To have a notion of dissipation, an additional energetic descriptor is needed. We postulate the existence of an additive real-valued *free energy* $\Psi(\mathcal{P}) = \int_{\mathcal{P}} \psi$, measuring the inner energy available to body-parts. Since we take integrals over body-parts with respect to the relaxed volume, ψ represents the free energy per unit *relaxed* volume. We call *power expended* along a process at time τ the *opposite* of the working expended by the inner force due to the process on the velocity *realized* at time τ. Hence, the power expended measures the working done by an *outer* force *balanced* with the *constitutively determined* inner force. The **dissipation principle** we enforce requires that the rate of energy dissipation—defined as the difference between the power expended along a process and the time derivative of the free energy—should be non-negative, for all body-parts, at all times.[3] This localizes into:

$$\dot{\psi} + \psi \mathrm{I} \cdot \mathbb{V} \leq \mathbb{S} \cdot \mathrm{Dv} + \mathbb{C} \cdot \mathbb{V}, \tag{10}$$

it being intended that v and $\mathbb{V}$ are given by (3), $\mathbb{S}$ and $\mathbb{C}$ by (9), and ψ has to be related to the process by an extra constitutive mapping. Our **main constitutive assumption** selects a rather special, but very interesting constitutive class, beautifully accounted for in Epstein, 1999. We posit that, at each body-point, the present value of the free energy per unit relaxed volume depends only on the *present* value of the *warp* at that point:

$$\psi(b, \tau) = \check{\psi}_b\,(\mathrm{F}(b, \tau))\,. \tag{11}$$

The requirement that (10) be satisfied along all processes is fulfilled if and only if for each b (which will be dropped from now on) the responses $\check{\mathbb{S}}$ and $\check{\mathbb{C}}$ satisfy (∂ denotes differentiation):

$$\check{\mathbb{S}} = \partial \check{\psi} + \overset{+}{\mathbb{S}}\,, \qquad \check{\mathbb{C}} = \mathbb{E} + \overset{+}{\mathbb{C}}\,, \tag{12}$$

with

$$\mathbb{E} := \check{\psi}\mathrm{I} - \mathrm{F}^{\mathsf{T}}\check{\mathbb{S}} \tag{13}$$

the **Eshelby coupling** between brute mechanics and remodelling, and the *extra-energetic responses* $\overset{+}{\mathbb{S}}$, $\overset{+}{\mathbb{C}}$ restricted by the **reduced dissipation inequality**

$$\overset{+}{\mathbb{S}} \cdot \dot{\mathrm{F}} + \overset{+}{\mathbb{C}} \cdot \mathbb{V} \geq 0\,, \tag{14}$$

to be abided by in the same sense as (10). As is seen, the Eshelbian coupling is mandatory (within the constitutive class we are considering) and independent of any special assumption on $\check{\mathbb{S}}$. Additional couplings—through the outer remodelling couple $\mathbb{B}$, in particular—are *not* ruled out.

It should be stressed that $\mathbb{B}$, while trivial in most—if not all—applications to "dead" engineering materials, plays a major role even in the simplest biomechanical applications, where it describes the mechanical feedback from the biochemical control system: think of so-called *stress-dependent growth laws* (Taber, 1995). Therefore, the idea—emphasized in Epstein and Maugin, 2000—that the Eshelby coupling is *the* "driving force of irreversible growth" is untenable for any smartly controlled material. The Eshelby coupling *by itself* does not drive any smart growth; rather, it drives a dull—though nontrivial—visco-plastic flow, as shown in DiCarlo, Nardinocchi and Teresi (forthcoming).

3. PIECEWISE COMPATIBLE BULK GROWTH

Let $\mathfrak{C}_{\mathrm{P}}$ be a collection of open disjoint *patches*, whose closures cover the body manifold $\mathcal{B}$, and $\mathfrak{C}_{\mathrm{F}}$ the corresponding collection of *interfaces*. In order to cover also the body boundary $\partial\mathcal{B}$ with (closures of) interfaces in $\mathfrak{C}_{\mathrm{F}}$, it is convenient to augment $\mathfrak{C}_{\mathrm{P}}$ with an idle patch $\mathcal{P}_{ext}$, the *exterior* of $\mathcal{B}$.

Now, enforce a *local compatibility constraint* on the relaxed stance $\mathbb{P}$, requiring that on each patch $\mathcal{P} \in \mathfrak{C}_{\mathrm{P}}$ there is a local placement

$$\boldsymbol{p}_{\mathcal{P}} : \mathcal{P} \to \mathcal{E}, \tag{15}$$

smooth up to $\partial\mathcal{P}$, such that

$$\mathbb{P}|\mathcal{P} = \nabla \boldsymbol{p}_{\mathcal{P}}. \tag{16}$$

Parallely, restrict the testing of the remodelling force to *piecewise compatible* growth velocities, i.e., on tensor fields $\mathbb{V}$ that, on each patch $\mathcal{P} \in \mathfrak{C}_{\mathrm{P}}$, are the relaxed gradient of a vector field $\mathbf{w}_{\mathcal{P}}$, smooth up to $\partial\mathcal{P}$:

$$\mathbb{V}|\mathcal{P} = \mathrm{D}\mathbf{w}_{\mathcal{P}} = (\nabla \mathbf{w}_{\mathcal{P}})\,\mathbb{P}^{-1}. \tag{17}$$

Then, denoting by $\mathbb{A} := \mathbb{B} - \mathbb{C}$ the total remodelling couple per unit relaxed volume ($\mathbb{A} = 0$ in $\mathcal{P}_{ext}$), the working expended on $\mathbb{V}$ is given by

$$\begin{aligned}\int_{\mathcal{B}} \mathbb{A}\cdot\mathbb{V} &= \sum_{\mathcal{P}\in\mathfrak{C}_{\mathrm{P}}} \int_{\mathcal{P}} \mathbb{A}\cdot\mathbb{V} = \sum_{\mathcal{P}\in\mathfrak{C}_{\mathrm{P}}} \int_{\mathcal{P}} \mathbb{A}\cdot(\mathrm{D}\mathbf{w}_{\mathcal{P}}) \\ &= \sum_{\mathcal{P}\in\mathfrak{C}_{\mathrm{P}}} \left(\int_{\partial\mathcal{P}} (\mathbb{A}\,\mathbf{n}_{\partial\mathcal{P}})\cdot\mathbf{w}_{\mathcal{P}} - \int_{\mathcal{P}} (\mathrm{Div}\,\mathbb{A})\cdot\mathbf{w}_{\mathcal{P}} \right),\end{aligned} \tag{18}$$

with $\mathbf{n}_{\partial\mathcal{P}}$ the *outward* unit normal to the relaxed shape of $\partial\mathcal{P}$. Hence, the principle of null working yields the following balances: for each $\mathcal{P} \in \mathfrak{C}_{\mathrm{P}}$,

$$\mathrm{Div}\,(\mathbb{E} + \overset{+}{\mathbb{C}} - \mathbb{B}) = 0 \quad \text{on } \mathcal{P}, \tag{19a}$$

$$(\mathbb{E} + \overset{+}{\mathbb{C}} - \mathbb{B})\,\mathbf{n}_{\partial\mathcal{P}} = 0 \quad \text{on } \partial\mathcal{P}. \tag{19b}$$

This set compares with, but is weaker than, (7) plus (9_2) and (12_2).

Eq. (19a) matches with the *configurational force balance* (5–10) on page 37 of Gurtin, 2000, provided that his *configurational stress* $\boldsymbol{C}$ is identified with $\mathbb{E}$—as implied by the correspondence between (13) and the *Eshelby relation* (6–9) on page 43—, and his *configurational body forces*, *internal* $\boldsymbol{g}$ and *external* $\boldsymbol{e}$, are related to my inner and outer remodelling couples by

$$\boldsymbol{g} \leftarrow \operatorname{Div} \overset{+}{\mathbb{C}}\,, \qquad \boldsymbol{e} \leftarrow -\operatorname{Div} \mathbb{B} \tag{20}$$

Here and in the following, *assignment statements* such as

$$q_{\,surf} \leftarrow T\, q_{\,bulk} \tag{21}$$

should be read as meaning that the constitutive assignment for quantity $q_{\,surf}$ in surface-growth theory is obtained as the T-image of the constitutive assignment for quantity $q_{\,bulk}$ in bulk-growth theory.

Analogously, (19b) agrees (modulo an obvious change in notation) with the *interfacial force balance* (7.7_2) given by Cermelli and Gurtin, 1994 (Part B: theory of *incoherent* interfaces *without* interfacial structure), provided that their *interfacial configurational forces*, *internal* $\mathbf{e}_{\partial\mathcal{P}}$ and *external* $\boldsymbol{f}_{\partial\mathcal{P}}$, are assigned according to

$$\mathbf{e}_{\partial\mathcal{P}} \leftarrow \overset{+}{\mathbb{C}}\,\mathbf{n}_{\partial\mathcal{P}}\,, \qquad \boldsymbol{f}_{\partial\mathcal{P}} \leftarrow -\mathbb{B}\,\mathbf{n}_{\partial\mathcal{P}} \tag{22}$$

Imitating Gurtin, 2000, I will now concentrate on coherent interfaces.

4. COHERENT INTERFACES

Interface coherency is obtained by enforcing the further constraint that the local relaxed placement in any two adjoining patches $\mathcal{P}_+$, $\mathcal{P}_-$ be *continuous* across the interface $\mathcal{S} := \partial\mathcal{P}_+ \cap \partial\mathcal{P}_-$: the jump across $\mathcal{S}$, $[\![\boldsymbol{p}]\!]_{\mathcal{S}} := \boldsymbol{p}_{\mathcal{S}}^+ - \boldsymbol{p}_{\mathcal{S}}^-$ (the limit from $\mathcal{P}_+$ minus that from $\mathcal{P}_-$), should vanish. If test velocities are analogously restricted by the condition $[\![\mathbf{w}]\!]_{\mathcal{S}} = 0$ on all $\mathcal{S} \in \mathfrak{C}_{\mathrm{F}}$, then the integrals over patch boundaries in (18) yield

$$\sum_{\mathcal{P}\in\mathfrak{C}_{\mathrm{P}}} \int_{\partial\mathcal{P}} (\mathbb{A}\,\mathbf{n}_{\partial\mathcal{P}}) \cdot \mathbf{w}_{\mathcal{P}} = -\sum_{\mathcal{S}\in\mathfrak{C}_{\mathrm{F}}} \int_{\mathcal{S}} ([\![\mathbb{A}]\!]_{\mathcal{S}}\mathbf{m}_{\mathcal{S}}) \cdot \mathbf{w}_{\mathcal{S}}\,, \tag{23}$$

where $\mathbf{w}_{\mathcal{S}} := \mathbf{w}_{\mathcal{S}}^+ = \mathbf{w}_{\mathcal{S}}^-$, $\mathbf{m}_{\mathcal{S}} := \mathbf{n}_{\partial\mathcal{P}_-} = -\mathbf{n}_{\partial\mathcal{P}_+}$ (pay attention to signs and draw a sketch!). Hence, on each $\mathcal{S} \in \mathfrak{C}_{\mathrm{F}}$,

$$[\![\mathbb{E} + \overset{+}{\mathbb{C}} - \mathbb{B}]\!]_{\mathcal{S}}\,\mathbf{m}_{\mathcal{S}} = 0\,, \tag{24}$$

in accord with the *configurational force balance at the interface* (11–8b) on page 68 of Gurtin, 2000, provided his *interface configurational forces, internal* $g^{\mathcal{S}}$ and *external* $e^{\mathcal{S}}$, be given by

$$g^{\mathcal{S}} \leftarrow [\![\overset{+}{\mathbb{C}}]\!]_{\mathcal{S}} \mathbf{m}_{\mathcal{S}}\,, \qquad e^{\mathcal{S}} \leftarrow -[\![\mathbb{B}]\!]_{\mathcal{S}} \mathbf{m}_{\mathcal{S}}\,. \tag{25}$$

To obtain a genuine surface-growth theory, the growth process should be further specialized, confining the *realized* growth velocity $\mathbb{V} = \dot{\mathbb{P}}\mathbb{P}^{-1}$ (cf. (3)) in thin layers around interfaces. For $\varepsilon > 0$ small enough, let

$$\mathcal{L}_\varepsilon := \{\, \boldsymbol{p}_{\mathcal{S}}(b) + \zeta\, \mathbf{m}_{\mathcal{S}}(b) \mid b \in \mathcal{S}\,, -\varepsilon < \zeta < \varepsilon \,\} \tag{26}$$

be the ε-thickening of the relaxed shape of an interface $\mathcal{S}$, and $\mathcal{L}_\varepsilon^+$, $\mathcal{L}_\varepsilon^-$ its upper ($\zeta > 0$) and lower ($\zeta < 0$) halves, respectively. Assume then (my definition of choice for tensor multiplication implying that $(\mathbf{m}\otimes\mathbf{w})\,\mathbf{r} = (\mathbf{m}\cdot\mathbf{r})\,\mathbf{w}$):

$$\mathbb{V} = \pm\varepsilon^{-1}\,\mathbf{m}_{\mathcal{S}} \otimes \mathbf{w}_{\mathcal{S}} \quad \text{on } \mathcal{L}_\varepsilon^{\mp}\,, \quad \mathbb{V} = 0 \quad \text{elsewhere.} \tag{27}$$

Notice that the minus sign prevails in $\mathcal{L}_\varepsilon^+$, and vice versa: the tensor field (27) is, to within $O(1)$ terms for $\varepsilon \downarrow 0$ on most of $\mathcal{L}_\varepsilon$, the relaxed gradient of the vector field given by $(b,\zeta) \mapsto (\varepsilon - |\zeta|\,)\,\mathbf{w}_{\mathcal{S}}(b)$ inside $\mathcal{L}_\varepsilon$ and vanishing outside it. The remodelling force, when evaluated on test fields having the structure (27), in the limit for $\varepsilon \downarrow 0$ yields a pure surface working:

$$\lim_{\varepsilon\downarrow 0} \int_{\mathcal{B}} \mathbb{A}\cdot\mathbb{V} = -\sum_{\mathcal{S}\in\mathfrak{C}_{\mathcal{F}}} \int_{\mathcal{S}} (\,[\![\mathbb{A}]\!]_{\mathcal{S}} \mathbf{m}_{\mathcal{S}}) \cdot \mathbf{w}_{\mathcal{S}}\,, \tag{28}$$

in agreement with (23) and (18), where the bulk integrals appearing in the last sum asymptotically vanish. Eqs. (27) and (28) explain how the remodelling *couples* dominating bulk growth induce configurational *forces*—in the sense of Gurtin—in the limit theory of surface growth (cf. (25)).

I now integrate the reduced dissipation inequality (14) over $\mathcal{L}_\varepsilon$ and take the limit for $\varepsilon \downarrow 0$, assuming that the brute dissipation per unit volume $\overset{+}{\mathbb{S}}\cdot\dot{\mathrm{F}}$ is $O(1)$. I thus obtain the **surface dissipation inequality**

$$(\,[\![\overset{+}{\mathbb{C}}]\!]_{\mathcal{S}} \mathbf{m}_{\mathcal{S}}) \cdot \mathbf{w}_{\mathcal{S}} = g^{\mathcal{S}} \cdot \mathbf{w}_{\mathcal{S}} \leq 0 \tag{29}$$

(recall (25_1)). Now, if *invariance under reparametrization* of the interface $\mathcal{S}$ is required, implying that the working depends only on the (scalar) *normal velocity* of the interface $V_S := \mathbf{w}_S \cdot \mathbf{m}_S$, then the internal configurational force on the interface is necessarily *normal:*

$$\boldsymbol{g}^S = g^S \, \mathbf{m}_S \,, \tag{30}$$

with g^S scalar-valued (cf. (11–11) on page 69 of Gurtin, 2000), and (29) coincides with the *interfacial dissipation inequality* (11–21) on page 71 of Gurtin, 2000: $g^S \, V_S \leq 0$.

5. CONCLUDING REMARKS

Lack of space prevents me from treating theories with interfacial structure, intersecting interfaces (junctions), or growing cracks (fracture). To allow for interfacial structure, the integral representation of the working (5) should be extended to include *measures* concentrated on interfaces, both for brute and remodelling forces; in (18), this would bring out an extra sum over $\mathfrak{C}_{\mathrm{F}}$ of area integrals over interfaces. Junctions and cracks require concentrations on subsets of higher *codimension:* the intersection of two or more interfaces, or the crack tip.

As a closing remark, let me quote the introduction to Gurtin, 2000: "Indetermination arises in the configurational system whenever there is no change in material structure." This is exactly why the theory of configurational forces, applied to changes in material structure confined to meagre subsets of the body manifold, is fraught with indeterminacy. The *bulk* balance (19a) (coincident with (5–10) on page 37 of Gurtin, 2000), while much weaker than (7), is altogether superfluous in *surface* growth, being identically satisfied by a *reactive* internal body force $\boldsymbol{g}$ (introduced in (20)), which is the reaction to the constraint of no structural change in the bulk. This does not help in advocating the independence and usefulness of the configurational force system. However, bulk growth gives us our revenge.

optional

Acknowledgments

I am grateful to the colloquium organisers, to the discussants at the colloquium and to the anonymous referee for prompting me, in various ways, to organize my ideas and to write them down. My work was supported by GNFM-INdAM (the Italian Group for Mathematical Physics), and by MIUR (the Italian Ministry of Education, University and Research) through the Project "Mathematical Models for Materials Science."

Notes

1. This is not at all obvious, since I can change my mind. But in this respect I did not. My coauthor did in Quiligotti, 2002, on the (not minor) issue of invariance requirements, convinced by an argument intimated by Green and Naghdi, 1971 and spelled out in Casey and Naghdi, 1980 and 1981, Casey, 1987, and elsewhere. For reasons I expound in DiCarlo (in preparation), I find that old argument faulty and misleading.

2. The physical dimensions of the working are God given: energy/time. Since the prototypal space of test velocities is $\mathcal{VE}$, with physical dimensions length/time, forces get typically identified with elements of $\mathcal{VE}$ itself, with physical dimensions energy/length (= "force" *by definition!*). As velocities have *not* dimensions length/time, in general (think of growth velocity $\mathbb{V}$), a general force has *not* dimensions "force". Another force related issue is that the space $\mathfrak{T}$ of test velocities should be endowed with the structure of a *topological* vector space, so as to make meaningful the physically essential requirement that force functionals be *continuous*. I leave gaps like this to be filled in by the mathematically conscious reader.

3. Due account is to be taken of the fact that the relaxed-volume form, let's say μ, *evolves* in time, as dictated by the growth velocity $\mathbb{V} = \dot{\mathbb{P}}\mathbb{P}^{-1}$: $\dot{\Psi}(\mathcal{P}) = (\int_{\mathcal{P}} \psi\, \mu)^{\cdot} = \int_{\mathcal{P}} (\psi\, \mu)^{\cdot} = \int_{\mathcal{P}} (\dot{\psi}\, \mu + \psi\, \dot{\mu}) = \int_{\mathcal{P}} (\dot{\psi} + \psi\, \mathrm{I}\cdot\mathbb{V})\, \mu$ (I denotes the identity on $\mathcal{VE}$; $\mathrm{I}\cdot\mathbb{V}$ is the trace of $\mathbb{V}$).

References

Cahn, John W. (1980). "Surface stress and the chemical equilibrium of small crystals. 1. The case of the isotropic surface," Act. Metall. **28**, 1333–1338.

Casey, James. (1987). "Invariance considerations in large strain elasto-plasticity," ASME J. Appl. Mech. **54**, 247.

Casey, James, and Paul M. Naghdi. (1980). "A remark on the use of the decomposition $\mathsf{F} = \mathsf{F}_e \mathsf{F}_p$ in plasticity," ASME J. Appl. Mech. **47**, 672–675.

Casey, James, and Paul M. Naghdi. (1981). "A correct definition of elastic and plastic deformation and its computational significance," ASME J. Appl. Mech. **48**, 683–984.

Cermelli, Paolo, and Morton E. Gurtin. (1994). "The dynamics of solid-solid phase transitions. 2. Incoherent interfaces," Arch. Rational Mech. Anal. **127**, 41–99.

DiCarlo, Antonio (in preparation). *Material Remodelling.*

DiCarlo, Antonio, Paola Nardinocchi, and Luciano Teresi (forthcoming). "Creep as passive growth."

DiCarlo, Antonio, and Sara Quiligotti. (2002). "Growth and balance," Mech. Res. Comm. **29**, 449–456.

Epstein, Marcelo. (1999). "On material evolution laws," in *Geometry, Continua & Microstructure* (G.A. Maugin ed.). Paris: Hermann, 1–9.

Epstein, Marcelo, and Gérard A. Maugin. (2000). "Thermomechanics of volumetric growth in uniform bodies," Int. J. Plast. **16**, 951–978.

Green, Albert E. and Paul M. Naghdi. (1971). "Some remarks on elastic-plastic deformation at finite strain," Int. J. Engng. Sci. **9**, 1219–1229.

Gurtin, Morton E. (2000). *Configurational Forces as Basic Concepts of Continuum Physics*. Berlin: Springer-Verlag.

Kröner, Ekkehart. (1960). "Allgemeine Kontinuumstheorie der Versetzungen und Eigenspannungen," Arch. Rational Mech. Anal. **4**, 273–334.

Lee, Erastus H. (1969). "Elastic-plastic deformations at finite strains," J. Appl. Mechanics **36**, 1–6.

Quiligotti, Sara. (2002). "On bulk growth mechanics of solid-fluid mixtures: kinematics and invariance requirements," Theor. Appl. Mech. **28/29**, 277–288.

Rodriguez, Edward K., Anne Hoger, and Andrew D. McCulloch. (1994). "Stress-dependent finite growth in soft elastic tissues," J. Biomechanics **27**, 455–467.

Taber, Larry A. (1995). "Biomechanics of growth, remodeling, and morphogenesis," Appl. Mech. Rev. **48**, 487–545.

Chapter 7

MECHANICAL AND THERMODYNAMICAL MODELLING OF TISSUE GROWTH USING DOMAIN DERIVATION TECHNIQUES

Jean Francois Ganghoffer
LEMTA – ENSEM, 2, Avenue de la Forêt de Haye, BP 160, 54054 Vandoeuvre CEDEX, France.
jfgangho@ensem.inpl-nancy.fr

1. INTRODUCTION

One outstanding problem in developmental biology is the understanding of the factors (genetic and epigenetic, such as strain and stress) that may promote the generation of biological form, involving growth (change of mass), remodeling (change of properties), and morphogenesis (shape changes), [1]. Contributions [2, 3, 4] analyze the problem of growth in terms of the evolution of a growth tensor, associated to a *natural configuration* of the living body. The growth of solids is one topic of configurational mechanics : mention for instance the 'accretive forces' introduced by Gurtin [6,7] to account for a microstructure, who complemented the more classical mechanical forces by *configurational forces*. Adopting herewith a different point of view, the thermomechanics of growth is analysed on the basis of the variation of the reference configuration, when mass is locally added to the material points.

2. KINEMATICS OF GROWTH

Considering the growing material as a single phase continuum, in which growth is envisaged as an increase of mass of the existing particles (at constant particle number) [4], the particles (thought as a population of cells in the case of biological tissues) can be labeled in any configuration of the body, thus a motion that connects all configurations can be defined. An intermediate configuration Ω_g due to growth (in between Ω_0 and Ω_t) is further introduced, so that a mapping between the material point $\underline{X}$ and its counterpart in Ω_g (after growth) $\underline{x}_g$ exists [3,6], given by

$$\underline{X} \in \Omega_0, t \in R^+ \mapsto \underline{x}_g := \underline{X} + \underline{U}_g(\underline{X}, t) \tag{2.1}$$

with $\underline{U}_g(\underline{X}, t)$ the growth displacement field and t a time-like parameter. The vector field $\underline{U}_g(\underline{X}, t)$ is in general non compatible at the macroscopic scale of the whole body, since a discontinuity or a superposition of material may occur due to the growth : thus, relation (2.1) establishes $\underline{x}_g$ as an *anholomic vector field.* However, since one can define the differential of the displacement between two non holonomic coordinate systems, the variation of the position in Ω_g is given by

$$\delta \underline{x}_{C_g} = \underline{\underline{F}}_g . \delta \underline{X} \tag{2.2}$$

introducing thereby the growth transformation gradient $\underline{\underline{F}}_g$. An additional growth accommodation tensor $\underline{\underline{F}}_a$ is needed to restore the continuity of the global displacement field over the whole body (see section 5), satisfying the multiplicative decomposition of the total deformation gradient $\underline{\underline{F}} := \nabla_X \underline{x}$:

$$\underline{\underline{F}} = \underline{\underline{F}}_a . \underline{\underline{F}}_g \tag{2.3}$$

The total, growth and accommodation strain tensors are respectively :

$$\underline{\underline{E}} := \frac{1}{2}\left(\underline{\underline{F}}^t . \underline{\underline{F}} - I\right) ; \quad \underline{\underline{E}}_g := \frac{1}{2}\left(\underline{\underline{F}}_g^t . \underline{\underline{F}}_g - I\right) ; \quad \underline{\underline{E}}_a := \frac{1}{2}\left(\underline{\underline{F}}_a^t . \underline{\underline{F}}_a - I\right)$$

The rate of the growth tensor is the gradient of the growth velocity field $\underline{V}_g$, being itself the material derivative of the growth displacement field $\underline{U}_g$. Considering domain derivation, the derivation of a functional (or a function) with respect to a domain Ω_t having its own motion at a velocity $\underline{W}$ consists in the domain derivative of the integral $B = \int_{\Omega_t} b(\underline{x}, t) dx$, as

$$\frac{\delta B}{\delta t} = \int_{\Omega_t} \frac{\partial b}{\partial t} dx + \int_{S_t} b \underline{W} . \underline{n} d\sigma = \int_{\Omega_t} \frac{db}{dt} dx - \int_{\Omega_t} \underline{\nabla b} . \underline{W} dx + \int_{S_t} b \underline{W} . \underline{n} d\sigma \tag{2.4}$$

with $\underline{n}$ the outward normal to the boundary $S_t = \partial \Omega_t$, see [10].

3. BALANCE LAWS

The growing solid consists of so-called germs of matter, the equivalent at a continuum level of the individual cells within a biological context. The germ is supposed to build a physical entity at a mesoscopic scale of description, exchanging work and matter with the surrounding germs, exerting contact forces on the boundary $\partial\Omega_g$. During growth, matter is added both in the bulk of Ω_g and on the portion S_g of $\partial\Omega_g$; thereby, the domain occupied by the germ changes. Assuming growth does not occur on the whole surface $\partial\Omega_g$, S_g is not closed, thus it has a boundary ∂S_g, fig. 1.

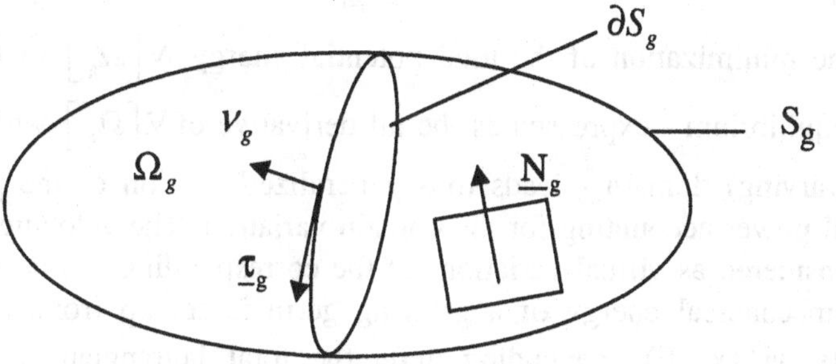

Fig. 1 : Germ of tissue under growth : configuration parameters

The impact of the domain variation on the balance laws shall be next assessed, focusing on the case of isothermal growth. The change of mass is written as the integral $\frac{dm}{dt} = \int_{\Omega_G} \Gamma_g \rho_g dx_g$, with ρ_g the density on the intermediate configuration, and Γ_g the rate of mass variation (having the dimension of the inverse of time). The local mass balance equation is :

$$\frac{\partial \rho_g}{\partial t} + \nabla.\left(\rho_g \underline{V}_g\right) = \Gamma_g \rho_g \tag{3.1}$$

For an open system undergoing chemical reactions, each of the r constituents satisfies the mass balance equation

$$\dot{\rho}_i + \rho_i \operatorname{div}(\underline{v}) + \operatorname{div}\underline{J}_i = \Phi_{pi} + \sum_{j=1}^{r} \upsilon_{ij} R_j \tag{3.2}$$

with $\underline{v}$ the barycentric velocity, $\underline{J}_i := \rho_i(\underline{v}_i - \underline{v})$ a diffusion flux due to the relative velocity $(\underline{v}_i - \underline{v})$ of the individual particles, Φ_{pi} a flux of the ith constituent, and R_j the rate of the jth chemical reaction. The υ_{ij} are the stoechiometric coefficients. Summing up the partial equations (3.2) gives the global mass balance :

$$\dot{\rho} + \rho \operatorname{div}(\underline{v}) = \Phi_\rho + \sum_{i=1}^{r}\sum_{j=1}^{r} \upsilon_{ij} R_j \tag{3.3}$$

Φ_ρ being the total flux of mass. Since $\frac{d\rho}{dt} = \frac{\partial \rho}{\partial t} + \underline{v}.\underline{\operatorname{grad}(\rho)}$, and comparing (3.3) and the eulerian counterpart of (3.1), the rate of mass change is $\Gamma = \frac{1}{\rho}\nabla.(\rho\underline{V}) = \Phi_\rho + \sigma_\rho$, with $\sigma_\rho = \sum_{i=1}^{r}\sum_{j=1}^{r} \upsilon_{ij} R_j$ the source of mass.

The minimization of the total potential energy $V\left[\Omega_g\right]$ of the germ at true equilibrium - expressed as the nil derivative of $V\left[\Omega_g\right]$ with respect to the (varying) domain - leads to a generalized version of the principle of virtual power accounting for the domain variation. The velocities shall here be considered as virtual variations of the corresponding displacements. The total mechanical energy of a growing germ is set up from a volumetric density $\psi^V(\underline{x}_g, \underline{\underline{E}})$, depending upon the total lagrangian strain $\underline{\underline{E}}$ and possibly on the position field $\underline{x}_g$, and a surface density $\psi^S\left(\tilde{\underline{x}}_g, \underline{N}_g, \tilde{\underline{\underline{E}}}, \tilde{\underline{\underline{E}}}_g\right)$, depending upon the surface lagrangian strain $\tilde{\underline{\underline{E}}}$, the normal $\underline{N}_g$ to the surface S_g, the surface growth deformation $\tilde{\underline{\underline{E}}}_g$, and possibly the position field $\tilde{\underline{x}}_g$ on S_g, envisioning the growth as an heterogeneous process occurring in a uniform material body [2]. The support of ψ^S is supposed to restrict to the sole growing surface S_g.

The total mechanical energy expresses in the various configurations as

$$E = \int_{\Omega_0} \psi^V dX + \int_{S_0} \psi^S d\Sigma = \int_{\Omega_g} \psi^V j_g dx_g + \int_{S_g} \psi^S j_{sg} d\sigma_g = \int_{\Omega} j\psi^V\left(\underline{x}, \underline{\underline{E}}\right) dx + \int_{S_g} j_s \psi^S\left(\tilde{\underline{x}}, \underline{n}, \tilde{\underline{\underline{E}}}, \tilde{\underline{\underline{E}}}_g\right) d\sigma \tag{3.4}$$

with $j_g := J_g^{-1}$ (resp. $j := J^{-1}$) and $j_{sg} := J_{sg}^{-1}$ (resp. $j_s := J_s^{-1}$) the inverses of the volume and the surface Jacobian, J_g and $J_{sg} := d\Sigma/d\sigma_g$ respectively. The surface Jacobian is $J_{sg} := J_g / \left(\underline{N}_g.\underline{\underline{F}}_g.\underline{N}_0\right)$, with $\underline{N}_0$ the unit normal to the surface $S_0 := \partial\Omega_0$. The work of the external forces is

$$W^{ext} := \int_{\Omega} \underline{f}_0 . \underline{u}(\underline{x}) dx + \int_{\partial\Omega} \underline{t}_0 . \underline{\tilde{u}}(\underline{\tilde{x}}) d\sigma + \int_{\partial S_{gt}} p_0 . \underline{\tilde{\tilde{u}}}\left(\underline{\tilde{\tilde{x}}}\right) \underline{\tau} dl =$$
$$\int_{\Omega_g} \underline{f}_0 . \underline{u}(\underline{x}) J_g dx_g + \int_{\partial\Omega_g} \underline{t}_0 . \underline{\tilde{u}}(\underline{\tilde{x}}_g) J_{sg} d\sigma_g + \int_{\partial S_g} p_0 . \underline{\tilde{\tilde{u}}} . \underline{\underline{F}}_a . \underline{\tau}_g dl_g \tag{3.5}$$

with $\underline{f}_0$ the volumetric body forces, $\underline{t}_0 = \underline{\underline{\sigma}}.\underline{n}$ the traction vector, product of the Cauchy stress tensor $\underline{\underline{\sigma}}$ with the normal $\underline{n}$ on the boundary $\partial\Omega$, and p_0 the (scalar valued function) density of line forces (acting on the close edge ∂S_{gt}). The tangent vector to the closed curve ∂S is noted $\underline{\tau}$ in the sequel (fig. 1). Easy calculations render the variation

$$\delta\underline{u} = \left(\underline{\underline{F}}_a . \underline{\underline{F}}_g - I\right).\delta\underline{X} = \left(\underline{\underline{F}}_a - \underline{\underline{F}}_g^{-1}\right).\delta\underline{x}_g = \left(\underline{\underline{F}}_a . \underline{\underline{F}}_g - I\right).\left(\underline{\underline{F}}_g - I\right)^{-1}.\delta\underline{U}_g \tag{3.6}$$

with the functional dependence $\underline{\underline{F}}_a = \underline{\underline{F}}_a(\underline{\underline{F}}_g)$, to be made explicit later on. The domain variation [6] of the total energy expresses as

$$\frac{\delta E}{\delta t} = \int_{\Omega} j\psi^V_{,x}\underline{V} dx + \int_{\Omega} j\psi^V_{,E} : \dot{\underline{\underline{E}}} dx + \left\{ -\int_{\Omega} j\psi^V \mathrm{div}\underline{V} dx - \int_{\Omega} \nabla_X \left(j\psi^V\right)^t . \underline{V} dx + \int_{S_{g_t}} j\psi^V \underline{V}.\underline{n} d\sigma \right\} +$$
$$+ \int_{S_{g_t}} j_s \left[\psi^S_{,x} . \underline{\tilde{V}} - \left(\psi^S_{,n}\right)^t . \underline{\underline{L}}\,\underline{\underline{P}}.\underline{\tilde{V}} + \underline{\underline{A}}^S : \dot{\underline{\underline{E}}}_g + \underline{\underline{\tilde{S}}} : \dot{\underline{\underline{\tilde{E}}}} \right] d\sigma + \int_{S_{g_t}} \psi^S \dot{j}_s d\sigma + \int_{S_{g_t}} j_s \psi^S \mathrm{div}_S \underline{V} d\sigma \tag{3.7}$$

The domain variation of the work of the external forces in (3.5) expresses as

$$\delta W^{ext} = \int_{\Omega} \underline{f}_0 . \delta\underline{U} dx - \int_{\Omega} \nabla_X \left(\underline{f}_0 . \underline{u}\right) . \delta\underline{U} dx + \int_{S_g} \underline{f}_0 . \underline{\tilde{u}} \delta\underline{\tilde{U}} \underline{n} d\sigma + \int_{\partial\Omega} \underline{t}_0 . \delta\underline{\tilde{u}} d\sigma - \int_{S_g} \nabla_S \left(\underline{t}_0 . \underline{\tilde{u}}\right)^t . \delta\underline{\tilde{U}} d\sigma +$$
$$+ \int_{\partial S_g} \underline{t}_0 . \underline{\tilde{\tilde{u}}}\, \delta\underline{\tilde{\tilde{U}}} \underline{\tau} dl + \int_{\partial S_g} p_0 \delta\underline{\tilde{\tilde{U}}} \underline{\tau} dl - \int_{\partial S_g} p_0 \underline{\tilde{\tilde{U}}} \underline{\tau} \underline{\underline{L}}\,\underline{\underline{P}} . \delta\underline{\tilde{\tilde{U}}} \underline{n} dl - \int_{\partial S_g} \mathrm{div}_S \left[\left(p_0 \underline{\tilde{\tilde{U}}} \underline{\tau}\right) . \underline{\tau} \otimes \underline{\tau} \right] . \delta\underline{\tilde{\tilde{U}}} dl \tag{3.8}$$

assuming the forces $\underline{f}_0, \underline{t}_0$ to act as dead loads. The operator div_S thereabove denotes the surface divergence [5,6], defined from the projection of the gradient by the tensorial operator $\underline{\underline{P}} := I - \underline{N} \otimes \underline{N}$, with $P \in L(N^{\perp}, R^3)$ the surface projection operator, such that one formally has $\nabla_S := P.\nabla$, [5, 10, 11]. The surface divergence is then formally defined as the following equality between operators :

$$\mathrm{div}_S (.) := \mathrm{Tr}\left(\nabla_S (.)\right)$$

From these definitions, one obtains the surface transformation gradient

$$\underline{\underline{\tilde{F}}} := \underline{\underline{F}}.P = \underline{\underline{F}}.(I - \underline{N}_g \otimes \underline{N}_g)$$

and furthermore its material derivative

$$\dot{\tilde{\underline{\underline{F}}}} = \dot{\underline{\underline{F}}} - \dot{\underline{\underline{F}}}.\left(\underline{N}_g \otimes \underline{N}_g\right) - \left(\underline{\underline{F}}.\underline{\underline{L}}.\underline{V}_{gT}\right) \otimes \underline{N}_g - \left(\underline{\underline{F}}.\underline{N}_g\right) \otimes \left(\underline{\underline{L}}_g.\underline{V}_{gT}\right)$$

which highlights the role of the curvature tensor of the surface S_g, viz $\underline{\underline{L}}_g := \nabla_S \underline{N}_g$

The additive decomposition of the total deformation rate is then postulated

$$\underline{\underline{D}} := \operatorname{sym}\left(\underline{\underline{L}}\right) = \underline{\underline{D}}_g + \underline{\underline{D}}_a \tag{3.9}$$

with $\underline{\underline{L}} := \dot{\underline{\underline{F}}}.\underline{\underline{F}}^{-1}$ the velocity gradient, sum of the growth deformation rate $\underline{\underline{D}}_g := \underline{\underline{F}}_g^{-t}.\dfrac{\delta \underline{\underline{E}}_g}{\delta t}.\underline{\underline{F}}_g^{-1}$, and $\underline{\underline{D}}_a$ the accommodation rate. The addition of matter occurring within the growing tissue is supposed smooth enough, so that the resulting accommodation strain can be neglected with regard to the growth deformation (case of *smooth growth*). The accommodation strain itself is supposed to be purely elastic, and described by lagrangian strain tensors, respectively the volumetric accommodation deformation tensor $\underline{\underline{E}}_a$, and the (elastic) surface accommodation deformation tensor $\tilde{\underline{\underline{E}}}_a$. The total lagrangian strain decomposes as

$$\underline{\underline{E}} = \underline{\underline{F}}_g^{\,t}.\underline{\underline{E}}_a.\underline{\underline{F}}_g + \underline{\underline{E}}_g\,.$$

Adopting a point of view similar to plasticity [14], one can write :

$$\int_\Omega j\underline{\underline{S}} : \dot{\underline{\underline{E}}} dx = \int_\Omega \underline{\underline{\sigma}} : \underline{\underline{D}} dx\,; \qquad \int_{S_{gt}} j_s \psi^S_{,E_g} : \dot{\tilde{\underline{\underline{E}}}}_g d\sigma = \int_{S_g} j_{sa}\, \underline{\underline{a}}^S : \tilde{\underline{\underline{d}}}_g d\sigma$$

introducing the following stress and strain measures :

$$\underline{\underline{\sigma}} = j\underline{\underline{F}}.\underline{\underline{S}}.\underline{\underline{F}}^t\ ; \underline{\underline{D}} = \underline{\underline{F}}^{-1}.\dot{\underline{\underline{E}}}.\underline{\underline{F}}^t\ ;\ \underline{\underline{a}}^S := j_{sg}\underline{\underline{F}}_g.\underline{\underline{A}}^S.\underline{\underline{F}}_g^t\ ;\ \tilde{\underline{\underline{d}}}_g := \underline{\underline{F}}_g^{-t}.\dot{\tilde{\underline{\underline{E}}}}_g.\underline{\underline{F}}_g^t$$

with $j_{sa} := d\sigma_g / d\sigma$ the surface accommodation Jacobian. The equilibrium equations of the growing germ are obtained from the annihilation of the Gateaux derivative of $V\left[\Omega_g\right] = E - W^{ext}$, in the direction of the admissible set of volumetric, surface and line growth directions $\left(\dot{\underline{x}}_g, \dot{\tilde{\underline{x}}}_g, \dot{\tilde{\tilde{\underline{x}}}}_g\right)$:

i) **Volume equilibrium:** The following vectorial equation is obtained

$$-\operatorname{div}_g\left[\frac{1}{2}j\left(\psi^{Y^t}_{,E} + \psi^{Y}_{,E}\right)\underline{\underline{F}}^t\right]\left(\underline{\underline{I}} - \underline{\underline{F}}^{-1}\right)^{-1} - \operatorname{div}_g\left[\frac{1}{2}j\left(\left(\underline{\underline{F}}_a^t.\underline{\underline{F}}.\psi^Y_{,E}\right)^t - \left(\underline{\underline{F}}_a^t.\underline{\underline{F}}.\psi^Y_{,E}\right)\right)\right] + \left(\underline{\underline{I}} - \underline{\underline{F}}^{-1}\right)^{-1}.\psi^Y_{,x_g} + \nabla_X\left(j\psi^Y\right) = \underline{f}_0 - \nabla_X\left(\underline{f}_0.\underline{u}\right)$$

ii) **Surface equilibrium:** the equilibrium is the vectorial equation

$$\frac{1}{2}j\left\{\left(\psi^{V t}_{,E}+\psi^{V}_{,E}\right)\underline{\underline{\tilde{F}}}.\left(I-\underline{\underline{\tilde{F}}}^{-1}\right)^{-1}+\left(\left(\underline{\underline{F}}^{t}_{a}.\underline{\underline{F}}.\psi^{V}_{,E}\right)^{t}-\left(\underline{\underline{F}}^{t}_{a}.\underline{\underline{F}}.\psi^{V}_{,E}\right)\right)-j\psi^{V}\right\}.\underline{n}+\left(I-\underline{\underline{\tilde{F}}}^{-1}\right)^{-1}.\psi^{S}_{,x_g}-$$

$$\mathrm{div}_S\left[\frac{1}{2}j_s\left(\left(\underline{\underline{\tilde{F}}}^{t}_{a}\underline{\underline{\tilde{F}}}\psi^{V}_{,E}\right)^{t}-\left(\underline{\underline{\tilde{F}}}^{t}_{a}\underline{\underline{\tilde{F}}}\psi^{V}_{,E}\right)\right)\right]-\mathrm{div}_S\left[\frac{1}{2}j_s\left(\psi^{S t}_{,E}+\psi^{S}_{,E}\right)\underline{\underline{\tilde{F}}}\right].\left(I-\underline{\underline{\tilde{F}}}^{-1}\right)^{-1}-\mathrm{div}_S\left\{j\left[\frac{\left(\psi^{S}_{,\tilde{E}_g}\right)^{t}+\psi^{S}_{,\tilde{E}_g}}{2}\underline{\underline{\tilde{F}}}^{t}_{g}\right]\right\}$$

$$+\underline{\underline{L}}.\underline{\underline{P}}.\psi^{S}_{,n}-jJ_s\underline{\underline{L}}.\underline{\underline{P}}.\left(\underline{\underline{F}}.\underline{N}_0\right)+\nabla_S\left(\psi^{S}\right)-jJ_s\left[\mathrm{div}_S\left(\underline{N}_0\otimes\underline{n}\right)\right]=\left(\underline{f}_0.\underline{\tilde{u}}\right)\left(I-P\right).\underline{n}+\underline{t}_0-\nabla_S\left(\underline{t}_0.\underline{\tilde{u}}\right)$$

iii) Line equilibrium: projecting on the direction $\underline{\upsilon}$ gives the scalar equation

$$j_{sg}\left[\frac{\left(\psi^{S}_{,\tilde{E}_g}\right)^{t}+\psi^{S}_{,\tilde{E}_g}}{2}\underline{\underline{\tilde{F}}}^{t}\right]+\frac{1}{2}j_s\underline{\upsilon}^{t}.\left(\left(\underline{\underline{\tilde{F}}}^{t}_{a}\underline{\underline{\tilde{F}}}\psi^{V}_{,E}\right)^{t}-\left(\underline{\underline{\tilde{F}}}^{t}_{a}\underline{\underline{\tilde{F}}}\psi^{V}_{,E}\right)\right).\underline{\upsilon}+\frac{1}{2}j_s\underline{\upsilon}^{t}.\left(\psi^{S t}_{,\tilde{E}}+\psi^{S}_{,\tilde{E}}\right)\underline{\underline{\tilde{F}}}^{t}.\left(I-\underline{\underline{\tilde{F}}}^{-1}\right)^{-1}.\underline{\upsilon}$$

$$-\psi^{S}+jJ_s\underline{\upsilon}^{t}\left(\underline{N}_0\otimes\underline{n}\right).\underline{\upsilon}=\underline{t}_0.\underline{\tilde{\tilde{u}}}-p_0\underline{\tilde{\tilde{U}}}.\underline{\tau}.\underline{\underline{L}}.\underline{\underline{P}}.\underline{\upsilon}-\underline{\upsilon}^{t}.\mathrm{div}_S\left[\left(p_0\underline{\tilde{\tilde{U}}}.\underline{\tau}\right).\underline{\tau}\otimes\underline{\tau}\right]$$

Projecting on the orthogonal direction $\underline{\tau}$ (fig. 1) gives the scalar equation

$$j_g J_s\underline{\upsilon}^{t}\left(\underline{N}_0\otimes\underline{n}\right).\underline{\tau}=p_0-p_0\underline{\tilde{\tilde{U}}}.\underline{\tau}.\underline{\underline{L}}.\underline{\underline{P}}.\underline{\tau}-\underline{\tau}^{t}.\mathrm{div}_S\left[\left(p_0\underline{\tilde{\tilde{U}}}.\underline{\tau}\right).\underline{\tau}\otimes\underline{\tau}\right]$$

The virtual power of the internal forces is then identified to the Gateaux derivative of the total mechanical energy of the germ. The *configurational forces* are defined from the domain variation of the work of external forces acting on the germ, expressed on the intermediate configuration, accounting for the vanishing of the domain variation of the total potential energy :

$$\delta W^{ext}=\int_{\Omega_g}J_g\underline{f}_0.\left(\underline{\underline{F}}-I\right).\left(\underline{\underline{F}}_g-I\right)^{-1}.\delta\underline{\tilde{U}}_g dx_g+\int_{S_g}J_{sg}\underline{t}_0.\left(\underline{\underline{\tilde{F}}}-I\right).\left(\underline{\underline{\tilde{F}}}_g-I\right)^{-1}.\delta\underline{\tilde{U}}_g d\sigma_g-\int_{S_g}\underline{t}_0.\underline{\tilde{u}}j_g J_{sg}{}^{2}\left(\underline{\underline{F}}_g\underline{N}_0\right)^{t}\underline{\underline{L}}.\underline{\underline{P}}.\delta\underline{\tilde{U}}_g d\sigma_g$$

$$-\int_{S_g}\nabla_S\left(J_{sg}\underline{t}_0.\underline{\tilde{u}}\right)^{t}.\delta\underline{\tilde{U}}_g d\sigma_g+\int_{\partial S_g}J_{sg}\underline{t}_0.\underline{\tilde{u}}\delta\underline{\tilde{U}}_g.\underline{\upsilon}_g dl_g+\int_{S_g}\nabla_S\left(j_g J_{sg}{}^{2}\underline{N}_0\otimes\underline{N}_0\right).\delta\underline{\tilde{U}}_g d\sigma_g+\int_{\partial S_g}\left(j_g J_{sg}{}^{2}\underline{N}_0\otimes\underline{N}_0\right).\delta\underline{\tilde{U}}_g.\underline{\upsilon}_g dl_g$$

$$-\int_{S_g}\nabla_S\left(J_{sg}\underline{t}_0.\underline{\tilde{u}}\right).\delta\underline{\tilde{U}}_g d\sigma_g+\int_{\partial S_g}J_{sg}\underline{t}_0.\underline{\tilde{u}}\,\delta\underline{\tilde{\tilde{U}}}_g.\underline{\upsilon}_g dl_g+\int_{\partial S_g}p_0\underline{\underline{F}}_a.\underline{\tau}_g.\left(\underline{\underline{\tilde{F}}}-I\right).\left(\underline{\underline{\tilde{F}}}_g-I\right)^{-1}.\delta\underline{\tilde{U}}_g dl_g-$$

$$\int_{\partial S_g}p_0\underline{\tilde{\tilde{u}}}\underline{n}\underline{\underline{L}}.\underline{\underline{P}}.\left(\underline{\underline{\tilde{F}}}-I\right).\left(\underline{\underline{\tilde{F}}}_g-I\right)^{-1}.\left(\delta\underline{\tilde{\tilde{U}}}_g.\underline{\tau}_g\right)\underline{\tau}_g dl_g-\int_{\partial S_g}\mathrm{div}_S\left(p_0\underline{\tilde{\tilde{U}}}_g\underline{\underline{F}}_a\underline{\tau}_g\right)\left(\delta\underline{\tilde{\tilde{U}}}_g.\underline{\tau}_g\right)\underline{\tau}_g\,dl_g$$

Although somewhat artificial, the balance of momentum expressed in the intermediate configuration (obtained as the vanishing of the gateaux derivative of $V\left[\Omega_g\right]$) has the merit to highlight the terms conjugated to the volume, surface and line growth velocity variations, that can be considered as *accretive forces*, that express the change of reference configuration due to the sole growth. These forces find alternative equivalent expressions when the virtual power of the external power is substituted by the gateaux derivative of the mechanical energy. The partial derivative $\underline{C}_s := \psi^S{}_{,N}$ defines an orientational configuration force of the growing surface [9], and the gradient $\underline{\underline{A}}_s := \psi^S{}_{,\underline{\underline{\tilde{E}}}_g}$ a surface growth driving force, both of which intervene in the resulting equilibrium equations [18].

4. 4. SECOND PRINCIPLE AND EVOLUTION LAWS

For material systems that exchange work and mass with their surrounding, the material derivative of the internal energy u (per unit volume) expresses as [12]

$$\dot{u} = \nabla.\left(\underline{J}_q\right) - p_i + \underline{J}_k.\underline{F}_k \tag{4.1}$$

with $\underline{J}_q$ the heat diffusion flux (that shall obey a constitutive relationship vs. the temperature T), $p_i = -\underline{\underline{\sigma}}:\underline{\underline{D}}(\underline{v})$ the volumetric density of the power of internal forces, $\underline{J}_k$ the diffusion flux of the k-specie, and $\underline{F}_k$ an external force acting on the k-specie. The entropy balance expresses in terms of the material derivative of the volumetric density of entropy s as

$$T\dot{s} = \dot{u} - \underline{\underline{\sigma}}:\underline{\underline{D}}^R - \mu_k \rho_k \dot{c}_k \tag{4.2}$$

considering $k = 1..N$ chemical species (the nutrients of the growth) each having the chemical potential μ_k and the concentration $c_k = \frac{\rho_k}{\rho}$ (ratio of the mass density of the kth constituent to the total density). The symmetric part of the velocity gradient $\underline{\underline{D}}$ is being additively decomposed into a reversible contribution $\underline{\underline{D}}^R$ and an irreversible contribution $\underline{\underline{D}}^{IR}$. The barycentric balance of mass for the kth constituent is

$$\rho\dot{c}_k = -\nabla.\left(\underline{J}_k\right) + \upsilon_{km} R_m \tag{4.3}$$

with R_m the velocity of the chemical reaction (a scalar quantity). Introducing (4.3) and (4.1) into (4.2) leads to

$$\rho T\dot{s} = \nabla.\left(\underline{J}_q\right) - p_i + \underline{J}_k.\underline{F}_k - \underline{\underline{\sigma}}:\underline{\underline{D}}^R + \mu_k \nabla.\left(\underline{J}_k\right) - \mu_i \upsilon_{ij} R_j \tag{4.4}$$

We further transform in 4.4 all terms having divergence multiplicative factors, thus giving

$$\rho T\dot{s} = \nabla.\left(\underline{J}_q\right) - p_i - \underline{\underline{\sigma}}:\underline{\underline{D}}^R + \nabla.\left(\mu_k \underline{J}_k\right) - \underline{J}_k.\nabla\mu_k + A_k R_k + \underline{J}_k.\underline{F}_k$$

with $A_k := \upsilon_{ik}\mu_i$ the chemical affinities. The identification with the conservation law of the entropy (4.4) gives (all terms having a divergence form are exchange contributions) the entropy flux and the entropy source σ_s

$$\underline{J}_s = \underline{J}_q + \mu_k \underline{J}_k \quad \rho T\sigma_s = -p_i - \underline{\underline{\sigma}}:\underline{\underline{D}}^R - \underline{J}_k.\nabla\mu_k + A_k R_k + \underline{J}_k.\underline{F}_k \tag{4.5$_{1,2}$}$$

The positivity of the entropy source $\sigma_s = \dot{s}_i$ implies

$$\underline{\underline{\sigma}}:\underline{\underline{D}} - \underline{\underline{\sigma}}:\underline{\underline{D}}^R - \underline{J}_k.\nabla\mu_k + A_k R_k + \underline{J}_k.\underline{F}_k = \underline{\underline{\sigma}}:\underline{\underline{D}}^{IR} - \underline{J}_k.\nabla\mu_k + A_k R_k + \underline{J}_k.\underline{F}_k \geq 0 \tag{4.6}$$

The global form of Clausius-Duhem inequality expresses the positiveness of the mechanical dissipation

$$\phi^{int} = -p_i + p_i^R \tag{4.7}$$

The reversible part of the power of internal forces is next identified, relying on the assumption that the terms involving the growth velocity field (arising from the domain variation tied to growth) are irreversible contributions to the total mechanical power ; for the same reason, so is also the term $\int j_{sa}\underline{\underline{a}}^S:\underline{\underline{\tilde{d}}}_g\,d\sigma$. Considering the decomposition (3.9), with $\underline{\underline{D}}_a$ supposed reversible, and the similar decomposition of the surface deformation rate $\underline{\underline{\tilde{d}}}$, viz $\underline{\underline{\tilde{d}}} = \underline{\underline{\tilde{d}}}_g + \underline{\underline{\tilde{d}}}_a$, thus giving the mechanical dissipation in the form

$$\phi^{int} = \int_{\Omega} \underline{F}^{IR}.\underline{V}_g\,dx + \int_{S_{gt}} \underline{\tilde{F}}^{IR}.\underline{\tilde{V}}_g\,d\sigma + \int_{\partial S_{gt}} \underline{\underline{\tilde{\tilde{F}}}}^{IR}.\underline{\tilde{\tilde{V}}}_g.\underline{\upsilon}\,dl \geq 0 \tag{4.8}$$

with $\underline{F}^{IR} = j\left(\psi^V_{,x}\right)^t - \text{div}\underline{\underline{\sigma}}$ the irreversible volumetric driving force for growth,

$$\underline{\tilde{F}}^{IR} = j_s \underline{\underline{LP}}\left(\psi^S_{,n}\right) - j_s\left(\psi^S_{,x}\right) - \left(\nabla_S\psi^S\right) + j\,\underline{\underline{I}}_s\underline{\underline{LP}}\left(\underline{\underline{F}}\underline{N}_0\right) - \nabla_s\left[j\,\underline{\underline{I}}_s\psi^S\left(\underline{n}\otimes\underline{N}_0\right)^t\right] + \nabla_s\left(j_s\psi^S\right) + \underline{\underline{\sigma}}\underline{n} - \text{div}_s\left(\underline{\underline{\tilde{\sigma}}} + j_{sa}\frac{1}{2}\left(\underline{\underline{a}}^S + \left(\underline{\underline{a}}^S\right)^t\right)\right)$$

the irreversible surface growth driving force, and

$$\underline{\underline{\tilde{\tilde{F}}}}^{IR} := (1+j_s)\psi^S I + jJ_s(\underline{n}\otimes\underline{N}_0)^t + \left(\underline{\underline{\tilde{\sigma}}} + j_{sa}\left(\underline{\underline{a}}^S + \left(\underline{\underline{a}}^S\right)^t\right)\right)$$

the irreversible line driving force for growth (normal to the closed contour ∂S_g). The evolution laws for the internal variables $\left(\underline{V}_g, \underline{\tilde{V}}_g, \underline{\tilde{V}}_g\right)$ that result from the positivity of the intrinsic dissipation have to be completed by the constitutive law of the tissue : this needs the specification of the form of the volumetric and surfacic potentials ψ^V and ψ^S, viz $\underline{\underline{S}} := \psi^V_{,E}$; $\underline{\underline{\tilde{S}}} := \psi^S_{,\tilde{E}}$.

5. GROWTH AND KINEMATIC COMPATIBILITY

Physically, the growth strain is the strain that would be observed at a point of the tissue if it could be insulated from the surrounding tissue during its growth under zero stress. Such isolated elements of matter (the germs) do not fit together into the whole body when fully grown, thus a residual stress field will be generated in order to maintain the kinematic integrity of the continuous body [4,16,17,18]. Considering growth to occur within a Riemanian geometry of an Euclidean space, the kinematic compatibility condition within a large transformations framework expresses as the Euclidean character of the tangent space to a differentiable manifold, thus the fourth order Riemann-Christoffel curvature tensor of the spatial body manifold (parameterized by curvilinear coordinates) vanishes.

We here assume that the growth develops as a mechanism not being influenced by the current stress state, and consider the instantaneous situation of a growth having developed into a portion of a cylindrical domain [18]. The material point in the initial configuration is assigned the coordinates $X = (R, \theta, Z)$; its coordinates in the intermediate configuration at the frozen time t_0 are given by $x_g = (\rho, \varphi, \xi)$. The growth model then expresses as the following mapping :

$$\rho = f(R) \quad \varphi = K_\theta(R)\theta \quad \xi = Z$$

The angular growth function $K_\theta(R)$ is supposed to depend upon the sole radial variable (and not upon the angle θ). The accommodation of growth is then modeled by the relationship between the coordinates in the intermediate and final configurations, viz

$$r = r(\rho) \quad \psi = \eta_\theta(\rho)\varphi \quad z = \varepsilon\xi$$

with ε a constant coefficient, corresponding to a plane strain deformation state. The linear growth model $\rho(R) = \dfrac{AR+B}{K_0}$ then solves the two differential equations resulting from (5.1), considering an homogenous angular growth, i.e. $K_\theta(R)=K_0$. The choice of a constant angular accommodation, viz. $\eta_\theta(\rho)=Cte=\eta_0$, then gives the two families of solutions :

* $r(\rho)=K_1\rho$, giving a linear accommodation of the (linear) growth;

$$r(\rho)=\pm(1-2\rho-2C)^{1/2}\pm\frac{1}{2}Ln\left(-1+(1-2\rho-2C)^{1/2}\right)\pm\frac{1}{2}Ln\left(1+(1-2\rho-2C)^{1/2}\right)-\frac{1}{2}Ln(\rho+C)$$

the constant C being determined from the boundary conditions. The set of mechanical balance equations have to be combined with both the constitutive equation of the solid growing material and the transport equations of the nutrients.

As a perspective, the modeling of the final goal reached by the system over its evolution shall be envisaged.

6. REFERENCES

1. L. Taber. Applied Mech. Rev., **48** (1995) 487-545.
2. M. Epstein, G.A. Maugin, Int. J. Plasticity 16 (2000) 951-978.
3. E.K. Rodriguez, A. Hoger, A. D. McCullogh. J. Biomechanics **27** n°4 (1994) 455-467.
4. D. Ambrosi, F. Mollica, Int. J Engng Sci. 40 (2002) 1297-1316.
5. M.E. Gurtin, Arch. Rational Mech. Anal. **131** (1995) 67-100.
6. G. Allaire, A. Henrot. C.R. Acad. Sci. Paris, t. 329 Série IIb (2001) 383-396.
7. C. Truesdell, W. Noll, The non-linear field theories of mechanics. Berlin, Springer-Verlag, 1992.
8. A.N. Norris, Int. J. Solids Struct. **35** N°36 (1998) 237-5252.
9. S.C. Cowin, Annals of Biomedical Engineering **11** (1983) 263-295.
10. H. Petryk, Z. Mroz, Arch. Mech., **38**, n° 5-6 (1986) 697-724.
11. P. Leo, R.F. Sekerka, Acta Metallurgica, **37** (1989) 3119-3138.
12. S.R. de Groot. Thermodynamics of irreversible processes. Amsterdam, 1952.
13. G.A. Maugin, The Thermomechanics of plasticity and fracture. Cambridge University Press, 1992.
14. A. Kahn, S. Huang, Continuum theory of plasticity, John Wiley, 1995.
15. R. Skalak, S. Zargaryan, R. K. Jain, P. A. Netti, A. Hoger, J. Math. Biol. 34 (1996) 889-914.
16. Y.C. Fung, Biomechanics : mechanical properties of living tissues. Springer, Berlin, 1993.
17. G. Helmlinger, P.A. Petti, H.C. Lichtenbeld, R.J. Melder, R.K. Jain, Nature Biotech. 15 (1997) 778-783.
18. J.F. Ganghoffer, B. Haussy. Mechanical modeling of biological tissue under growth. Part I : constitutive framework. Submited to Int. J. Engng Sci.

Chapter 8

MATERIAL FORCES IN THE CONTEXT OF BIOTISSUE REMODELLING

Krishna Garikipati, Harish Narayanan, Ellen M. Arruda, Karl Grosh and Sarah Calve
krishna@umich.edu
University of Michigan, Ann Arbor, USA
Keywords: Microstructure, configurational stress, equilibrium, self-assembly

Abstract Remodelling of biological tissue, due to changes in microstructure, is treated in the continuum mechanical setting. Microstructural change is expressed as an evolution of the reference configuration. This evolution is expressed as a point-to-point map from the reference configuration to a remodelled configuration. A "preferred" change in configuration is considered in the form of a globally incompatible tangent map. This field could be experimentally determined, or specified from other insight. Issues of global compatibility and evolution equations for the resulting configurations are addressed. It is hypothesized that the tissue reaches local equilibrium with respect to changes in microstructure. A governing differential equation and boundary conditions are obtained for the microstructural changes by posing the problem in a variational setting. The Eshelby stress tensor, a separate configurational stress, and thermodynamic driving (material) forces arise in this formulation, which is recognized as describing a process of self-assembly. An example is presented to illustrate the theoretical framework.

1. INTRODUCTION

The development of biological tissue consists of distinct processes of growth, remodelling and morphogenesis—a classification suggested by Taber, 1995. In our treatment of the problem, growth is defined as the addition or depletion of mass through processes of transport and reaction coupled with mechanics. As a result there is an evolution of the concentrations of the various species that make up the tissue. Nominally, these include the solid phase (cells and extra cellular matrix), the fluid phase (interstitial fluid), various amino acids, enzymes, nutrients, and byproducts of reactions between them. The stress and deformation state of the tissue also evolve due to mechanical loads, and the coupling between transport, reaction and mechanics.

Remodelling is the process of microstructural reconfiguration within the tissue. It can be viewed as an evolution of the reference configuration to a "remodelled" configuration. While it usually occurs simultaneously with growth, it is an independent process. For the purpose of conceptual clarity we will ignore growth in this paper, and focus upon a continuum mechanical treatment of remodelling.

The microstructural reconfiguration that underlies remodelling is a motion of material points in *material space.* An example of remodelling driven by stress is provided by the micrographs of Figure 8.1 from Calve et al., 2003.

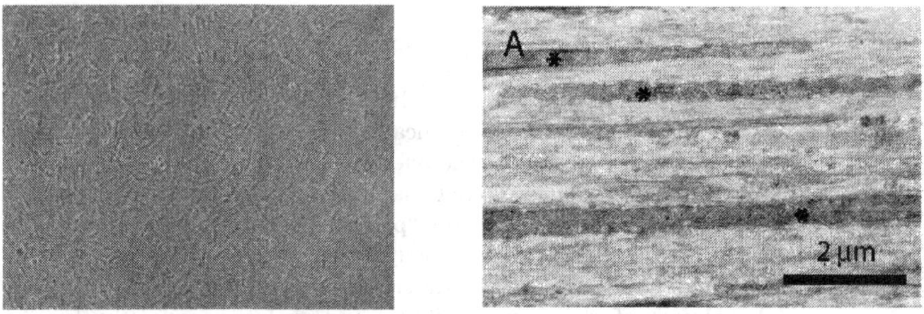

(a) (b)

Figure 8.1. (a) Micrograph taken 3 days after plating of cells, shows a random distribution and orientation of tendon fibroblast cells in engineered tendon. (b) As growth occurs the cells organize into a more ordered microstructure seen in this micrograph taken about a month after plating of cells. The horizontal alignment of cells corresponds to the orientation of a uniaxial stress that was imposed externally on the growing tendon construct. The alignment of cells along the stress axis is evidence of remodelling due to stress in the engineered tendon during growth.

Remodelling also can be driven by the local density of the tissue's solid or fluid phases, availability of various chemical factors, temperature, etc. We assume that it is possible (through experiments or other approaches) to define a phenomenological law that specifies this evolution. Since any conditions that could drive remodelling vary pointwise through the tissue, such a "preferred" remodelled state will, in general, be globally incompatible. However, in its final remodelled state, the tissue is virtually always free of such incompatibilities (see Figure 8.1). We therefore propose that a further, compatibilty-restoring material motion occurs, carrying material points to the remodelled configuration.

Taber and Humphrey, 2001 and Ambrosi and Mollica, 2002 have previously referred to remodelling. However, the treatments in these papers are based upon concentration (or density) changes in growing tissue and the mechanics—mainly internal stress—that is associated with them. By our definitions, these papers describe growth rather than remodelling. In a largely descriptive paper, Humphrey and Rajagopal, 2002 have proposed the evolution of "natural configurations" that seems closest to our ideas. To our knowledge, however, no quantitative treatment exists paralleling the ideas of microstructural reconfiguration, material motion, material/configurational forces and their relation to remodelling, as described in the present paper.

2. VARIATIONAL FORMULATION: MATERIAL MOTION AND MATERIAL FORCES

Figure 8.2 depicts the kinematics associated with remodelling. The preferred remodelled state is given by a tangent map of material motion $\boldsymbol{K}^{\mathrm{r}}: \Omega_0 \times [0,T] \mapsto \mathbb{GL}^3$, where $\mathbb{GL}^3$ is the space of 3×3 matrices. The reference configuration is $\Omega_0 \subset \mathbb{R}^3$. Our first assumption is that $\boldsymbol{K}^{\mathrm{r}}$ is given. Future communications will address the derivation of such an evolution law from our experiments (see Calve et al., 2003 for preliminary results). Since $\boldsymbol{K}^{\mathrm{r}}$ is generally incompatible (see Section 1), a further tangent map of material motion, $\boldsymbol{K}^{\mathrm{c}}: \Omega_0 \times [0,T] \mapsto \mathbb{R}^3$, acts to render the tissue of interest, $\mathcal{B}$, compatible in its remodelled configuration, Ω_t^*. The point-to-point map $\boldsymbol{\kappa}: \Omega_0 \times [0,T] \mapsto \mathbb{R}^3$ carries material points from Ω_0 to $\Omega_t^* \subset \mathbb{R}^3$, the remodelled configuration. It is a material motion, and its *compatible* tangent map, $\boldsymbol{K} = \partial\boldsymbol{\kappa}/\partial\boldsymbol{X}$ satisfies $\boldsymbol{K} = \boldsymbol{K}^{\mathrm{c}}\boldsymbol{K}^{\mathrm{r}}$. The placement of material points in Ω_t^* is $\boldsymbol{X}^* = \boldsymbol{\kappa}(\boldsymbol{X}, t)$. Further deformation, brought about by the displacement, $\boldsymbol{u}^*$, carries material points from Ω_t^* to the spatial configuration Ω_t. The deformation gradient is $\boldsymbol{F}^* = \mathbf{1} + \partial\boldsymbol{u}^*/\partial\boldsymbol{X}^*$. In this initial treatment we do not consider any further decompositions of $\boldsymbol{F}^*$. The overall motion of a point is $\boldsymbol{\varphi}(\boldsymbol{X}, t) = \boldsymbol{\kappa}(\boldsymbol{X}, t) + \boldsymbol{u}^*(\boldsymbol{X}^*, t) \circ \boldsymbol{\kappa}(\boldsymbol{X}, t)$, and the corresponding tangent map is $\boldsymbol{F} = \mathbf{1} + \partial\boldsymbol{\varphi}/\partial\boldsymbol{X}$. It admits the multiplicative decomposition $\boldsymbol{F} = \boldsymbol{F}^*\boldsymbol{K}^{\mathrm{c}}\boldsymbol{K}^{\mathrm{r}}$. To reiterate upon the foregoing distinction between the material motion and deformation components of the kinematics, we emphasize that $\boldsymbol{u}^*$ is a displacement, while $\boldsymbol{\kappa}$ is a motion in material space. The corresponding tangent maps are $\boldsymbol{F}^*$ (a classical deformation gradient) and $\boldsymbol{K}$ (a material motion gradient).

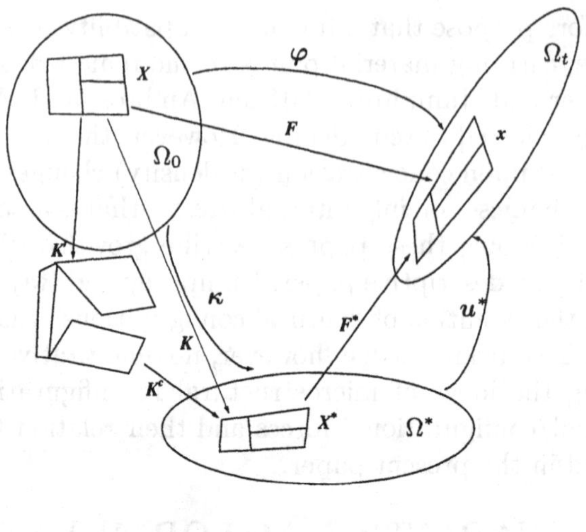

Figure 8.2. The kinematics of remodelling

2.1. A VARIATIONAL FORMULATION

We consider the following energy functional:

$$\Pi[\boldsymbol{u}^*, \boldsymbol{\kappa}] := \int_{\Omega_t^*} \hat{\psi}^*(\boldsymbol{F}^*, \boldsymbol{K}^{\mathrm{c}}, \boldsymbol{X}^*)\mathrm{d}V^* - \int_{\Omega_t^*} \boldsymbol{f}^* \cdot (\boldsymbol{u}^* + \boldsymbol{\kappa})\mathrm{d}V^* - \int_{\partial\Omega_t^*} \bar{\boldsymbol{t}}^* \cdot (\boldsymbol{u}^* + \boldsymbol{\kappa})\mathrm{d}A^*, \quad (1)$$

where $\psi^* = \hat{\psi}^*(\boldsymbol{F}^*, \boldsymbol{K}^{\mathrm{c}}, \boldsymbol{X}^*)$ is the stored energy function. Observe that ψ^* is assumed to depend upon the compatibility-restoring material motion, $\boldsymbol{K}^{\mathrm{c}}$, in addition to the usual dependence on $\boldsymbol{F}^*$. Furthermore, material heterogeneity is allowed. The body force per unit volume in Ω_t^* is $\boldsymbol{f}^*$, and the surface traction per unit area on $\partial\Omega_t^*$ is $\bar{\boldsymbol{t}}^*$. Since the total motion of a material point is $\boldsymbol{\kappa} + \boldsymbol{u}^*$, the potential energy of the external loads is as seen in the second and third terms.

Recall that the Euler-Lagrange equations obtained by imposing equilibrium with respect to $\boldsymbol{u}^*$ (stationarity of Π with respect to variations in $\boldsymbol{u}^*$) represent the quasistatic balance of linear momentum in Ω_t^*.

$$\left.\frac{\mathrm{d}}{\mathrm{d}\varepsilon}\Pi[\boldsymbol{u}_\varepsilon^*, \boldsymbol{\kappa}]\right|_{\varepsilon=0} = 0 \text{ where } \boldsymbol{u}_\varepsilon^* = \boldsymbol{u}^* + \varepsilon\delta\boldsymbol{u}^* \quad (2)$$

$$\Longrightarrow \mathrm{Div}^*\boldsymbol{P}^* + \boldsymbol{f}^* = \boldsymbol{0}, \text{ in } \Omega^*;\ \boldsymbol{P}^*\boldsymbol{N}^* = \bar{\boldsymbol{t}}^* \text{ on } \partial\Omega^*;\ \boldsymbol{P}^* := \frac{\partial\psi^*}{\partial\boldsymbol{F}^*} \quad (3)$$

Observe that the definition of $\boldsymbol{P}^*$ resembles the constitutive relation for the first Piola-Kirchhoff stress if Ω_t^* were the reference configuration.

A final assumption is that the tissue also reaches local equilibrium with respect to $\boldsymbol{\kappa}$ (stationarity of Π with respect to variations in $\boldsymbol{\kappa}$). The variational statement is:

$$\frac{\mathrm{d}}{\mathrm{d}\varepsilon}\Pi[\boldsymbol{u}^*, \boldsymbol{\kappa}_\varepsilon]\bigg|_{\varepsilon=0} = 0, \text{ where } \boldsymbol{\kappa}_\varepsilon = \boldsymbol{\kappa} + \varepsilon\delta\boldsymbol{\kappa}, \text{ and } \boldsymbol{\varphi} \text{ is fixed.} \tag{4}$$

The calculations are lengthy, but entirely standard, and yield the following Euler-Lagrange equations:

$$-\mathrm{Div}^*\left[\psi^*\mathbf{1} - \boldsymbol{F}^{*^{\mathrm{T}}}\boldsymbol{P}^* + \frac{\partial\psi^*}{\partial\boldsymbol{K}^{\mathrm{c}}}\boldsymbol{K}^{\mathrm{c}^{\mathrm{T}}}\right] + \frac{\partial\psi^*}{\partial\boldsymbol{X}^*} = \mathbf{0} \text{ in } \Omega^* \tag{5}$$

$$\left[\psi^*\mathbf{1} - \boldsymbol{F}^{*^{\mathrm{T}}}\boldsymbol{P}^* + \frac{\partial\psi^*}{\partial\boldsymbol{K}^{\mathrm{c}}}\boldsymbol{K}^{\mathrm{c}^{\mathrm{T}}}\right]\boldsymbol{N}^* = \mathbf{0} \text{ on } \partial\Omega^*. \tag{6}$$

Observe that the Eshelby stress $\psi^*\mathbf{1} - \boldsymbol{F}^{*^{\mathrm{T}}}\boldsymbol{P}^*$ makes its appearance. Hereafter, it will be denoted $\boldsymbol{\mathcal{E}}$. The term $\frac{\partial\psi^*}{\partial\boldsymbol{K}^{\mathrm{c}}}\boldsymbol{K}^{\mathrm{c}^{\mathrm{T}}}$ is a thermodynamic driving quantity giving the change in stored energy, ψ^*, corresponding to a change in configuration, $\boldsymbol{K}^{\mathrm{c}}$. It is stress-like in its physical dimensions and tensorial form, and we therefore refer to it as a configurational stress. Hereafter we will write $\boldsymbol{\Sigma}^* = \frac{\partial\psi^*}{\partial\boldsymbol{K}^{\mathrm{c}}}\boldsymbol{K}^{\mathrm{c}^{\mathrm{T}}}$.

Remark 1: A distinct class of variations can be considered than those in (4). Specifically, consider

$$\frac{\mathrm{d}}{\mathrm{d}\varepsilon}\Pi[\boldsymbol{u}^*, \boldsymbol{\kappa}_\varepsilon]\bigg|_{\varepsilon=0} = 0, \text{ where } \boldsymbol{\kappa}_\varepsilon = \boldsymbol{\kappa} + \varepsilon\delta\boldsymbol{\kappa}, \text{ and } \boldsymbol{\varphi}_\varepsilon = \boldsymbol{X} + \boldsymbol{\kappa} + \varepsilon\delta\boldsymbol{\kappa} + \boldsymbol{u}^*, \tag{7}$$

which is distinct from (4) in that variations on the material motion result in variations on the final placement as well. In this case too, the Euler-Lagrange equations (5) and (6) are arrived at. This is an important property: The system of equations governing the evolution of the microstructural configuration must be independent of the particular class of variations considered.

2.2. RESTRICTIONS FROM THE DISSIPATION INEQUALITY

The dissipation inequality written per unit volume in the reference configuration takes the familiar form,

$$\boldsymbol{\tau}\colon(\dot{\boldsymbol{F}}\boldsymbol{F}^{-1}) - \frac{\partial}{\partial t}\left(\det[\boldsymbol{K}]\psi^*\right) \geq 0, \tag{8}$$

where $\boldsymbol{\tau}$ is the Kirchhoff stress defined in Ω_t. Observe that $\det[\boldsymbol{K}]\psi^*$ is the stored energy per unit volume in Ω_0. Using the multiplicative decomposition $\boldsymbol{F} = \boldsymbol{F}^*\boldsymbol{K}^{\mathrm{c}}\boldsymbol{K}^{\mathrm{r}}$, and the defining relation for the configurational stress $\boldsymbol{\Sigma}^* = \frac{\partial\psi^*}{\partial\boldsymbol{K}^{\mathrm{c}}}\boldsymbol{K}^{\mathrm{c}^{\mathrm{T}}}$, standard manipulations result in the following equivalent form:

$$\left(\boldsymbol{\tau}\boldsymbol{F}^{*^{-\mathrm{T}}} - \det[\boldsymbol{K}]\frac{\partial\psi^*}{\partial\boldsymbol{F}^*}\right)\dot{\boldsymbol{F}}^* - \det[\boldsymbol{K}]\left(\boldsymbol{\mathcal{E}} + \boldsymbol{\Sigma}^*\right) : \left(\dot{\boldsymbol{K}}^{\mathrm{c}}\boldsymbol{K}^{\mathrm{c}-1}\right)$$
$$-\det[\boldsymbol{K}]\boldsymbol{\mathcal{E}} : \left(\boldsymbol{K}^{\mathrm{c}}\dot{\boldsymbol{K}}^{\mathrm{r}}\boldsymbol{K}^{-1}\right) - \det[\boldsymbol{K}]\frac{\partial\psi^*}{\partial\boldsymbol{X}^*}\cdot\dot{\boldsymbol{\kappa}} \geq 0. \qquad (9)$$

Adopting the constitutive relation, $\boldsymbol{\tau} = \det[\boldsymbol{K}]\frac{\partial\psi^*}{\partial\boldsymbol{F}^*}\boldsymbol{F}^{*^{\mathrm{T}}}$ (this is consistent with the observation that $\boldsymbol{P}^* = \frac{\partial\psi^*}{\partial\boldsymbol{F}^*}$ is related to the first Piola-Kirchhoff stress with Ω_t^* as the reference configuration), results in the reduced dissipation inequality

$$-\det\boldsymbol{K}\left(\boldsymbol{\mathcal{E}} + \boldsymbol{\Sigma}^*\right) : \left(\dot{\boldsymbol{K}}^{\mathrm{c}}\boldsymbol{K}^{\mathrm{c}-1}\right)$$
$$-\det[\boldsymbol{K}]\boldsymbol{\mathcal{E}} : \left(\boldsymbol{K}^{\mathrm{c}}\dot{\boldsymbol{K}}^{\mathrm{r}}\boldsymbol{K}^{-1}\right) - \det[\boldsymbol{K}]\frac{\partial\psi^*}{\partial\boldsymbol{X}^*}\cdot\dot{\boldsymbol{\kappa}} \geq 0, \qquad (10)$$

which places restrictions on the evolution law for $\boldsymbol{K}^{\mathrm{r}}$, and on the functional dependencies $\hat{\psi}^*(\bullet, \boldsymbol{K}^{\mathrm{c}}, \bullet)$ through $\boldsymbol{\Sigma}^*$, and $\hat{\psi}^*(\bullet, \bullet, \boldsymbol{X}^*)$.

3. REMODELLING OF ONE-DIMENSIONAL BARS

In general, the examples of remodelling encountered in soft and hard biological tissue involve complex microstructural changes. Evolution laws for $\boldsymbol{K}^{\mathrm{r}}$ and the functional form, $\hat{\psi}^*(\bullet, \boldsymbol{K}^{\mathrm{c}}, \bullet)$, to model these complexities are critical components for the successful application of the theoretical framework outlined in this paper. In future communications we will describe our experimental program to extract such constitutive information. However, the working of the formulation can be demonstrated by academic, but illuminating, examples. In the interest of brevity we restrict ourselves to a single example in this paper.

Consider two parallel bars, that may represent adjacent strips of a long bone (Figure 8.3). We wish to consider a scenario in which each bar undergoes a preferred material motion to change its length from L to L_i^{r}, $i = 1, 2$. In general, this configuration is incompatible as the bars can attain different lengths. If they are required to remain of the same length in the remodelled configuration, further material motion occurs, resulting in a length L^* for each bar. This is the remodelled

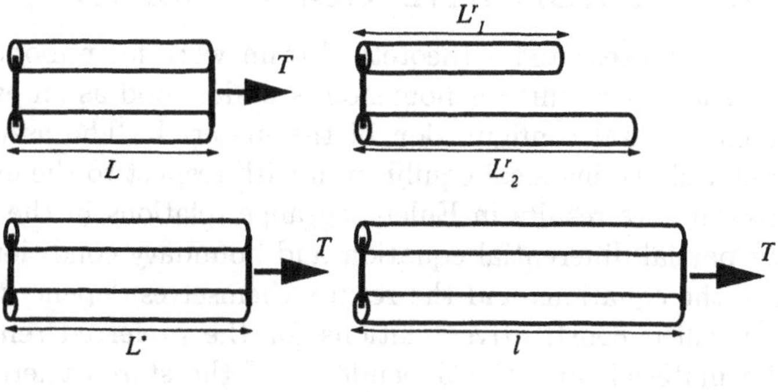

Figure 8.3. One-dimensional remodelling of bars.

configuration, Ω_t^*, in which the total material motion of each bar is $\kappa = L^* - L$. If the remodelling takes place under an external load, T, the bars each stretch to a final length l. The deformation is $u^* = l - L^*$. We examine the equations that govern the deformation and material motion by considering the following energy functional:

$$\Pi[u^*, \kappa] = \frac{1}{2}k^*(\kappa + L - L_1^r)^2 + \frac{1}{2}k^*(\kappa + L - L_2^r)^2 + 2\cdot\frac{1}{2}ku^{*2} - T(u^* + \kappa), \quad (11)$$

where k^* and k are spring constants for the material motion- and stretch-dependent portions of the stored energy, respectively. These portions are assumed to be separable. The theory of Section 2 results in the following relations:

$$\frac{\partial \Pi}{\partial u^*} = 0 \quad \Rightarrow 2ku^* = T, \quad (12)$$

$$\frac{\partial \Pi}{\partial \kappa} = 0 \quad \Rightarrow \kappa = \frac{k}{k^*}u^* - \left(L - \frac{L_1^r + L_2^r}{2}\right). \quad (13)$$

In (12) the standard relation is seen for the stretch of a linear spring with effective stiffness $2k$. The more interesting result is (13). Observe that when $L = \frac{1}{2}(L_1^r + L_2^r)$, material motion can occur, driven by stress, since $u^* = T/2k$ from (12). In this case remodelling can be incompatible if $L_1^r \neq L_2^r$. However, remodelling does not drive material motion, κ in this case. Instead there is stress-driven remodelling as described in Section 1. On the other hand, in the absence of an external load, material motion is obtained when $L \neq \frac{1}{2}(L_1^r + L_2^r)$. In this case the compatibility-restoring remodelling, motivated in Section 1 and described by the tangent map, $\boldsymbol{K}^c$, also leads to overall material motion, $\boldsymbol{\kappa}$.

4. DISCUSSION AND CONCLUSION

This paper has presented a theoretical framework for remodelling in biological tissue, where this phenomenon is understood as an evolution of the microstructural configuration of the material. The assumption that the material attains local equilibrium with respect to the evolution of its microstructure results in Euler-Lagrange relations in the form of a governing partial differential equation and boundary conditions. The final form of the equations and the results themselves depend critically upon the specified constitutive relations for the preferred remodelled state of the material, and the dependence of the stored energy upon the compatibility-restoring component of remodelling. These are open problems, and will be addressed in future papers by our group. The following points are noteworthy at this stage of the development of the theory:

- This work does not deal with the approach to local equilibrium, which may take time on the order of days in biological tissue. Furthermore, the equilibrium state, being defined by the external loads, evolves upon perturbation of these conditions: Additional remodelling occurs in biological tissue when the load is altered.

- The energy functionals in (1) and (11) generalize to the Gibbs free energy of the body under constant loads and isothermal conditions. Further contributions that drive the process, such as chemistry or electrical stimuli, can be encompassed by the Gibbs energy. In such a setting the process we have described here would be termed a "self-assembly" in the realms of materials science or physics.

References

Ambrosi, D. and Mollica, F. (2002). On the mechanics of a growing tumor. *Int. J. Engr. Sci.*, 40:1297–1316.

Calve, S., Dennis, R., Kosnik, P., Baar, K., Grosh, K., and Arruda, E. (2003). Engineering of functional tendon. Submitted to *Tissue Engineering*.

Humphrey, J. D. and Rajagopal (2002). A constrained mixture model for growth and remodeling of soft tissues. *Math. Meth. Mod. App. Sci.*, 12(3):407–430.

Taber, L. A. (1995). Biomechanics of growth, remodelling and morphogenesis. *Applied Mechanics Reviews*, 48:487–545.

Taber, L. A. and Humphrey, J. D. (2001). Stress-modulated growth, residual stress and vascular heterogeneity. *J. Bio. Mech. Engrg.*, 123:528–535.

IV

NUMERICAL ASPECTS

Chapter 9

ERROR-CONTROLLED ADAPTIVE FINITE ELEMENT METHODS IN NONLINEAR ELASTIC FRACTURE MECHANICS

Marcus Rüter and Erwin Stein

Institute of Mechanics and Computational Mechanics (IBNM)

University of Hannover, Appelstraße 9a, 30167 Hannover, Germany

rueter/stein@ibnm.uni-hannover.de

Abstract Goal-oriented a posteriori error estimators are presented in this contribution for the error obtained while approximately evaluating the J-integral, i.e. the material force acting at the crack tip, in nonlinear elastic fracture mechanics using the finite element method. The error estimators rest upon the strategy of solving an auxiliary dual problem and can be classified as equilibrated residual error estimators based on the solutions of Neumann boundary value problems on the element level. Finally, an illustrative numerical example is presented.

Keywords: A posteriori error estimation, J-integral, material forces

Introduction

A key issue in elastic fracture mechanics is the evaluation of the J-integral or an equivalent fracture criterion. However, its main feature of independence from the integration path or area cannot be maintained by an approximate numerical evaluation using the finite element method. As a consequence, adaptive finite element methods are required to control the error of the J-integral. In this contribution we therefore follow Rüter and Stein, 2003 to derive appropriate *goal-oriented* error estimators, as introduced by Eriksson et al., 1995 and Becker and Rannacher, 1996, in elastic fracture mechanics which are based on equilibrated Neumann problems on the element level.

1. THE MODEL PROBLEM

Let the elastic body be given by the closure of a bounded open set $\Omega \subset \mathbb{R}^d$ with a piecewise smooth, polyhedral and Lipschitz continuous boundary $\Gamma = \partial\Omega$ such that $\Gamma = \bar{\Gamma}_D \cup \bar{\Gamma}_N$ and $\Gamma_D \cap \Gamma_N = \emptyset$, where Γ_D and Γ_N are the portions of the boundary Γ where Dirichlet and Neumann boundary conditions are imposed, respectively. The variational form of the nonlinear model problem of finite elasticity then reads: find a solution $\boldsymbol{u} \in \mathcal{V} = \{\boldsymbol{v} \in [W^{1,r}(\Omega)]^3 \; ; \; \boldsymbol{v}|_{\Gamma_D} = \boldsymbol{0}\}$, with $r \geq 2$, such that

$$a(\boldsymbol{u}, \boldsymbol{v}) = F(\boldsymbol{v}) \quad \forall \boldsymbol{v} \in \mathcal{V}. \tag{1}$$

Here, $a : \mathcal{V} \times \mathcal{V} \to \mathbb{R}$ denotes a semi-linear form (i.e. a is linear w.r.t. its second argument only) and $F : \mathcal{V} \to \mathbb{R}$ is a linear functional defined as

$$a(\boldsymbol{u}, \boldsymbol{v}) = \int_\Omega \boldsymbol{\tau}(\boldsymbol{u}) : \operatorname{grad} \boldsymbol{v} \, \mathrm{d}V \quad \text{and} \quad F(\boldsymbol{v}) = \int_{\Gamma_N} \bar{\boldsymbol{t}} \cdot \boldsymbol{v} \, \mathrm{d}S, \tag{2}$$

respectively. In the above, $\boldsymbol{\tau}$ denotes the Kirchhoff stress tensor, and $\bar{\boldsymbol{t}} \in [L_2(\Gamma_N)]^3$ are prescribed tractions on Γ_N. Note that body forces are omitted in this formulation. The discrete counterpart of (1) consists in seeking a solution $\boldsymbol{u}_h$ in a finite dimensional subspace $\mathcal{V}_h \subset \mathcal{V}$ satisfying

$$a(\boldsymbol{u}_h, \boldsymbol{v}_h) = F(\boldsymbol{v}_h) \quad \forall \boldsymbol{v}_h \in \mathcal{V}_h. \tag{3}$$

In the finite element method the finite dimensional subspace $\mathcal{V}_h$ is usually constructed by subdividing the domain $\bar{\Omega}$ into n_e elements $\bar{\Omega}_e$, such that $\bar{\Omega} = \bigcup_{n_e} \bar{\Omega}_e$, and defining piecewise polynomials on the elements $\bar{\Omega}_e$.

2. ELASTIC FRACTURE MECHANICS

In this contribution we aim at investigating the accuracy of approximately determined fracture criteria for nonlinear elastic materials using the finite element method. More precisely, we shall confine our attention to the J-integral concept (Rice, 1968).

Within the framework of Eshelbian mechanics (see Eshelby, 1951, Maugin, 1993, Steinmann, 2000 and others), the J-integral can be conveniently derived as a material force acting at the crack tip. For the sake of simplicity, let the material be isotropic, and let the geometry and the loads be symmetric with respect to the x-z-plane (in 3D) or w.r.t. the x-axis (in 2D). Then, in the absence of crack face tractions the domain expression of the J-integral (Shih et al., 1986) is defined as

$$J(\boldsymbol{u}) = -\int_{\Omega_J} \operatorname{Grad}(q\boldsymbol{e}_{||}) : \tilde{\boldsymbol{\Sigma}}(\boldsymbol{u}) \, \mathrm{d}V \tag{4}$$

with arbitrary but continuously differentiable weighting function $q = q(x, y, z)$ (or $q = q(x, y)$ in 2D), where $q = 1$ at the crack tip and $q|_{\Gamma_J = \partial\Omega_J} = 0$. Furthermore, $\boldsymbol{e}_{||}$ is a unit vector in the direction of crack propagation, and $\tilde{\boldsymbol{\Sigma}} = W_s\mathbf{1} - \boldsymbol{H}^T \cdot \boldsymbol{P}$ is termed the Newton-Eshelby stress tensor with specific strain-energy function W_s, identity tensor $\mathbf{1}$, displacement gradient $\boldsymbol{H}$ and first Piola-Kirchhoff stress tensor $\boldsymbol{P}$. Note that the value of J represents the length of the material force vector. For elaborations on error control of material forces we refer to Rüter and Stein, 2003.

3. GENERAL FRAMEWORK FOR GOAL-ORIENTED ERROR ANALYSIS

3.1. THE DISCRETIZATION ERROR

In this section we turn our attention to a general framework for goal-oriented error analysis. We shall begin our study by introducing the finite element discretization error $\boldsymbol{e}_u = \boldsymbol{u} - \boldsymbol{u}_h$. Furthermore, let us define the weak form of the residual $R : \mathcal{V} \to \mathbb{R}$ by

$$R(\boldsymbol{v}) = F(\boldsymbol{u}_h) - a(\boldsymbol{u}_h, \boldsymbol{v}). \tag{5}$$

In particular, with (3) it turns out that $R(\boldsymbol{v}_h) = 0$ for all $\boldsymbol{v}_h \in \mathcal{V}_h$ which is also referred to as the *Galerkin orthogonality* for general nonlinear problems. Recalling (1), we may rewrite (5) as follows

$$R(\boldsymbol{v}) = a(\boldsymbol{u}, \boldsymbol{v}) - a(\boldsymbol{u}_h, \boldsymbol{v}) = \int_0^1 a'(\boldsymbol{\xi}(s); \boldsymbol{e}_u, \boldsymbol{v}) \, \mathrm{d}s \quad \forall \boldsymbol{v} \in \mathcal{V} \tag{6}$$

with $\boldsymbol{\xi}(s) = \boldsymbol{u}_h + s\boldsymbol{e}_u$, $s \in [0, 1]$, where we made use of the fundamental theorem of calculus. Here, $a' : \mathcal{V} \times \mathcal{V} \to \mathbb{R}$ is the tangent bilinear form of a, i.e. the Gâteaux derivative of a with respect to $\boldsymbol{u}$. Now, we may construct a bilinear form $a_S : \mathcal{V} \times \mathcal{V} \to \mathbb{R}$ defined by

$$a_S(\boldsymbol{u}, \boldsymbol{u}_h; \boldsymbol{e}_u, \boldsymbol{v}) = \int_0^1 a'(\boldsymbol{\xi}(s); \boldsymbol{e}_u, \boldsymbol{v}) \, \mathrm{d}s \, . \tag{7}$$

It is important to note that a_S represents an *exact* linearization of the semi-linear form a and is therefore a secant form. Combining (7) and (6) finally results in the *exact* error representation

$$a_S(\boldsymbol{u}, \boldsymbol{u}_h; \boldsymbol{e}_u, \boldsymbol{v}) = R(\boldsymbol{v}) \quad \forall \boldsymbol{v} \in \mathcal{V}. \tag{8}$$

Since a_S involves the exact solution $\boldsymbol{u}$, we replace $\boldsymbol{u}$ with $\boldsymbol{u}_h$. In other words, we replace a_S with the tangent form $a_T(\cdot, \cdot) = a'(\boldsymbol{u}_h; \cdot, \cdot) \approx$

$a_S(\boldsymbol{u}, \boldsymbol{u}_h; \cdot, \cdot)$ at $\boldsymbol{u}_h$ which yields the approximate error representation

$$a_T(\boldsymbol{e}_u, \boldsymbol{v}) = R(\boldsymbol{v}) \quad \forall \boldsymbol{v} \in \mathcal{V}. \tag{9}$$

Note that this error representation holds for small errors $\boldsymbol{e}_u$ only.

3.2. ERROR MEASURES

In order to control the error of the J-integral, let us introduce the *goal-oriented error measure* $E : \mathcal{V} \times \mathcal{V} \to \mathbb{R}$ that can be any differentiable (non)linear functional satisfying the condition $E(\boldsymbol{v}, \boldsymbol{v}) = 0$ for all $\boldsymbol{v} \in \mathcal{V}$, see Larsson et al., 2002. The error measure E can be either defined as a quantity of the error, such as an error norm, or as the error of a quantity, such as the nonlinear J-integral, i.e. $E(\boldsymbol{u}, \boldsymbol{u}_h) = J(\boldsymbol{u}) - J(\boldsymbol{u}_h)$. As in (6), we may now introduce the relation

$$E(\boldsymbol{u}, \boldsymbol{u}_h) = E(\boldsymbol{u}, \boldsymbol{u}_h) - E(\boldsymbol{u}_h, \boldsymbol{u}_h) = \int_0^1 E'(\boldsymbol{\xi}(s), \boldsymbol{u}_h; \boldsymbol{e}_u)\, \mathrm{d}s \tag{10}$$

with tangent form E'. Next, we may construct a linear functional $E_S : \mathcal{V} \to \mathbb{R}$ in such a fashion that

$$E_S(\boldsymbol{u}, \boldsymbol{u}_h; \boldsymbol{e}_u) = \int_0^1 E'(\boldsymbol{\xi}(s), \boldsymbol{u}_h; \boldsymbol{e}_u)\, \mathrm{d}s\,. \tag{11}$$

Note that E_S represents an *exact* linearization and is therefore a secant form. Combining (11) and (10) we end up with the important relation

$$E_S(\boldsymbol{u}, \boldsymbol{u}_h; \boldsymbol{e}_u) = E(\boldsymbol{u}, \boldsymbol{u}_h). \tag{12}$$

Again, for small errors $\boldsymbol{e}_u$, we may replace E_S with the tangent $E_T(\cdot) = E'(\boldsymbol{u}_h, \boldsymbol{u}_h; \cdot) \approx E_S(\boldsymbol{u}, \boldsymbol{u}_h; \cdot)$ at $\boldsymbol{u}_h$. In the case of the J-integral, E_T is given by

$$E_T(\boldsymbol{v}) = -\int_{\Omega_J} \mathrm{Grad}(q\boldsymbol{e}_{||}) : \tilde{\mathbb{C}}(\boldsymbol{u}_h) : (\boldsymbol{F}^T(\boldsymbol{u}_h) \cdot \mathrm{Grad}\, \boldsymbol{v})\, \mathrm{d}V \tag{13}$$

with the deformation gradient $\boldsymbol{F}$ and $\tilde{\mathbb{C}} = 2\partial\tilde{\boldsymbol{\Sigma}}/\partial \boldsymbol{C}$, with $\boldsymbol{C} = \boldsymbol{F}^T \cdot \boldsymbol{F}$, being the tensor of elastic tangent moduli associated with $\tilde{\boldsymbol{\Sigma}}$.

3.3. DUALITY TECHNIQUES

In order to derive an error representation formula for the error measure E, we follow the well-established strategy of solving an auxiliary dual problem which is based on the dual bilinear form $a_T^* : \mathcal{V} \times \mathcal{V} \to \mathbb{R}$ of a_T. As we saw earlier, for small errors $\boldsymbol{e}_u$ it is sufficient to use the approximations $a_T^*(\cdot,\cdot) = a_S^*(\boldsymbol{u}_h, \boldsymbol{u}_h; \cdot,\cdot) \approx a_S^*(\boldsymbol{u}, \boldsymbol{u}_h; \cdot,\cdot)$ and E_T. Thus, in the (approximate) dual problem we are seeking a solution $\boldsymbol{g} \in \mathcal{V}$ such that

$$a_T^*(\boldsymbol{g}, \boldsymbol{v}) = E_T(\boldsymbol{v}) \quad \forall \boldsymbol{v} \in \mathcal{V}. \tag{14}$$

Substituting $\boldsymbol{v}$ with $\boldsymbol{e}_u$ and making use of the Galerkin orthogonality, the (approximate) error representation for the error measure E reads

$$E(\boldsymbol{u}, \boldsymbol{u}_h) = E_T(\boldsymbol{e}_u) = a_T^*(\boldsymbol{g} - \boldsymbol{g}_h, \boldsymbol{e}_u) \quad \forall \boldsymbol{g}_h \in \mathcal{V}_h. \tag{15}$$

Note that if $\boldsymbol{g}_h \in \mathcal{V}_h$ is chosen to be the finite element approximation of $\boldsymbol{g}$, then $\boldsymbol{e}_g = \boldsymbol{g} - \boldsymbol{g}_h$ is referred to as the discretization error of the dual problem with associated (approximate) error representation

$$a_T^*(\boldsymbol{e}_g, \boldsymbol{v}) = \bar{R}_g(\boldsymbol{v}) \quad \forall \boldsymbol{v} \in \mathcal{V}. \tag{16}$$

Here, $\bar{R}_g(\boldsymbol{v}) = E_T(\boldsymbol{v}) - a_T^*(\boldsymbol{g}_h, \boldsymbol{v})$ is the (approximate) weak form of the residual of the dual problem.

4. RESIDUAL GOAL-ORIENTED A POSTERIORI ERROR ESTIMATORS

The objective of this section is to present strategies to obtain a posteriori estimators of the error measure E. To begin with, we introduce Ritz projections $\bar{\boldsymbol{e}}_u$ and $\bar{\boldsymbol{e}}_g$ of $\boldsymbol{e}_u$ and $\boldsymbol{e}_g$, respectively, by

$$b(\bar{\boldsymbol{e}}_u, \boldsymbol{v}) = a_T(\boldsymbol{e}_u, \boldsymbol{v}) \quad \forall \boldsymbol{v} \in \mathcal{V} \tag{17}$$
$$b(\bar{\boldsymbol{e}}_g, \boldsymbol{v}) = a_T(\boldsymbol{e}_g, \boldsymbol{v}) \quad \forall \boldsymbol{v} \in \mathcal{V}, \tag{18}$$

since $a_T(\boldsymbol{v}, \boldsymbol{v})$ may become negative for some $\boldsymbol{v} \in \mathcal{V}$. In the above, the bilinear form $b : \mathcal{V} \times \mathcal{V} \to \mathbb{R}$ denotes an arbitrary but continuous, $\mathcal{V}$-elliptic and symmetric bilinear form inducing the norm $|\!|\!|\cdot|\!|\!|_b = b(\cdot,\cdot)^{\frac{1}{2}}$ on $\mathcal{V}$. One possible option is therefore to choose b as the bilinear form in linear elasticity.

In order to derive a posteriori error estimators, let us suppose that we have computed suitable approximations $\boldsymbol{\psi}_{u,e}, \boldsymbol{\psi}_{g,e}, \bar{\boldsymbol{\psi}}_{u,e}, \bar{\boldsymbol{\psi}}_{g,e} \in \mathcal{V}_e = \{\boldsymbol{v}|_{\bar{\Omega}_e} : \boldsymbol{v} \in \mathcal{V}\}$ of $\boldsymbol{e}_u|_{\bar{\Omega}_e}, \boldsymbol{e}_g|_{\bar{\Omega}_e}, \bar{\boldsymbol{e}}_u|_{\bar{\Omega}_e}, \bar{\boldsymbol{e}}_g|_{\bar{\Omega}_e} \in \mathcal{V}_e$, respectively, on the element level. These approximations can be obtained by the solutions of the following local Neumann problems that correspond to local forms of the error representation formulas (9) and (16):

$$a_{T,e}(\boldsymbol{\psi}_{u,e}, \boldsymbol{v}_e) = \tilde{R}_e(\boldsymbol{v}_e) \quad \forall \boldsymbol{v}_e \in \mathcal{V}_e \tag{19a}$$
$$a_{T,e}(\boldsymbol{\psi}_{g,e}, \boldsymbol{v}_e) = \tilde{\bar{R}}_{g,e}(\boldsymbol{v}_e) \quad \forall \boldsymbol{v}_e \in \mathcal{V}_e \tag{19b}$$
$$b_e(\bar{\boldsymbol{\psi}}_{u,e}, \boldsymbol{v}_e) = \tilde{R}_e(\boldsymbol{v}_e) \quad \forall \boldsymbol{v}_e \in \mathcal{V}_e \tag{19c}$$

$$b_e(\bar{\psi}_{g,e}, v_e) = \tilde{\tilde{R}}_{g,e}(v_e) \quad \forall v_e \in \mathcal{V}_e. \tag{19d}$$

Here, $a_{T,e} : \mathcal{V}_e \times \mathcal{V}_e \to \mathbb{R}$, $b_e : \mathcal{V}_e \times \mathcal{V}_e \to \mathbb{R}$, $\tilde{R}_e : \mathcal{V}_e \to \mathbb{R}$ and $\tilde{\tilde{R}}_{g,e} : \mathcal{V}_e \to \mathbb{R}$ are the restrictions of a_T, b and the equilibrated residuals $\tilde{R}$ and $\tilde{\tilde{R}}_g$ (involving equilibrated tractions), respectively, to an element $\bar{\Omega}_e$.

Upon recalling (15), we then get the error estimator

$$E(u, u_h) = \sum_{n_e} a_{T,e}(e_u|_{\bar{\Omega}_e}, e_g|_{\bar{\Omega}_e}) \approx \sum_{n_e} a_{T,e}(\psi_{u,e}, \psi_{g,e}), \tag{20}$$

henceforth designated as "Neumann 1", without error bounds but good approximation properties if the approximations $\psi_{u,e}, \psi_{g,e}$ are good enough. An upper bound, that is not guaranteed, can be obtained as follows

$$|E(u, u_h)| \leq \sum_{n_e} |a_{T,e}(e_u|_{\bar{\Omega}_e}, e_g|_{\bar{\Omega}_e})| \approx \sum_{n_e} |a_{T,e}(\psi_{u,e}, \psi_{g,e})| \tag{21}$$

("Neumann 2"). Next, we employ the Cauchy-Schwarz inequality on the element level which yields (under certain assumptions) the error estimator

$$|E(u, u_h)| \leq \sum_{n_e} \|\bar{e}_u|_{\bar{\Omega}_e}\|_{b,\Omega_e} \|\bar{e}_g|_{\bar{\Omega}_e}\|_{b,\Omega_e} \approx \sum_{n_e} \|\bar{\psi}_{u,e}\|_{b,\Omega_e} \|\bar{\psi}_{g,e}\|_{b,\Omega_e} \tag{22}$$

("Neumann 3"), cf. Prudhomme and Oden, 1999 for the case of linear problems. Note that the upper bound is still not guaranteed, since we cannot estimate the local norms due to the pollution error. Finally, upon applying the Cauchy-Schwarz inequality globally, due to the properties of the Neumann problems (19) (see for example Ainsworth and Oden, 2000) an upper bound error estimator can be obtained, which reads

$$|E(u, u_h)| \leq \|\bar{e}_u\|_b \|\bar{e}_g\|_b \leq \left(\sum_{n_e} \|\bar{\psi}_{u,e}\|^2_{b,\Omega_e} \right)^{\frac{1}{2}} \left(\sum_{n_e} \|\bar{\psi}_{g,e}\|^2_{b,\Omega_e} \right)^{\frac{1}{2}} \tag{23}$$

("Neumann 4"). The bounds, however, might not be very sharp.

5. NUMERICAL EXAMPLE

In the numerical example we investigate the fracture behavior of a pre-cracked foam rubber cross specimen in plane-strain state under biaxial loading. The foam rubber is modeled by a Blatz-Ko material ($\mu = 75$ N/mm^2, $\nu = 0.225$, $f = 0.868$). Due to symmetry conditions only one quarter of the system is modeled as shown in Fig. 9.1. For the

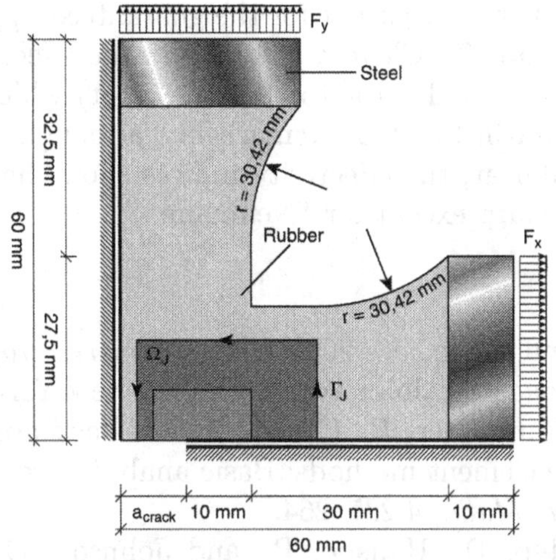

Figure 9.1. Modeled system and primal loading.

discretization we choose Q_1-elements. The experiment is carried out by a biaxial tensile/compression device that is modeled by the steel parts of the specimen for which a neo-Hooke material is chosen ($\lambda = 121153.85$ N/mm^2, $\mu = 80769.23$ N/mm^2). The loads are chosen as $F_x = 30$ N/mm^2 and $F_y = 70$ N/mm^2, and the weighting function q is defined as a modified plateau function. The reference solution $J/2 = 601.2914$ kJ/m^2 is obtained by using Q_2-elements with 76176 degrees of freedom, based on adaptive mesh refinements.

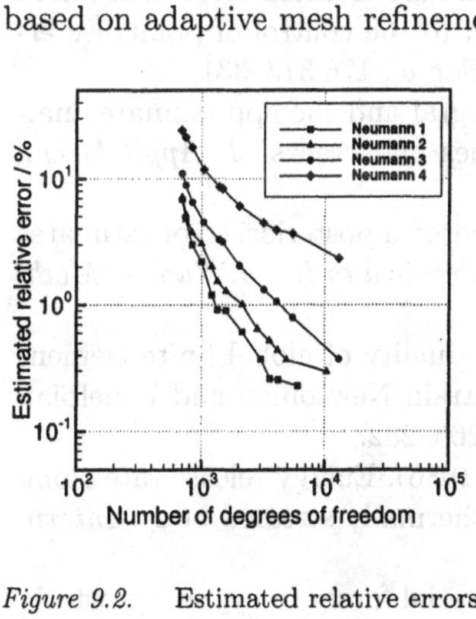

Figure 9.2. Estimated relative errors.

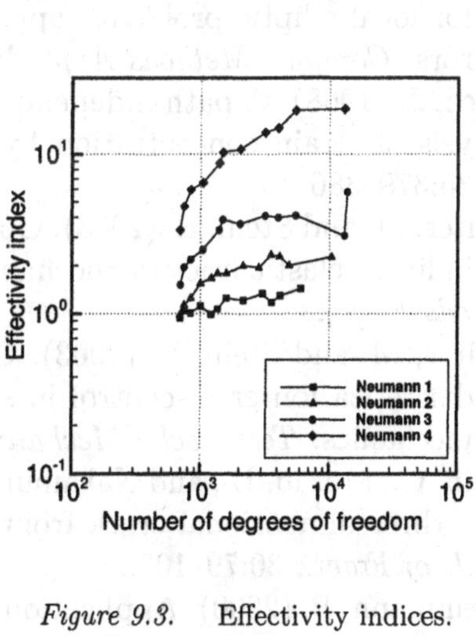

Figure 9.3. Effectivity indices.

The convergence behavior of the goal-oriented error estimators presented is depicted in Fig. 9.2. As can be seen, all error estimators show good convergence behavior except for the guaranteed upper bound error estimator "Neumann 4". Clearly, this affects the effectivity index (i.e. the ratio of the estimated error to the true error) which is plotted for each error estimator in Fig. 9.3. Remarkably, although the model problem is highly nonlinear, the effectivity indices show that the estimated errors are rather sharp except for "Neumann 4".

References

Ainsworth, M. and Oden, J. (2000). *A posteriori error estimation in finite element analysis.* John Wiley & Sons, New York.

Becker, R. and Rannacher, R. (1996). A feed-back approach to error control in finite element methods: Basic analysis and examples. *East-West J. Numer. Math.*, 4:237–264.

Eriksson, K., Estep, D., Hansbo, P., and Johnson, C. (1995). Introduction to adaptive methods for differential equations. *Acta Numer.*, pages 106–158.

Eshelby, J. (1951). The force on an elastic singularity. *Philos. Trans. Roy. Soc. London Ser. A*, 244:87–112.

Larsson, F., Hansbo, P., and Runesson, K. (2002). On the computation of goal-oriented a posteriori error measures in nonlinear elasticity. *Int. J. Numer. Methods Engrg.*, 55:879–894.

Maugin, G. (1993). *Material Inhomogeneities in Elasticity.* Chapman & Hall, London.

Prudhomme, S. and Oden, J. (1999). On goal-oriented error estimation for local elliptic problems: application to the control of pointwise errors. *Comput. Methods Appl. Mech. Engrg.*, 176:313–331.

Rice, J. (1968). A path independent integral and the approximate analysis of strain concentration by notches and cracks. *J. Appl. Mech.*, 35:379–386.

Rüter, M. and Stein, E. (2003). Goal-oriented a posteriori error estimates in linear elastic fracture mechanics. *Submitted to Int. J. Numer. Methods Engrg.*

Rüter, M. and Stein, E. (2003). On the duality of global finite element discretization error-control in small strain Newtonian and Eshelbian mechanics. *Technische Mechanik*, 23:265–282.

Shih, C., Moran, B., and Nakamura, T. (1986). Energy release rate along a three-dimensional crack front in a thermally stressed body. *Intern. J. of Fract.*, 30:79–102.

Steinmann, P. (2000). Application of material forces to hyperelastostatic fracture mechanics. I. Continuum mechanical setting. *Int. J. Solids Structures*, 37:7371–7391.

Chapter 10

MATERIAL FORCE METHOD. CONTINUUM DAMAGE & THERMO–HYPERELASTICITY

Ralf Denzer, Tina Liebe, Ellen Kuhl, Franz Josef Barth and Paul Steinmann

Chair of Applied Mechanics, University of Kaiserslautern

67663 Kaiserslautern, Germany

denzer@rhrk.uni-kl.de, liebe@rhrk.uni-kl.de, kuhl@rkrk.uni-kl.de, bath@rhrk.uni-kl.de, ps@rhrk.uni-kl.de

Abstract

The numerical analysis of material forces in the context of continuum damage and thermo–hyperelasticity constitutes the central topic of this work. We consider the framework of geometrically non–linear spatial and material settings that lead to either spatial or material forces, respectively. Thereby material forces essentially represent the tendency of material defects to move relative to the ambient material. Material forces are thus important in the context of damage mechanics and thermo–elasticity, where an evolving damage variable or thermal effects can be understood as a potential source of heterogeneity. Thus the appearance of distributed material volume forces that are due to the damage or temperature gradient necessitates the discretization of the damage or temperature variable as an independent field in addition to the deformation field. Consequently we propose a monolithic solution strategy for the corresponding coupled problem. As a result in particular global discrete nodal quantities, the so–called material node point (surface) forces, are obtained and are studied for a number of computational examples.

Keywords: Material Force Method, damage mechanics, thermo-hyperelasticity, finite element method

Introduction

The concern of this work is to establish a theoretical and computational link between defect mechanics and dissipative materials by the use of the Material

Force Method. Our developments are essentially based on the exposition of the continuum mechanics of inhomogeneities as comprehensively outlined by Maugin [8, 9], Gurtin [5] and our own recent contributions by Steinmann [15], Steinmann et al. [17] and Denzer et al. [2]. Material (configurational) forces are concerned with the response to variations of material placements of 'physical particles' with respect to the ambient material. Material forces as advocated by Maugin [10, 11] are especially suited for the assessment of general defects as inhomogeneities, interfaces, dislocations and cracks, where the material forces are directly related to the classical J-Integral in fracture mechanics. First numerical concepts of material forces within the FE-method date back to Braun [1], who derived for the hyperelastic case node point forces from the discretized potential energy with respect to the material node point positions, that contain the material stress in the spirit of Eshelby [3, 4]. Thereby the algorithmic representation of the material balance of momentum resulting in the notion of discrete material forces is proposed as the so called Material Force Method, see Steinmann [14, 17]. The Material Force Method is especially appealing in the numerical treatment of problems in fracture mechanics, see also the applications in Müller et al. [12] and Müller and Maugin [13]. In the present work we focus on continuum damage and thermo–hyperelasticity. This results in distributed material volume forces that are due to the damage or the temperature gradient, respectively. Thus the Galerkin discretization of the damage or temperature variable as an independent field becomes necessary in addition to the deformation field. The coupled problem will be solved using a monolithic solution strategy. The resulting material node point quantities, which we shall denote discrete material node point forces are demonstrated to be closely related to the classical J-integral in fracture mechanic problems. In particular we investigate the behavior of the Material Force Method in the case of cracked specimen while the damage zone or the temperature field evolves. Thereby e.g. a shielding effect of the distributed damage field w.r.t. the macroscopic crack is clearly demonstrated.

1. KINEMATICS AND KINETICS

To set the stage, we review the underlying basic geometrically non–linear kinematics of the quasi–static material motion problem.

In the material motion problem $\mathcal{B}_t$ denotes the spatial configuration occupied by the body of interest at time t. Then $\boldsymbol{\Phi}(\boldsymbol{x})$ denotes the non–linear deformation map assigning the spatial placements $\boldsymbol{x} \in \mathcal{B}_t$ of a 'physical particle' to the material placements $\boldsymbol{X} = \boldsymbol{\Phi}(\boldsymbol{x}) \in \mathcal{B}_0$ of the same 'physical particle'. Thus, the material placements are followed through the ambient material at fixed spatial position, i.e. the observer takes essentially the Eulerian viewpoint.

Next, the material motion linear tangent map is given by the deformation gradient $\boldsymbol{f} = \nabla_x \boldsymbol{\Phi}$ transforming line elements from the tangent space $T\mathcal{B}_t$ to line elements from the tangent space $T\mathcal{B}_0$, see Fig. 10.1. The spatial Jacobian, i.e. the determinant of $\boldsymbol{f}$ is denoted by $j = \det \boldsymbol{f}$ and relates volume elements $dv \in \mathcal{B}_t$ to volume elements $dV \in \mathcal{B}_0$.

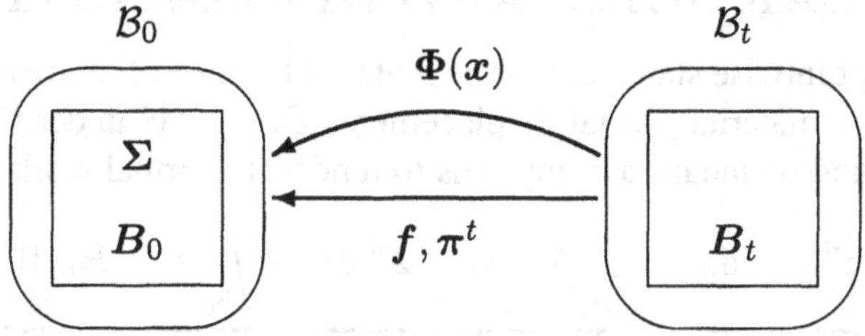

Figure 10.1. Kinematics and Kinetics of the Material Motion Problem

For the material motion problem the quasi–static balance of momentum reads

$$-\text{div}\boldsymbol{\pi}^t = \boldsymbol{B}_t \quad \longrightarrow \quad -\text{Div}\boldsymbol{\Sigma}^t = \boldsymbol{B}_0 \tag{1}$$

It involves the momentum flux $\boldsymbol{\pi}^t$, a two–point tensor that we shall call the material motion Piola stress, see Fig. 10.1, and the momentum source $\boldsymbol{B}_t$, a vector in material description with spatial reference called the material motion volume force density.

The Piola transformation of $\boldsymbol{\pi}^t$ is called the Eshelby stress $\boldsymbol{\Sigma}^t = J\boldsymbol{\pi}^t \cdot \boldsymbol{f}^t$, alternatively the terminology energy-momentum tensor or configurational stress tensor is frequently adopted. The spatial motion volume force density with material reference is given by $\boldsymbol{B}_0 = J\boldsymbol{B}_t$.

2. CONSTITUTIVE EQUATION

For the material motion problem the free energy density $\psi_t = j\psi_0$ with spatial reference is expressed in terms of the material motion deformation gradient $\boldsymbol{f}$ (or its inverse $\boldsymbol{F}$) and additional internal variable α for a generic scalar or tensorial quantity. The explicit dependence on the material placement is captured by the field $\boldsymbol{X} = \boldsymbol{\Phi}(\boldsymbol{x})$

$$\psi_t = \psi_t(\boldsymbol{f}, \alpha, \boldsymbol{\Phi}(\boldsymbol{x})) \tag{2}$$

The conjugated counterpart to the internal variable α is given by $A_t = jA_0 = -d_\alpha \psi_t$. Then the familiar constitutive equations for the so-called Eshelby stress in $\mathcal{B}_0$ are given as $\boldsymbol{\Sigma}^t = j\boldsymbol{\pi}^t \cdot \boldsymbol{f}^t = \psi_0 \boldsymbol{I} - \boldsymbol{F}^t \cdot \boldsymbol{\Pi}^t$ with $\boldsymbol{\Pi}^t = \partial_F \Psi_0$. Note that the distributed volume forces now take the following particular format with respect to the additional internal variable α

$$\boldsymbol{B}_t = A_t \nabla_X \alpha - \partial_{\boldsymbol{\Phi}} \psi_t + \boldsymbol{B}_t^{ext} \tag{3}$$

where the first part is related to material heterogeneities and the second to material inhomogeneities.

3. WEAK FORM AND DISCRETIZATION

Here, the pointwise statement of the material balance of momentum, see Eq. 1, is tested by material virtual displacements $\delta \boldsymbol{\Phi} = \boldsymbol{W}$ under the necessary smoothness and boundary assumptions to render the virtual work expression

$$\underbrace{\int_{\partial \mathcal{B}_0} \boldsymbol{W} \cdot \boldsymbol{\Sigma}^t \cdot \boldsymbol{N} \, \mathrm{d}A}_{\mathfrak{W}^{sur}} = \underbrace{\int_{\mathcal{B}_0} \nabla_X \boldsymbol{W} : \boldsymbol{\Sigma}^t \, \mathrm{d}V}_{\mathfrak{W}^{int}} - \underbrace{\int_{\mathcal{B}_0} \boldsymbol{W} \cdot \boldsymbol{B}_0 \, \mathrm{d}V}_{\mathfrak{W}^{vol}} \quad \forall \boldsymbol{W} \tag{4}$$

For a conservative system the different energetic terms $\mathfrak{W}^{sur}$, $\mathfrak{W}^{int}$ and $\mathfrak{W}^{vol}$ may be interpreted by considering the material variation at fixed $\boldsymbol{x}$ of the free energy density ψ_t. As a result, the contribution $\mathfrak{W}^{sur}$ denotes the material variation of ψ_t due to its complete dependence on the material position, whereas the contributions $\mathfrak{W}^{int}$ and $\mathfrak{W}^{vol}$ denote the material variations of ψ_t due to its implicit and explicit dependence on the material position, respectively.

By expanding the geometry $\boldsymbol{x}$ elementwise with shape functions N_x^k in terms of the positions $\boldsymbol{x}_k$ of the node points and by using a Bubnov-Galerkin finite element method based on the iso–parametric concept, we end up with discrete algorithmic material node point (surface) forces at the global node point K given by

$$\mathfrak{F}^h_{sur,K} = \mathop{\mathbf{A}}_{e} \int_{\mathcal{B}_0^e} \boldsymbol{\Sigma}^t \cdot \nabla_X N^k - N_{\boldsymbol{\Phi}}^k \boldsymbol{B}_0 \, \mathrm{d}V, \tag{5}$$

whereby we denote the material surface forces $\mathfrak{F}^h_{sur,K}$ by 'SUR' in the diagrams later in the example section. Furthermore we separate them into an internal ('INT') and a volume part ('VOL')

$$\mathfrak{F}^h_{int,K} = \mathop{\mathbf{A}}_{e} \int_{\mathcal{B}_0^e} \boldsymbol{\Sigma}^t \cdot \nabla_X N^k \, \mathrm{d}V \quad \text{and} \quad \mathfrak{F}^h_{vol,K} = \mathop{\mathbf{A}}_{e} \int_{\mathcal{B}_0^e} N_{\boldsymbol{\Phi}}^k \boldsymbol{B}_0 \, \mathrm{d}V \tag{6}$$

Thus we have in summary the obvious result

$$\mathfrak{F}^h_{sur,K} = \mathfrak{F}^h_{int,K} - \mathfrak{F}^h_{vol,K} \tag{7}$$

Based on these results we advocate the Material Force Method with the notion of global discrete material node point (surface) forces, that (in the sense of Eshelby) are generated by variations relative to the ambient material at fixed spatial positions. Such forces corresponding to the material motion problem are trivially computable once the spatial motion problem has been solved.

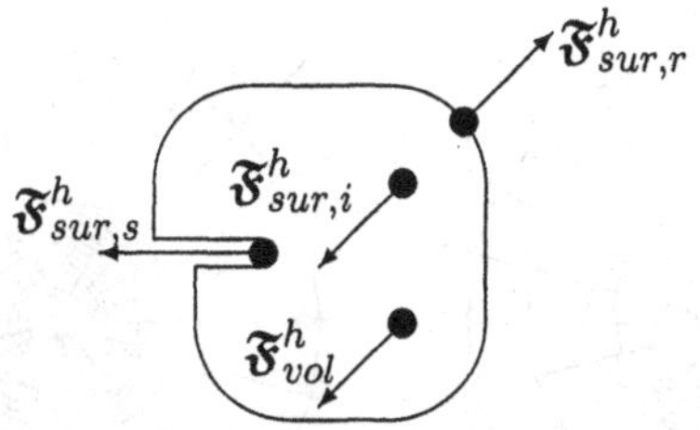

Figure 10.2. Balance of discrete material node point forces

3.1. DISCRETIZED FORMAT OF J-INTEGRAL

Consider the resulting discrete material node point (surface) force $\mathfrak{F}^h_{sur,s}$ acting on a crack tip, see Fig. 10.2. The exact value $\mathfrak{F}_{sur,s}$ can be approximated by the discrete regular surface part $\mathfrak{F}^h_{sur,r}$ and the discrete volume part $\mathfrak{F}^h_{vol}$ of the discrete material node point (surface) forces $\mathfrak{F}_{sur,s} \approx -\mathfrak{F}^h_{sur,r} - \mathfrak{F}^h_{vol}$. These in turn are balanced by discrete singular material surface forces $\mathfrak{F}^h_{sur,s}$ and (spurious) discrete internal material surface forces $\mathfrak{F}^h_{sur,i}$, which stem from an insufficient discretization accuracy as $-\mathfrak{F}^h_{sur,r} - \mathfrak{F}^h_{vol} = \mathfrak{F}^h_{sur,s} + \mathfrak{F}^h_{sur,i}$. Note thus, that the sum of all discrete algorithmic material node point surface forces $\mathfrak{F}^h_{sur,K}$ corresponds according to Eq. 5 to the resulting value

$$\mathfrak{F}_{sur,s} \approx \sum_{K \in \mathcal{V}^h_0 \backslash \partial \mathcal{V}^{r,h}_0} \mathfrak{F}^h_{sur,K} = \mathfrak{F}^h_{sur,s} + \mathfrak{F}^h_{sur,i} \tag{8}$$

Thus an improved value for $\mathfrak{F}_{sur,s}$ is obtained by summing up all discrete material node point surface forces in the vicinity of the crack tip, see also Denzer et al. [2].

4. NUMERICAL EXAMPLES

As a first example we look at a hyperelastic material coupled to isotropic damage. Thereby isotropic damage is characterized by a degradation measure in terms of a scalar damage variable $0 \leq d \leq 1$ that acts as a reduction factor of the local stored energy density of the virgin material $W_0 = JW_t$ per unit volume in $\mathcal{B}_0$ (or $W_t = jW_0$ per unit volume in $\mathcal{B}_t$, respectively), which is supposed to be an objective and isotropic function in $\boldsymbol{F}$ (or $\boldsymbol{f}$, respectively), see Liebe et al. [7]. The free energy density then reads $\psi_t = \psi_t(d, \boldsymbol{f}, \boldsymbol{\Phi}(\boldsymbol{x})) = [1 - d] W_t(\boldsymbol{f}, \boldsymbol{\Phi}(\boldsymbol{x}))$ and $Y_0 = -\partial_d \psi_0 = W_0$. In this case the distributed volume forces take the particular format $\boldsymbol{B}_0 = Y_0 \nabla_X d - \partial_{\boldsymbol{X}} \psi_0 - \boldsymbol{F}^t \cdot \boldsymbol{b}^{ext}_0$.

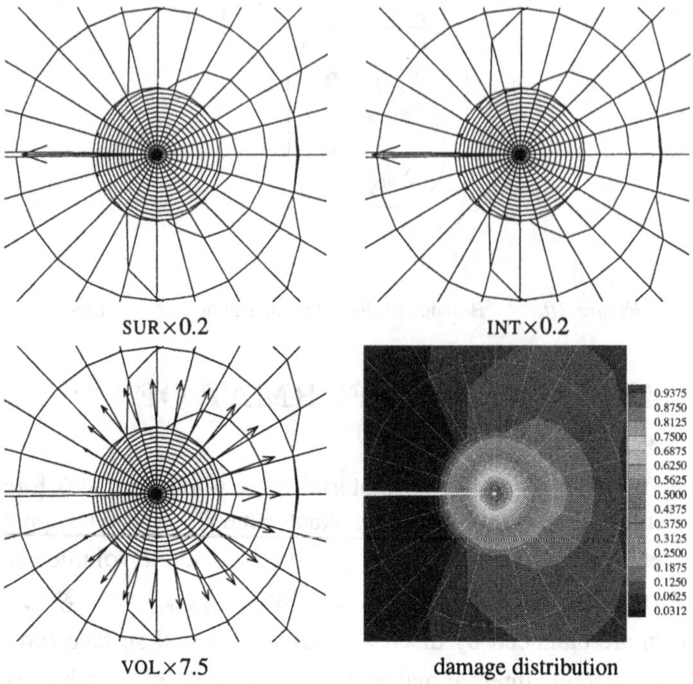

Figure 10.3. Damage distribution at crack tip for different damage states

The second example consists of thermo-hyperelastic material with the free energy function $\psi_0 = \psi_0(\boldsymbol{F}, \theta; \boldsymbol{X})$ depending on the absolute temperature θ with the entropy density given as $S_0 = -D_\theta \psi_0$. Here the distributed volume forces results in $\boldsymbol{B}_0 = S_0 \nabla_X \theta - \partial_{\boldsymbol{X}} \psi_0 - \boldsymbol{F}^t \cdot \boldsymbol{b}_0^{ext}$, see Kuhl et al. [6].

For the case of the hyperelastic material with isotropic damage we modeled the virgin material as a compressible Neo-Hookean formulation $W_0 = \mu\,[[I_1 - \ln J]/2 - 3] + \lambda \ln^2 J/2$ with the Lamé parameters μ, λ. We consider a 'Modified Boundary Layer' formulation (MBL-formulation) of a straight, traction free crack. The MBL-formulation is based on an isolated treatment of the crack tip region which is independent of the surrounding specimen. We prescribe a given material force at the boundary of the MBL-region by Cauchy stresses $\boldsymbol{\sigma}$ gained from the first term of the asymptotic linear elastic stress series near a crack tip given by Williams [18] as $\boldsymbol{\sigma} = K_I/\sqrt{2\pi r}\,\boldsymbol{f}(\theta)$. In the case of small external loads the linear elastic relation $J_{pre} = K_I^2/E'$ with $E' = E/[1 - \nu^2]$ for plane strain between the J-integral and the mode I stress intensity factor K_I holds. The resulting discrete material forces for a damaged crack tip and a contour plot of the damage variable are depicted in Fig. 10.3.

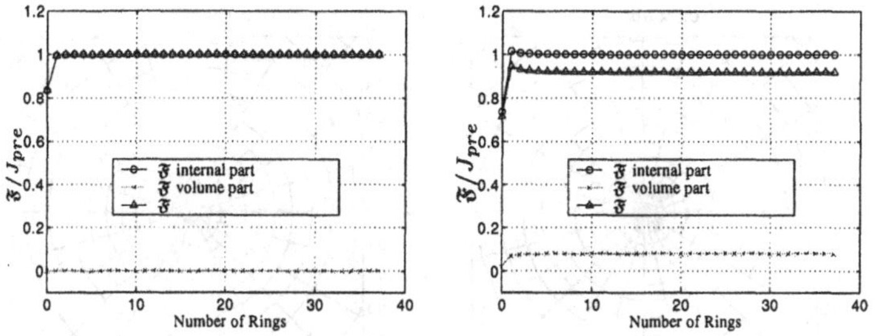

Figure 10.4. Purely elastic state vs. advanced damage state

To investigate the influence of the damage zone on the resulting material force acting on the crack tip we compare a purely elastic state and an advanced damaged state, see Fig. 10.4.

According to Eq. 8 the sum of all discrete algorithmic material node point surface forces renders an improved value for the material force at the crack tip. The sum is taken over a varying number of rings of elements around the crack tip. After only a few number of rings the internal part of the material node point surface force is converged to the prescribed material load and remains constant. Due to the increasing damage zone around the crack tip the volume part of the discrete material node point forces increases accordingly with the gradient of the damage field. Therefore the discrete material node point surface force on the crack tip is decreased due to the evolving damage around the crack tip compared to the purely elastic state. Thus the crack tip might be considered as being shielded by the distributed damage field. This 'shield' is formed closely around the crack tip and converges after a few rings to a constant value.

For thermo–hyperelastic materials, we introduce the following free energy function

$$\begin{aligned} \Psi_0 \;=\; & \tfrac{\lambda}{2}\ln^2 J + \tfrac{\mu}{2}\,[\boldsymbol{b} - \boldsymbol{I}] : \boldsymbol{I} - \mu \ln J \\ & -3\alpha\kappa\,[\theta - \theta_0]\tfrac{\ln J}{J} \\ & +c_0\,[\theta - \theta_0 - \theta \ln \tfrac{\theta}{\theta_0}] - [\theta - \theta_0] S^\circ \end{aligned} \tag{9}$$

whereby the first three terms represent the classical free energy function of Neo–Hooke type characterized through the two Lamé constants λ and μ. The fourth term introduces a thermo–mechanical coupling in terms of the thermal expansion coefficient α weighting the product of the bulk modulus κ and the difference between the current temperature θ and the reference temperature θ_0. The fifth term finally accounts for the purely thermal behavior in terms of the specific heat capacity c_0 and the last term defines the absolute entropy density S° at the reference temperature θ_0. Moreover, we assume the material heat flux $\boldsymbol{Q}$ to obey Fourier's law $\boldsymbol{Q} = -K_0\,\boldsymbol{G} \cdot \nabla_X \theta$, introducing a materially

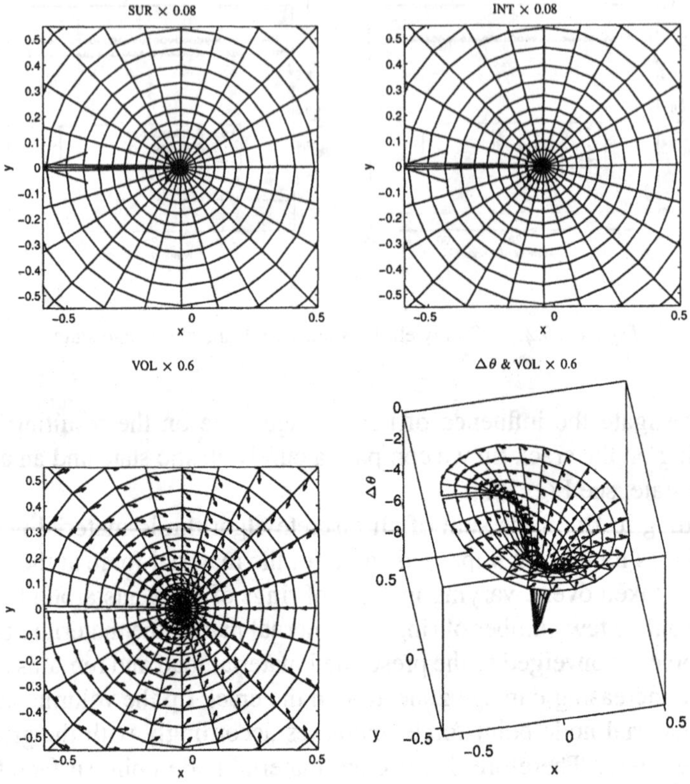

Figure 10.5. Discrete material node point surface, internal and volume forces and temperature distribution in the vicinity of the crack tip at time $t = 0.01$ and $t = 1.0$

isotropic behavior in terms of the conductivity K_0, the material metric $\boldsymbol{G}$ and the material temperature gradient $\nabla_X \theta$.

As a example we want to discuss a single edged tension specimen typically used in fracture mechanics. The height to width ratio is set to $H/W = 3$ and the ratio of crack length to width is $a/W = 0.5$. The specimen is discretized by bilinear Q1-elements and the mesh is heavily refined around the crack tip. The elements which are connected to the crack tip are P1-elements. The material is modeled with the parameter given in the previous section which roughly corresponds to an Aluminium alloy. A constant symmetric elongation of totally 0.1667% of the height W is applied at the top and bottom of the specimen within the time step $\Delta t = 0.01$. The computed discrete material node point /surface) forces 'SUR', the internal part 'INT', the volume part 'VOL' and the temperature distribution $\Delta\theta$ in the vicinity of the crack tip are shown in Fig. 10.5 at the time state $t = 0.01$.

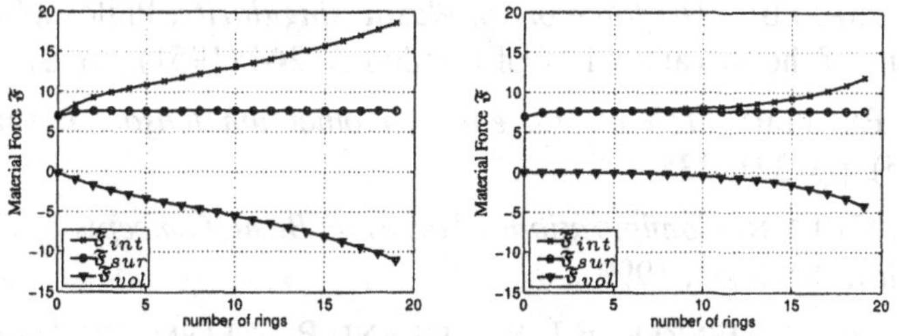

Figure 10.6. Material Forces at time $t = 0.01$

We now apply our improvement of the Material Force Method for the vectorial J-Integral evaluation, as given by Eq. 8 The resulting material forces in the vicinity of the crack tip are shown in Fig. 10.6.

Although the internal part $\boldsymbol{F}_{int}$ and the volume part $\boldsymbol{F}_{vol}$ of the material surface force are domain dependent due to the temperature gradient $\nabla_X \theta$ the resulting material surface force $\boldsymbol{F}_{sur}$ behaves domain independent.

5. CONCLUSION

The objective of this work was to exploit the notion of material forces within the framework of isotropic geometrically non-linear continuum damage and thermo–hyperelasticity. Thereby the Material Force Method, see e.g. Steinmann et al. [17] was combined with an internal variable formulation of computational continuum damage mechanics and thermo–hyperelasticity. This particularly leads to the notion of distributed material volume forces that are conjugated to the damage or the temperature gradient, respectively. To this end, it was necessary to set up a two field formulation, i.e. the additional discretization of the damage variable or the temperature as an independent field next to the deformation field. With regard to fracture mechanics it could be shown that the evolving damage zone shields the crack tip. Hence the driving material surface force at the crack tip is significantly less than the applied external material load. This is the crucial difference with the hyperelastic case, where both are equal.

References

[1] M. Braun, *Configurational forces induced by finite-element discretization*, Proc. Estonian Acad. Sci. Phys. Math., **46** (1997), pp. 24–31.

[2] R. Denzer, F.J. Barth, and P. Steinmann, *Studies in elastic fracture mechanics based on the material force method*, Int. J. Num. Meth. Eng., in press (2003).

[3] J.D. ESHELBY, *The force on an elastic singularity*, Philosophical transactions of the Royal Society of London A, 244 (1951), pp. 87–112.

[4] J.D. ESHELBY, *The elastic energy-momentum tensor*, J. Elasticity, **5** (1975), pp. 321–335.

[5] M.E. GURTIN, *Configurational Forces as Basic Concepts of Continuum Physics*, (Springer, 1999).

[6] E. KUHL, R. DENZER, F.J. BARTH AND P. STEINMANN *Application of the material force method to thermo–hyperelasticity*, Comp. Meth. Appl. Mech. Eng., **193** (2004), pp. 3303–3325.

[7] T. LIEBE, R. DENZER AND P. STEINMANN *Application of the material force method to isotropic continuum damage*, Computational Mechanics, **30** (2003), pp. 171–184.

[8] G.A. MAUGIN, *Material Inhomogeneities in Elasticity*, (Chapman & Hall, London, 1st ed., 1993).

[9] G.A. MAUGIN, *Material forces: Concepts and applications*, Appl. Mech. Rev., **48** (1995), pp. 213–245.

[10] G.A. MAUGIN, *Canonical momentum and energy in elastic systems with additional state variables*, C. R. Acad. Sci. Paris, **323IIb** (1996), pp. 407–412.

[11] G.A. MAUGIN, *On the universality of the thermomechanics of forces driving singular sets*, Arch. Appl. Mech., **70** (2000), pp. 31–45.

[12] R. MÜLLER, S. KOLLING, AND D. GROSS, *On configurational forces in the context of the Finite Element method*, Int. J. Num. Meth. Eng., **53** (2002), pp. 1557–1574.

[13] R. MÜLLER AND G.A. MAUGIN, *On material forces and Finite Element discretization*, Comp. Mech., **29** (2002), pp. 52–60.

[14] P. STEINMANN, *Application of material forces to hyperelastostatic fracture mechanics. Part I: continuum mechanical setting*, Int. J. Solids Struct., **37** (2000), pp. 7371–7391.

[15] P. STEINMANN, *On spatial and material settings of hyperelastodynamics*, Acta Mechanica, **156** (2002), pp. 193–218.

[16] P. STEINMANN, *On spatial and material settings of thermo-hyperelastodynamics*, J. Elasticity, **66** (2002), pp. 109–157.

[17] P. STEINMANN, D. ACKERMANN, AND F.J. BARTH, *Application of material forces to hyperelastostatic fracture mechanics. Part II: computational setting*, Int. J. Solids Struct., **38** (2001), pp. 5509–5526.

[18] M.L. WILLIAMS, *On the stress distribution at the base of a stationary crack*, Journal of Applied Mechanics, **24** (1957), pp. 109–114.

Chapter 11

DISCRETE MATERIAL FORCES IN THE FINITE ELEMENT METHOD

Ralf Mueller and Dietmar Gross

Institute of Mechanics, TU Darmstadt, D-64289 Darmstadt, GERMANY

r.mueller@mechanik.tu-darmstadt.de

Abstract This paper discusses the calculation of discrete material forces within the Finite Element Method. The possible use of these discrete material forces in r- and h-adaptive procedures is demonstrated by two short examples from non-linear elasticity.

Keywords: Material forces, configurational forces, FEM, adaptivity

1. INTRODUCTION

The intention of this paper is not to elaborate on the wide possible use of material/configurational forces in the context of inhomogeneous materials or fracture and defects mechanics, but to concentrate on the possible use of material forces in the Finite Element Method (FEM). The central part in the theory of material forces (see Maugin, 1993) or of configurational forces (see Gurtin, 1995; Gurtin, 2000) is the energy-momentum tensor. This quantity was introduced by Eshelby (see Eshelby, 1951; Eshelby, 1970) in continuum mechanics and is therefore also termed Eshelby-stress tensor. The use of material forces within the FEM was first mentioned (to the knowledge of the authors) in Braun, 1997. This line of research was continued in Steinmann, 2000; Steinmann et al., 2001; Mueller et al., 2002; Mueller and Maugin, 2002.

2. MATERIAL FORCE BALANCE

For a hyper-elastic material a strain energy density

$$W = \hat{W}(\boldsymbol{F}, \boldsymbol{X}) \tag{1}$$

per unit volume of the reference configuration exists. It is assumed that W depends on the deformation gradient $\boldsymbol{F}$ and explicitly on the position $\boldsymbol{X}$ (in the reference configuration). The first Piola-Kirchhoff stress tensor is given by

$$\boldsymbol{P} = \frac{\partial \hat{W}}{\partial \boldsymbol{F}}. \tag{2}$$

In static equilibrium the Piola-Kirchhoff stress satisfies the (local) equilibrium condition

$$\mathrm{Div}\boldsymbol{P} + \boldsymbol{f} = \boldsymbol{0}. \tag{3}$$

The vector $\boldsymbol{f}$ represents the physical volume forces (defined per unit volume of the reference configuration).
The gradient of the strain energy W with respect to the reference configuration in conjunction with compatibility and the equilibrium condition (3) yields after rearrangement of terms the material or configurational force balance

$$\mathrm{Div}\boldsymbol{\Sigma} + \boldsymbol{g} = \boldsymbol{0}. \tag{4}$$

The configurational stress tensor, the Eshelby stress tensor or the energy-momentum tensor $\boldsymbol{\Sigma}$ and the configurational force $\boldsymbol{g}$, given by

$$\boldsymbol{\Sigma} = W\boldsymbol{1} - \boldsymbol{F}^{\mathrm{T}}\boldsymbol{P} \quad \text{and} \quad \boldsymbol{g} = -\boldsymbol{F}^{\mathrm{T}}\boldsymbol{f} - \left.\frac{\partial \hat{W}}{\partial \boldsymbol{X}}\right|_{\text{expl.}}, \tag{5}$$

are introduced to obtain a formula that resembles the structure of equation (3). Introducing the symmetric second Piola-Kirchhoff stress tensor $\boldsymbol{S}$, and the symmetric right Cauchy-Green tensor $\boldsymbol{C}$, defined by

$$\boldsymbol{S} = \boldsymbol{F}^{-1}\boldsymbol{P} \quad \text{and} \quad \boldsymbol{C} = \boldsymbol{F}^{\mathrm{T}}\boldsymbol{F} \tag{6}$$

respectively, $(5)_1$ can be written as

$$\boldsymbol{\Sigma} = W\boldsymbol{1} - \boldsymbol{C}\boldsymbol{S}, \quad \text{thus} \quad \boldsymbol{\Sigma}\boldsymbol{C} = \boldsymbol{C}\boldsymbol{\Sigma}^{\mathrm{T}}. \tag{7}$$

The last expression is the symmetry of the energy-momentum tensor $\boldsymbol{\Sigma}$ with respect to right Cauchy-Green tensor $\boldsymbol{C}$. From (4) an important observation can be made: If the body is homogeneous and no body forces are applied, the divergence of the energy-momentum tensor vanishes, i.e.

$$\mathrm{Div}\boldsymbol{\Sigma} = \boldsymbol{0}. \tag{8}$$

Thus within the body the energy-momentum tensor satisfies a strict conservation law, see for example Kienzler and Herrmann, 2000. This is an important property, that will be used in the subsequent applications.

3. NUMERICAL IMPLEMENTATION

For a FE scheme the balance equation (4) is discretized in a way consistent with the approximation of the other field equations. Starting from the weak form of (4) multiplied by a test function $\boldsymbol{\eta}$,

$$\int_{\mathcal{B}_0} (\mathrm{Div}\boldsymbol{\Sigma} + \boldsymbol{g}) \cdot \boldsymbol{\eta} \, dV = 0 \tag{9}$$

integration by parts yields:

$$\int_{\partial\mathcal{B}_0} (\boldsymbol{\Sigma}\boldsymbol{N}) \cdot \boldsymbol{\eta} \, dA - \int_{\mathcal{B}_0} \boldsymbol{\Sigma} : \mathrm{Grad}\boldsymbol{\eta} \, dV + \int_{\mathcal{B}_0} \boldsymbol{g} \cdot \boldsymbol{\eta} \, dV = 0. \tag{10}$$

It is assumed that the test function $\boldsymbol{\eta}$ vanishes on the boundary $\partial\mathcal{B}_0$, therefore the first integral in (10) is zero. This assumption corresponds to a stationary boundary, which does not change its (material/referential) position $\boldsymbol{X}$.
Here attention is restricted to a 2D formulation (plane strain). The test function in (10) is approximated in every element $\mathcal{B}_e$ by node values $\boldsymbol{\eta}^I$ and shape functions N^I:

$$\underline{\eta} = \sum_I N^I \underline{\eta}^I, \quad \text{where} \quad \underline{\eta} = [\,\eta_1 \quad \eta_2\,]^{\mathrm{T}} \quad \text{and} \quad \underline{\eta}^I = [\,\eta_1^I \quad \eta_2^I\,]^{\mathrm{T}}. \tag{11}$$

Expressing the gradient of the test function and the energy-momentum tensor in the following matrix notation

$$\begin{aligned} \underline{\mathrm{Grad}\eta} &= \textstyle\sum_I \underline{D}^I \underline{\eta}^I, \quad \text{where} \quad \underline{D}^I = \begin{bmatrix} N^I_{,X_1} & 0 \\ 0 & N^I_{,X_2} \\ N^I_{,X_2} & 0 \\ 0 & N^I_{,X_1} \end{bmatrix}, \\ \underline{\Sigma} &= [\,\Sigma_{11} \quad \Sigma_{22} \quad \Sigma_{12} \quad \Sigma_{21}\,]^{\mathrm{T}} \quad \text{and} \quad \underline{g} = [\,g_1 \quad g_2\,]^{\mathrm{T}} \end{aligned} \tag{12}$$

yields the discretized version of (10)

$$\sum_I \underline{\eta}^{I\,\mathrm{T}} \cdot \left[-\int_{\mathcal{B}_e} \underline{D}^{I\,\mathrm{T}} \cdot \underline{\Sigma} \, dV + \int_{\mathcal{B}_e} N^I \underline{g} \, dV \right] = 0. \tag{13}$$

It is mentioned that the standard Voigt-notation for symmetric tensors can not be used, as the energy-momentum tensor is in general not symmetric, see (7). The energy momentum tensor $\boldsymbol{\Sigma}$ is only symmetric if the material is isotropic, i.e. $\boldsymbol{C}$ and $\boldsymbol{S}$ are coaxial. As (13) has to be satisfied for arbitrary node values $\boldsymbol{\eta}^I$, the term in square brackets must vanish. This naturally introduces the discrete material forces as

$$\begin{aligned} \underline{G}_e^I &= \begin{Bmatrix} G_{e1}^I \\ G_{e2}^I \end{Bmatrix} = \int_{\mathcal{B}_e} N^I \underline{g}^I \, dV = \int_{\mathcal{B}_e} \underline{D}^{I\,\mathrm{T}} \cdot \underline{\Sigma} \, dV \\ &= \int_{\mathcal{B}_e} \begin{Bmatrix} N^I_{,X_1}\Sigma_{11} + N^I_{,X_2}\Sigma_{12} \\ N^I_{,X_1}\Sigma_{21} + N^I_{,X_2}\Sigma_{22} \end{Bmatrix} dV. \end{aligned} \tag{14}$$

The material forces G_e^I of all n_e elements adjacent to node K are then assembled to a total material force $\underline{G}^K = \bigcup_{e=1}^{n_e} \underline{G}_e^I$. This does not pose a new boundary value problem, as the nodal values G^K are obtained purely from quantities that are already known (strain energy, stress and deformation measures) from the solution of displacement field.

4. ADAPTIVITY

4.1. *R*-ADAPTIVITY

As was discussed in section 2 the material volume force g has to vanish for a homogeneous body without physical volume forces. In the discretized boundary value problem it is equivalent to demand that the material node forces G^I vanish. Due to the requirement that the test function for the material force balance vanishes on the boundary, discrete material node forces occur on the boundary as reaction forces to this constraint. Therefore only discrete material forces for nodes in the interior can be expected to vanish. It is well understood that the approximative character of the FE method introduces some spurious (numerical) material forces in the interior. This originates from the fact that material forces are energetic quantities, as the energy-momentum tensor is computed from strain energy, stresses and strains. Using a standard FE discretization only the displacements are approximated continuously (C^0), not their derivatives and therefore strains, stresses and strain energy are not continuous. The discontinuities between the elements lead to numerically caused material forces.
It is reasonable to ask if an internal arrangement of nodes exists, which forces the material forces to vanish. This question is a finite dimensional (probably non-convex) optimization problem, which will be illustrated in more detail in the example 5.1. Allowing all interior nodes to be movable, we can propose a simple updated rule

$$\boldsymbol{X}^K_{\text{new}} \to \boldsymbol{X}^K_{\text{old}} - c\boldsymbol{G}^K, \tag{15}$$

where K represents all interior nodes. The constant c has to be sufficiently small to avoid "unhealthy" mesh distortions. This technique can be understood as a steepest descent method, see Mueller et al.,

2002; Mueller and Maugin, 2002 which reduces the total potential of the system. In Sussman and Bathe, 1985 a similar strategy is used without relation to the concept of material forces.
The r-adaptivity leads to an energetically optimized mesh, where the connectivity remains unaltered. If the connectivity constraint is relaxed more degrees of freedom can be introduced in a h- or p-adaptive scheme. A way based on the material force balance is discussed in the following.

4.2. *H*-ADAPTIVITY

In order to derive an h-adaptive scheme the magnitude of the numerically caused material forces will be used to check the discretization and assign new mesh sizes. The mesh quality is given by a local mesh size h^I, which is defined in every node I. In the following examples the quantity h^I is the smallest edge length of all elements adjacent to the node I. In an adaptive scheme a discretization has to be generated from a new distribution of mesh sizes. The refinement factors r^I are introduced to assign the new nodal values of the mesh size by

$$h^I_{\text{new}} \to h^I_{\text{old}} r^I. \tag{16}$$

The rule to obtain the refinement factor from the discrete material force is sketched in fig. 11.1a), where $G^I = \left|\boldsymbol{G}^I\right|$ is the absolute value of $\boldsymbol{G}^I$. As mentioned earlier, some consideration must be given to the difference between points on the boundary and interior points. The nodes are therefore split into two sets $\mathcal{N}_{\text{int}}$ and $\mathcal{N}_{\text{boun}}$, which contain the points in the interior and on the boundary, respectively. The maximum and minimum value of G^I for all interior nodes $I \in \mathcal{N}_{\text{int}}$ is denoted by $G_{\max}$ and $G_{\min}$. The corresponding node numbers are $I_{\max}$ and $I_{\min}$. At node $I_{\max}$, the mesh is strongly densified by assigning a refinement fac-

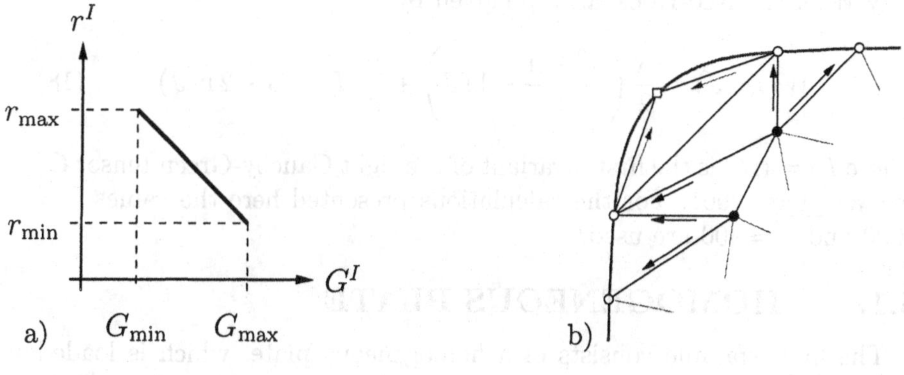

Figure 11.1. a) Refinement strategy, b) extrapolation scheme for nodes on the boundary

tor r_{min}, see(16). In this region the material force balance is only crudely approximated. Because at node I_{min} the approximation is better, only a mild refinement (or a possible coarsening) of the mesh is possible by assigning r_{max}. For all other points the refinement is linearly interpolated according to fig. 11.1a. The refinement process is controlled by the two parameters r_{max} and r_{min}. Practical experience showed that the following ranges

$$r_{\text{max}} = 0.1, \ldots, 0.5 \qquad \text{and} \qquad r_{\text{min}} = 0.5, \ldots, 0.9 \tag{17}$$

are useful. For points on the boundary ($I \in \mathcal{N}_{\text{boun}}$)an interpolation strategy is used: The refinement factors on the boundary are extrapolated from the interior of the domain. In a first step boundary nodes, which are connected by element edges to interior nodes, are assigned a refinement factor by the arithmetic mean over all connected interior nodes. For boundary nodes which are not connected to interior nodes a second extrapolation cycle is applied. The average is now taken over all neighboring boundary nodes. A schematic sketch of this procedure is shown in fig. 11.1b). In the first step the extrapolation assigns refinement factors to the nodes marked by open circles, the second step then concerns the nodes marked by open squares. With this new size distribution a new mesh is generated. We use the mesh generator GiD, which allows the assignment of so called background meshes. This strategy is used in conjuction with simple triangular elements, with a linear displacement interpolation and constant element stresses (CST: constant stress triangle).

5. EXAMPLES

We consider a non-linear isotropic elastic material with a strain energy W of the Neo-Hookean type given by

$$W(I_C, J) = \frac{\lambda}{2}\left(\frac{J^2-1}{2} - \ln J\right) + \frac{\mu}{2}\left(I_C - 3 - 2\ln J\right), \tag{18}$$

where $I_C = \mathrm{tr}\boldsymbol{C}$ is the first invariant of the right Cauchy-Green tensor $\boldsymbol{C}$, see Wriggers, 2001. For the calculations presented here the values $\lambda = 1000$ and $\mu = 400$ are used.

5.1. HOMOGENEOUS PLATE

The first example consists of a homogeneous plate, which is loaded by a displacement w on its top side. The situation is sketched in fig. 11.2a). The calculated discrete material forces at the nodes are shown in fig. 11.2b). Large material forces occur at the boundary. As discussed in the theory section, no material forces should appear in the interior, see remark after eqn. (7). Spurious material node forces originate from the fact that the FE approximation is not smooth with respect to strains and stresses. Thus the numerical value of the discrete material forces does not vanish. As material points on the boundary are not allowed to change their position in the reference configuration, large material forces appear as reaction forces to this constraint. As a first illustrative investigation we consider the following: Is it possible to find a X_2-position for the marked point in fig. 11.2c) and its symmetrical partner, so that the material forces on this points vanish? The mesh changes associated with this node movement are also sketched in fig. 11.2c). The material force G_2 is plotted as a function of the X_2-position in fig. 11.3a). For the considered node the position that causes a zero material force is

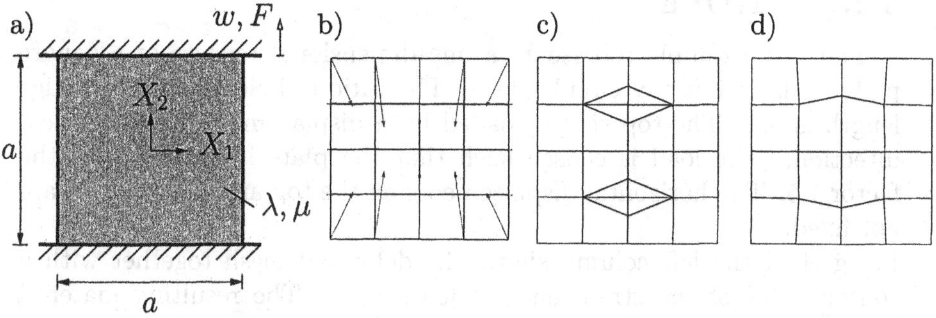

Figure 11.2. Homogenous block: a) Sketch of the problem, b) discrete material forces, c) simple mesh changes, d) optimized mesh

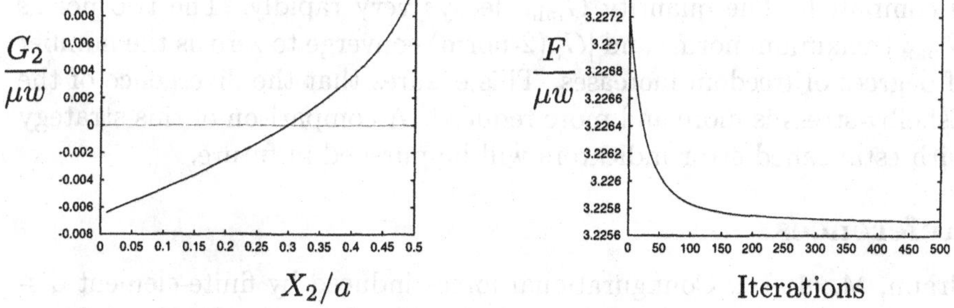

Figure 11.3. Change in a) material force, b) reaction force

at $X_2/a \approx 0.3$.
The mesh evolution of the updating strategy of eqn. (15) is presented in fig. 11.2d). The mesh modification softens the system, as can be seen from fig. 11.3b), where the evolution of the physical reaction force F resulting from the displacement of the top surface is plotted. However, the change in reaction force is relatively small, as only moderate mesh modifications are encountered.

5.2. HOLE

The first example represents a one dimensional tension of a square plate with an initially circular hole. The ratio of hole diameter to edge length is 0.4. The top side is loaded by a displacement in the vertical direction. The load is chosen such that the plate is stretched by the factor 1.5. The horizontal displacements on the top and bottom face are not fixed.
In fig. 11.4 the left column shows the deformed mesh together with a contour plot of the strain energy density W. The resulting material forces are plotted in the right column. Of course material forces occur on the boundary. In the initial mesh rather large material forces occur at interior nodes located near the diagonal. This area is refined in the first adaptive step. The refinement process continues to refine this region. In the 2. refinement a smooth distribution of element sizes develops on the hole boundary. The number of nodes and elements is reported in figure 11.5a). The plot 11.5b) verifies that during the refinement the material force balance is satisfied with increasing accuracy. Therefore the already introduced quantities $G_{\max}$ and $G_{\min}$, see section 4, are analyzed. In addition the norm of all discrete material forces in the interior, defined by

$$|G| = \Big(\sum_{I \in \mathcal{N}_{\text{int}}} G_I^2 \Big)^{1/2} \tag{19}$$

is computed. The quantity $G_{\min}$ decays very rapidly. The two norms $G_{\max}$ (maximum norm) and $|G|$ (2-norm) converge to zero as the number of degrees of freedom increases. This ensures that the divergence of the Eshelby-stress is more and more reduced. A comparison of this strategy with established error indicators will be pursued in future.

References

Braun, M. (1997). Configurational forces induced by finite-element discretization. *Proc. Estonian Acad. Sci. Phys. Math.*, 46(1/2):24–31.
Eshelby, J. D. (1951). The force on an elastic singularity. *Phil. Trans. Roy. Soc. London A*, 244:87–112.

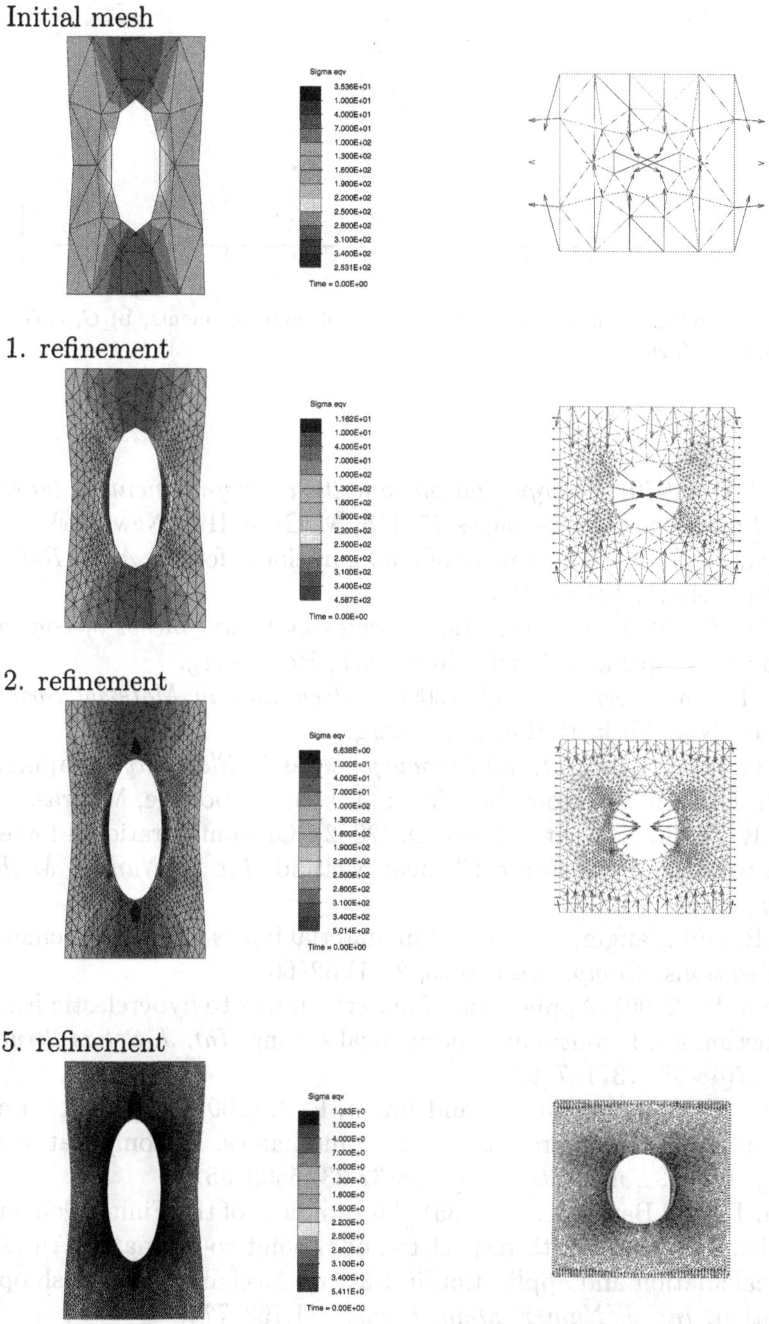

Figure 11.4. Adaptivity with $r_{\min} = 0.1$, $r_{\max} = 0.8$

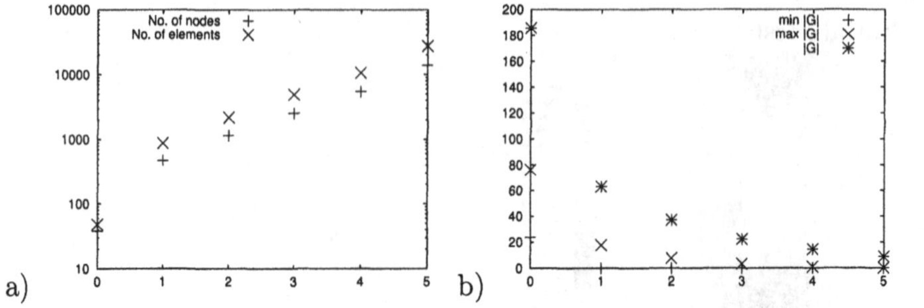

Figure 11.5. Refinement statistics: a) Number of nodes/elements, b) $G_{\min}$,$G_{\max}$ and $|G|$ during refinement

Eshelby, J. D. (1970). *Energy relations and the energy-momentum tensor in continuum mechanics*, pages 77–115. McGraw Hill, New York.

Gurtin, M. E. (1995). The nature of configurational forces. *Arch. Rational Mech. Anal.*, 131:67–100.

Gurtin, M. E. (2000). *Configurational forces as basic concept of continuum physics.* Springer, Berlin, New York, Heidelberg.

Kienzler, R. and Herrmann, G. (2000). *Mechanics in Material Space.* Springer, New York, Berlin, Heidelberg.

Maugin, G. A. (1993). *Material Inhomogeneities in Elasticity.* Chapman & Hall, London, Glasgow, New York, Tokyo, Melbourne, Madras.

Mueller, R., Kolling, S., and Gross, D. (2002). On configurational forces in the context of the Finite Element Method. *Int. J. Numer. Meth. Engng.*, 53:1557–1574.

Mueller, R. and Maugin, G. (2002). On material forces and finite element discretizations. *Comp. Mechanics*, 29(1):52–60.

Steinmann, P. (2000). Application of material forces to hyperelastic fracture mechanics. I. continuum mechanical setting. *Int. J. Solids Structures*, 37(48-50):7371–7391.

Steinmann, P., Ackermann, D., and Barth, F. J. (2001). Application of material forces to hyperelastic fracture mechanics. II. computational setting. *Int. J. Solids Structures*, 38(32-33):5509–5526.

Sussman, T., and Bathe, K. J. (1985). The gradient of the Finite Element variational indicator with respect to nodal point co-ordinates: An explicit calculation and application in fracture mechanics and mesh optimization. *Int. J. Numer. Meth. Engng.*, 21:763–774.

Wriggers, P. (2001). *Nichtlineare Finite-Element-Methoden.* Springer, Berlin, Heidelberg, New York, Barcelona, HongKong, London, Mailand, Paris, Singapur, Tokio.

Chapter 12

COMPUTATIONAL SPATIAL AND MATERIAL SETTINGS OF CONTINUUM MECHANICS. AN ARBITRARY LAGRANGIAN EULERIAN FORMULATION

Ellen Kuhl
Chair of Applied Mechanics, Faculty of Mechanical Engineering
University of Kaiserslautern, Germany
ekuhl@rhrk.uni-kl.de

Harm Askes
Computational Mechanics, Faculty of Civil Engineering and Geosciences
Delft University of Technology, Netherlands
h.askes@citg.tudelft.nl

Paul Steinmann
Chair of Applied Mechanics, Faculty of Mechanical Engineering
University of Kaiserslautern, Germany
ps@rhrk.uni-kl.de

Abstract

The present contribution aims at deriving a generic hyperelastic Arbitrary Lagrangian Eulerian formulation embedded in a consistent variational framework. The governing equations follow straightforwardly from the Dirichlet principle for conservative mechanical systems. Thereby, the key idea is the reformulation of the total variation of the potential energy at fixed referential coordinates in terms of its variation at fixed material and at fixed spatial coordinates. The corresponding Euler–Lagrange equations define the spatial and the material motion version of the balance of linear momentum, i.e. the balance of spatial and material forces, in a consistent dual format. In the discretised setting, the governing equations are solved simultaneously rendering the spatial and the material configuration which minimise the overall potential energy of the system. The remeshing strategy of the ALE formulation is thus no longer user–defined but objective in the sense of energy minimisation. As the governing equations are derived from a potential, they are inherently symmetric, both in the continuous case and in the discrete case.

Keywords: Arbitrary Lagrangian Eulerian formulation, spatial and material settings, variational principle, spatial and material forces, finite element technology.

Introduction

The deformation of a body is essentially characterised through the balance of linear momentum. Typically, this balance of momentum is formulated in terms of the spatial deformation map mapping the material placements of particles in the material configuration to their spatial placements in the spatial configuration. Alternatively, we could formulate the balance of momentum in terms of the material motion map, see [Maugin, 1993; Gurtin, 2000; Kienzler and Herrmann, 2000; Steinmann, 2002a; Steinmann, 2002b; Kuhl and Steinmann, 2003]. Being the inverse of the spatial motion map, the material motion map characterises the mapping of particles in the spatial configuration to their material motion counterparts. While the former approach relates to the equilibration of the classical spatial forces in the sense of Newton, the latter lends itself to the equilibration of material forces in the sense of Eshelby.

The finite element discretisation renders the discrete residual statement of the balance of momentum. In the classical sense, the solution of the finite element method corresponds to the vanishing residual of the spatial force balance which is related to the minimum of the potential energy. This minimum, however, is only a minimum with respect to fixed material placements.

In the discrete setting, a vanishing residual of discrete spatial forces does not necessarily imply that the discrete material forces vanish equivalently, see [Braun, 1997; Mueller and Maugin, 2002; Kuhl et al., 2003; Askes et al., 2003; Thoutireddy, 2003]. Non–vanishing discrete material forces indicate that the underlying discretisation is not yet optimal. An additional release of energy can take place when finite element nodes are moved in the direction opposite to the resulting discrete material node point forces. In this respect, the proposed strategy can be interpreted as a particular version of an Arbitrary Lagrangian Eulerian formulation, see e.g. [Donea, 1980; Hughes et al., 1981] for ALE formulations in gen-

eral and [Askes, 2000; Kuhl et al., 2003] for their particular application to adaptive remeshing. In contrast to the existing ALE formulations, however, the selection of the mesh is no longer user–defined in the proposed strategy. Rather, the optimal mesh follows straightforwardly from a variational principle. Unlike classical ALE formulations, the formulation derived herein is thus inherently symmetric.
The present work is motivated by a one–dimensional model problem introduced in section 1. After reviewing the relevant ALE kinematics in section 2, we illustrate the variational framework of our ALE formulation based on the ALE Dirichlet principle in section 3. Section 4 finally treats its spatial discretisation parameterised in terms of the material and the spatial reference mapping. We close with a brief discussion in section 5.

1. MOTIVATION

Let us motivate our work by considering a simple one–dimensional bar, clamped on both sides and loaded by a constant line load $b_0 = \text{const}$. The resulting displacement field can be determined analytically through the classical Dirichlet principle as the minimum of the potential energy $\mathcal{I}(\varphi)$. Hereby, the potential energy can be understood as the total potential energy density U integrated over the entire bar length as $\mathcal{I} = \int_{\mathcal{B}} U \, dV$ whereby $U = W + V$ typically consists of an internal contribution $W = \frac{1}{4}E\,[\,F^2 - 1 - 2\,\ln(F)\,]$ and an external contribution $V = -b\varphi$. The potential energy is thus a function of the spatial placement φ, its gradient $F = \nabla_X \varphi$, Young's modulus E and of the amount of loading b. Its minimum is characterised through its vanishing variation with respect to the independent variables, in this case, the spatial placement φ.

$$\mathcal{I}(\varphi) = \int_{\mathcal{B}} U \, dV \rightarrow \inf \qquad \delta\mathcal{I}(\varphi) = \int_{\mathcal{B}} \delta U \, dV \doteq 0 \tag{1}$$

The above equations define the spatial placement φ or rather the deformation u as the difference between material and spatial placement $u = X - \varphi$. The analytical solution is indicated through the curves in figure 12.1. However, typically, the analytical solution is not easily available such that a numerical approximation has to be carried out instead. In the simplest case, we could discretise the clamped bar with two linear finite elements as illustrated in figure 12.1, left. We thus introduce one single degree of freedom for the midpoint node which is located right in the middle of the bar at $X = 0.5\,L$. The discrete potential energy $\mathcal{I}(\varphi^h)$ can then be expressed as follows.

$$\mathcal{I}(\varphi^h) = X\,U_1^h + [\,L - X\,]\,U_2^h \rightarrow \inf_{\varphi^h} \tag{2}$$

The black lines in figure 12.1, left, illustrate the corresponding discrete solution for the displacement $u^h = X - \varphi^h$ of this classical Lagrangian analysis. The area between the curve and the lines represents a measure for the discretisation error. A significant improvement of the discrete solution can obviously be obtained by shifting the midnode to the left as illustrated in figure 12.1, right. Strictly speaking, we introduce the node point position Φ^h as an additional degree of freedom next to the spatial placement φ^h. The corresponding discrete potential energy $\mathcal{I}(\varphi^h, \Phi^h)$ can then be expressed in the following form.

$$\mathcal{I}(\Phi^h, \varphi^h) = \Phi^h U_1^h + [\, L - \Phi^h \,] U_2^h \rightarrow \inf_{\Phi^h, \varphi^h} \tag{3}$$

Figure 12.1, right, shows the solution for the material placement Φ^h and the displacement $u^h = \varphi^h - \Phi^h$ for this Arbitrary Lagrangian Eulerian analysis. The optimal node point position is found at $\Phi = 0.2800L$. The corresponding potential energy $\mathcal{I} = -12541$ indicates, that the potential energy $\mathcal{I} = -10405$ of the classical Lagrangian analysis has been reduced by more than 20 % through the relaxation of the node point position. Figure 12.2, which shows the potential energy $\mathcal{I}$ as a function of the material placements X and the displacements u supports these observations. For the classical Lagrangian analysis, we look for the minimum of the potential energy at fixed material position. It is obvious, that the potential energy can be reduced considerably upon relaxation of this node point position. In what follows, we will present a general theoretical and numerical strategy which allows to determine

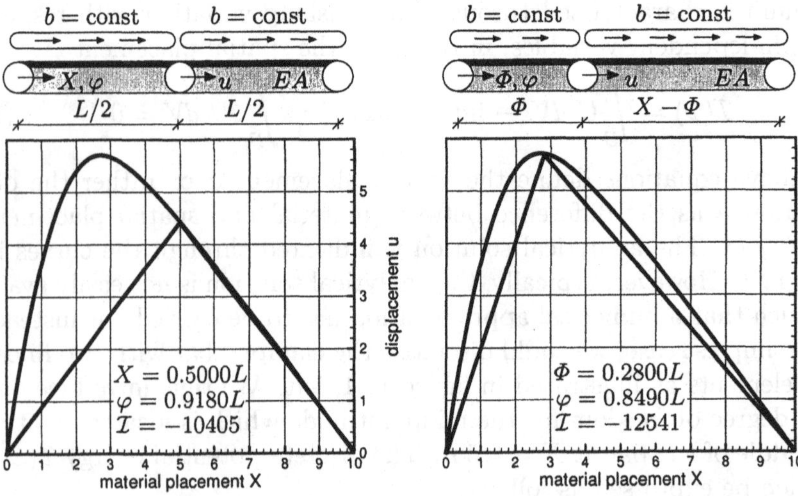

Figure 12.1. Model problem – continuous vs. discrete Lagrangian and discrete ale solution

not only the spatial but also the material placements within a consistent variational framework.

2. ALE KINEMATICS

Let us briefly summarise the fundamental kinematic relations of the present ALE formulation which is basically characterised through the introduction of an independent fixed reference domain $\mathcal{B}_\square$ next to the classical material and spatial domain $\mathcal{B}_0$ and $\mathcal{B}_t$. The following considerations are essentially based on two independent mappings, i.e. the referential maps from the fixed referential configuration $\mathcal{B}_\square$ to the material configuration $\mathcal{B}_0$ and from the referential configuration $\mathcal{B}_\square$ to the spatial configuration $\mathcal{B}_t$ as illustrated in figure 12.3. Let $\tilde{\boldsymbol{\Phi}}$ denote the

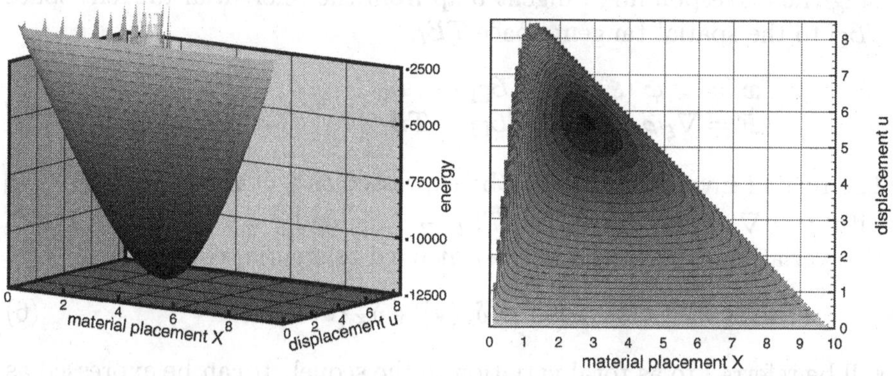

Figure 12.2. Potential energy $\mathcal{I}$ in terms of material placements X and displacements u

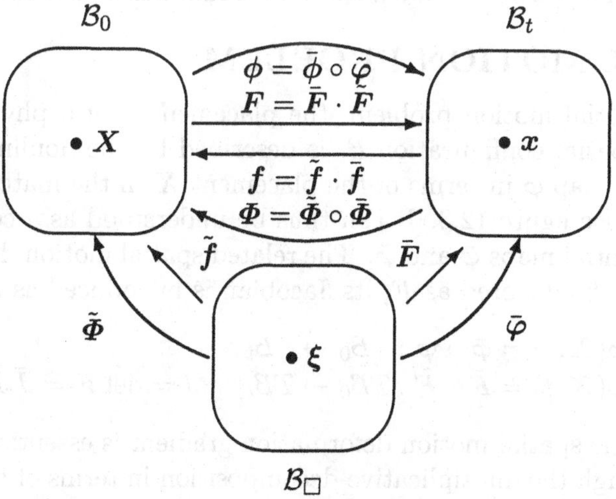

Figure 12.3. Reference domain $\mathcal{B}_\square$, material domain $\mathcal{B}_0$ and spatial domain $\mathcal{B}_t$

mapping of physical particles from a fixed position $\boldsymbol{\xi}$ in the referential domain $\mathcal{B}_\square$ to their material position $\boldsymbol{X}$ in the material domain $\mathcal{B}_0$. The related tangent map from the referential tangent space $T\mathcal{B}_\square$ to the material tangent space $T\mathcal{B}_0$ with the related Jacobian $\tilde{j}$ is denoted as $\tilde{\boldsymbol{f}}$.

$$\begin{aligned} \boldsymbol{X} &= \quad \tilde{\boldsymbol{\Phi}}\,(\boldsymbol{\xi},t): \quad \mathcal{B}_\square \rightarrow \mathcal{B}_0 \\ \tilde{\boldsymbol{f}} &= \nabla_{\xi}\tilde{\boldsymbol{\Phi}}\,(\boldsymbol{\xi},t): \ T\mathcal{B}_\square \rightarrow T\mathcal{B}_0 \qquad \tilde{j} = \det \tilde{\boldsymbol{f}} > 0 \end{aligned} \tag{4}$$

With $\tilde{j} > 0$, we assure the existence of the inverse map $\tilde{\varphi}$ with $\boldsymbol{\xi} = \tilde{\varphi}\,(\boldsymbol{X},t)\ : \mathcal{B}_0 \rightarrow \mathcal{B}_\square$ whereby $\tilde{\boldsymbol{F}} = \nabla_X \tilde{\varphi}(\boldsymbol{X},t) = \tilde{\boldsymbol{f}}^{-1}\ : T\mathcal{B}_0 \rightarrow T\mathcal{B}_\square$ with $\tilde{J} = \det \tilde{\boldsymbol{F}} = 1\,/\,\tilde{j} > 0$. Moreover, we introduce the second independent mapping $\bar{\varphi}$ from the referential domain $\mathcal{B}_\square$ to the spatial domain $\mathcal{B}_0$ with the related referential gradient $\bar{\boldsymbol{F}}$ and its Jacobian $\bar{J}$ characterising the corresponding tangent map from the referential tangent space $T\mathcal{B}_\square$ to the spatial tangent space $T\mathcal{B}_t$.

$$\begin{aligned} \boldsymbol{x} &= \quad \bar{\varphi}\,(\boldsymbol{\xi},t): \quad \mathcal{B}_\square \rightarrow \mathcal{B}_t \\ \bar{\boldsymbol{F}} &= \nabla_{\xi}\bar{\varphi}\,(\boldsymbol{\xi},t): \ T\mathcal{B}_\square \rightarrow T\mathcal{B}_t \qquad \bar{J} = \det \bar{\boldsymbol{F}} > 0 \end{aligned} \tag{5}$$

The related inverse map $\bar{\boldsymbol{\Phi}}$ can be expressed as $\boldsymbol{\xi} = \bar{\boldsymbol{\Phi}}\,(\boldsymbol{x},t)\ : \mathcal{B}_t \rightarrow \mathcal{B}_\square$ with $\bar{\boldsymbol{f}} = \nabla_x \bar{\boldsymbol{\Phi}}(\boldsymbol{x},t) = \bar{\boldsymbol{F}}^{-1}\ : T\mathcal{B}_t \rightarrow T\mathcal{B}_\square$ and $\bar{j} = \det \bar{\boldsymbol{f}} = 1\,/\,\bar{J} > 0$. The variation of any function $\{\bullet\}$ at fixed referential coordinate $\boldsymbol{\xi}$

$$\delta\{\bullet\} = \delta_x\{\bullet\} + \delta_X\{\bullet\} \tag{6}$$

will be referred to as total variation in the sequel. It can be expressed as the variation with respect to the spatial coordinates $\boldsymbol{x}$ at fixed material coordinates $\boldsymbol{X}$ denoted as $\delta_x\{\bullet\}$ plus a contribution with respect to the material coordinates $\boldsymbol{X}$ at fixed spatial coordinates $\boldsymbol{x}$ denoted as $\delta_X\{\bullet\}$.

SPATIAL MOTION PROBLEM

In the spatial motion problem, the placement $\boldsymbol{x}$ of a physical particle in the spatial configuration $\mathcal{B}_t$ is described by the nonlinear spatial deformation map φ in terms of the placement $\boldsymbol{X}$ in the material configuration $\mathcal{B}_0$, see figure 12.3. It can thus be understood as a composition of the referential maps $\bar{\varphi}$ and $\tilde{\varphi}$. The related spatial motion deformation gradient will be denoted as $\boldsymbol{F}$, its Jacobian is introduced as J.

$$\begin{aligned} \boldsymbol{x} &= \quad \varphi(\boldsymbol{X},t) = \bar{\varphi} \circ \tilde{\varphi}: \quad \mathcal{B}_0 \rightarrow \mathcal{B}_t \\ \boldsymbol{F} &= \nabla_X \varphi(\boldsymbol{X},t) = \bar{\boldsymbol{F}} \cdot \tilde{\boldsymbol{F}}: T\mathcal{B}_0 \rightarrow T\mathcal{B}_t \qquad J = \det \boldsymbol{F} = \bar{J}\,\tilde{J} > 0 \end{aligned} \tag{7}$$

Note, that the spatial motion deformation gradient is essentially characterised through the multiplicative decomposition in terms of the referential gradients $\bar{\boldsymbol{F}}$ and $\tilde{\boldsymbol{F}}$. In what follows, the variation of a function $\{\bullet\}$

with respect to the spatial coordinates $\boldsymbol{x}$ at fixed material coordinates $\boldsymbol{X}$ will be denoted as $\delta_{\boldsymbol{x}}\{\bullet\} = \delta\{\bullet\}|_{\boldsymbol{X}\text{fixed}}$. Consequently, the variation of the spatial motion deformation map and its gradient can be expressed as $\delta_{\boldsymbol{x}}\boldsymbol{\varphi} = \delta\bar{\boldsymbol{\varphi}}$ and $\delta_{\boldsymbol{x}}\boldsymbol{F} = \nabla_{\boldsymbol{X}}\delta\bar{\boldsymbol{\varphi}}$ defining the following essential relation $\delta_{\boldsymbol{x}}\{\bullet\}_0(\boldsymbol{F},\boldsymbol{\varphi};\boldsymbol{X}) = \mathrm{D}_{\boldsymbol{F}}\{\bullet\}_0 : \nabla_{\boldsymbol{X}}\delta\bar{\boldsymbol{\varphi}} + \partial_{\boldsymbol{x}}\{\bullet\}_0 \cdot \delta\bar{\boldsymbol{\varphi}}$.

MATERIAL MOTION PROBLEM

In complete analogy to the spatial motion problem, we can introduce a material motion map $\boldsymbol{\Phi}$ defining the mapping of the placement $\boldsymbol{X}$ of physical particles in the material configuration $\mathcal{B}_0$ in terms of the related placement $\boldsymbol{x}$ in the spatial configuration $\mathcal{B}_t$. As illustrated in figure 12.3, the material deformation map can thus be interpreted as a composition of the referential mappings $\tilde{\boldsymbol{\Phi}}$ and $\bar{\boldsymbol{\Phi}}$. The related linear tangent map from the spatial tangent space $T\mathcal{B}_t$ to the material tangent space $T\mathcal{B}_0$ is defined through the material motion deformation gradient $\boldsymbol{f}$ and its Jacobian j,

$$\begin{aligned} \boldsymbol{X} &= \quad \boldsymbol{\Phi}(\boldsymbol{x},t) = \tilde{\boldsymbol{\Phi}} \circ \bar{\boldsymbol{\Phi}} : \quad \mathcal{B}_t \rightarrow \quad \mathcal{B}_0 \\ \boldsymbol{f} &= \nabla_{\boldsymbol{x}}\boldsymbol{\Phi}(\boldsymbol{x},t) = \tilde{\boldsymbol{f}} \cdot \bar{\boldsymbol{f}} : T\mathcal{B}_t \rightarrow T\mathcal{B}_0 \qquad j = \det \boldsymbol{f} = \tilde{j}\,\bar{j} > 0 \end{aligned} \tag{8}$$

whereby $\boldsymbol{f}$ can be decomposed multiplicatively in the referential gradients $\tilde{\boldsymbol{f}}$ and $\bar{\boldsymbol{f}}$. The variation of any function $\{\bullet\}$ with respect to the material coordinates $\boldsymbol{X}$ at fixed spatial coordinates $\boldsymbol{x}$ denoted as $\delta_{\boldsymbol{X}}\{\bullet\} = \delta\{\bullet\}|_{\boldsymbol{x}\text{fixed}}$ defines the variation of the material motion deformation map $\delta_{\boldsymbol{X}}\boldsymbol{\Phi} = \delta\tilde{\boldsymbol{\Phi}}$ and its gradient as $\delta_{\boldsymbol{X}}\boldsymbol{f} = \nabla_{\boldsymbol{x}}\delta\tilde{\boldsymbol{\Phi}}$. We thus obtain the following essential relation $\delta_{\boldsymbol{X}}\{\bullet\}_t(\boldsymbol{f},\boldsymbol{\Phi};\boldsymbol{x}) = \mathrm{d}_{\boldsymbol{f}}\{\bullet\}_t : \nabla_{\boldsymbol{x}}\delta\tilde{\boldsymbol{\Phi}} + \partial_{\boldsymbol{X}}\{\bullet\}_t \cdot \delta\tilde{\boldsymbol{\Phi}}$.

3. ALE DIRICHLET PRINCIPLE

Restricting ourselves to hyperelastostatic conservative systems for which the Dirichlet principle defines the appropriate variational setting, we introduce the total potential energy density $U_\square$ per unit volume in $\mathcal{B}_\square$ as the sum of the internal and external potential energy density $W_\square$ and $V_\square$, thus $U_\square = W_\square + V_\square$. Then, the conservative mechanical system is essentially characterised through the infimum of the total energy $\mathcal{I}$, the integration of $U_\square$ over the reference domain $\mathcal{B}_\square$, which corresponds to its vanishing total variation $\delta\mathcal{I}$, i.e. the variation at fixed referential coordinates $\boldsymbol{\xi}$.

$$\mathcal{I}(\bar{\boldsymbol{\varphi}},\tilde{\boldsymbol{\Phi}}) = \int_{\mathcal{B}_\square} U_\square \,\mathrm{d}V_\square \rightarrow \inf \qquad \delta\mathcal{I}(\bar{\boldsymbol{\varphi}},\tilde{\boldsymbol{\Phi}}) = \int_{\mathcal{B}_\square} \delta U_\square \,\mathrm{d}V_\square \doteq 0 \tag{9}$$

According to equation (6), this total variation consists of a variation with respect to the spatial coordinates $\boldsymbol{x}$ at fixed material coordinates

$\boldsymbol{X}$ plus variation with respect to the material coordinates $\boldsymbol{X}$ at fixed spatial coordinates $\boldsymbol{x}$, $\delta\mathcal{I} = \delta_{\boldsymbol{x}}\mathcal{I} + \delta_{\boldsymbol{X}}\mathcal{I}$, as illustrated in figure 12.4.

SPATIAL MOTION PROBLEM

The variation of the total energy $\mathcal{I}$ with respect to the spatial coordinates $\boldsymbol{x}$ at fixed material coordinates $\boldsymbol{X}$ can be expressed in the following form.

$$\begin{aligned} \delta_{\boldsymbol{x}}\mathcal{I} &= \int_{\mathcal{B}_0} \nabla_{\boldsymbol{X}}\, \delta\bar{\boldsymbol{\varphi}} : \mathrm{D}_{\boldsymbol{F}} U_0 + \delta\bar{\boldsymbol{\varphi}} \cdot \partial_{\boldsymbol{x}} U_0 \;\, \mathrm{d}V_0 \\ &= \int_{\mathcal{B}_0} \nabla_{\boldsymbol{X}}\, \delta\bar{\boldsymbol{\varphi}} : \;\; \boldsymbol{\Pi}^t \;\; + \delta\bar{\boldsymbol{\varphi}} \cdot \;\; \boldsymbol{b}_0 \;\;\, \mathrm{d}V_0 \end{aligned} \tag{10}$$

Hereby, we have made use of the following definitions of the spatial motion momentum flux and source $\boldsymbol{\Pi}^t$ and $\boldsymbol{b}_0$,

$$\boldsymbol{\Pi}^t = \mathrm{D}_{\boldsymbol{F}} U_0 \qquad \boldsymbol{b}_0 = -\, \partial_{\boldsymbol{x}} U_0 \tag{11}$$

whereby the former corresponds to the first Piola–Kirchhoff stress tensor while the latter denotes the spatial volume force density per unit volume in $\mathcal{B}_0$.

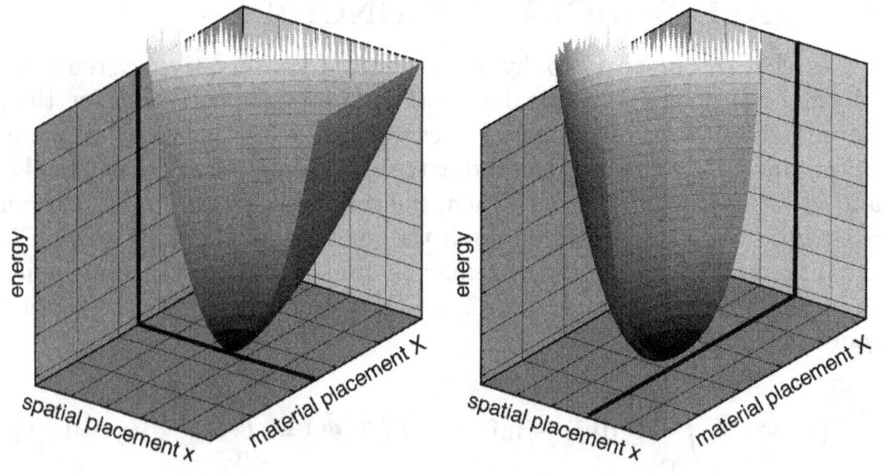

Figure 12.4. Variation of potential energy $\mathcal{I}$ with respect to spatial and material coordinates

MATERIAL MOTION PROBLEM

The variation of total energy $\mathcal{I}$ with respect to the material coordinates $\boldsymbol{X}$ at fixed spatial coordinates $\boldsymbol{x}$ can be expressed as follows.

$$\begin{aligned} \delta_{\boldsymbol{X}}\mathcal{I} &= \int_{\mathcal{B}_t} \nabla_{\boldsymbol{x}}\, \delta\tilde{\boldsymbol{\Phi}} : \mathrm{d}_{\boldsymbol{f}} U_t + \delta\tilde{\boldsymbol{\Phi}} \cdot \partial_{\boldsymbol{X}} U_t \;\mathrm{d}V_t \\ &= \int_{\mathcal{B}_t} \nabla_{\boldsymbol{x}}\, \delta\tilde{\boldsymbol{\Phi}} : \;\boldsymbol{\pi}^t \;+ \delta\tilde{\boldsymbol{\Phi}} \cdot \;\boldsymbol{B}_t \;\;\mathrm{d}V_t \end{aligned} \tag{12}$$

In complete analogy to the spatial motion case, we have introduced the material motion momentum flux $\boldsymbol{\pi}^t$ and the corresponding momentum source $\mathcal{B}_t$.

$$\boldsymbol{\pi}^t = \mathrm{d}_{\boldsymbol{f}} U_t \qquad \boldsymbol{B}_t = -\,\partial_{\boldsymbol{X}} U_t \tag{13}$$

4. ALE FINITE ELEMENT DISCRETISATION

In the spirit of the finite element method, the reference domain $\mathcal{B}_\square$ is discretised in n_{el} elements which are characterised through the corresponding element domain $\mathcal{B}^e_\square$ such that $\mathcal{B}_\square = \bigcup_{e=1}^{n_{el}} \mathcal{B}^e_\square$. The underlying geometry $\boldsymbol{\xi}$ is interpolated elementwise by the shape functions N^i_ξ in terms of the discrete node point positions $\boldsymbol{\xi}_i$ of the $i = 1, \ldots, n_{en}$ element nodes as $\boldsymbol{\xi}^h = \sum_{i=1}^{n_{en}} N^i_\xi\, \boldsymbol{\xi}_i$. Following the isoparametric concept, the unknowns $\bar{\varphi}$ and $\tilde{\boldsymbol{\Phi}}$ are interpolated on the element level with the same shape functions $N^i_{\bar{\varphi}}$ and $N^j_{\tilde{\Phi}}$ as the element geometry $\boldsymbol{\xi}$. Similar shape functions are applied to interpolate the test functions $\delta\bar{\varphi}$ and $\delta\tilde{\boldsymbol{\Phi}}$ according to the classical Bubnov–Galerkin technique.

$$\begin{aligned} \delta\bar{\varphi}^h &= \sum_{i=1}^{n_{en}} N^i_{\bar{\varphi}}\, \delta\bar{\varphi}_i \in H^0_1(\mathcal{B}_\square) \qquad \bar{\varphi}^h = \sum_{k=1}^{n_{en}} N^k_{\bar{\varphi}}\, \bar{\varphi}_k \in H_1(\mathcal{B}_\square) \\ \delta\tilde{\boldsymbol{\Phi}}^h &= \sum_{j=1}^{n_{en}} N^j_{\tilde{\Phi}}\, \delta\tilde{\boldsymbol{\Phi}}_j \in H^0_1(\mathcal{B}_\square) \qquad \tilde{\boldsymbol{\Phi}}^h = \sum_{l=1}^{n_{en}} N^l_{\tilde{\Phi}}\, \tilde{\boldsymbol{\Phi}}_l \in H_1(\mathcal{B}_\square) \end{aligned} \tag{14}$$

The discrete residual of the balance of momentum of the spatial motion problem $\mathbf{R}^I_{\bar{\varphi}}$ and of the material motion problem $\mathbf{R}^J_{\tilde{\Phi}}$ can thus be expressed as

$$\begin{aligned} \mathbf{R}^I_{\bar{\varphi}} &= \mathop{\mathbf{A}}_{e=1}^{n_{el}} \int_{\mathcal{B}^e_0} \nabla_{\boldsymbol{X}} N^i_{\bar{\varphi}} \cdot \boldsymbol{\Pi}\, \mathrm{d}V_0 - \int_{\partial\mathcal{B}^{te}_0} N^i_{\bar{\varphi}}\, \boldsymbol{t}_0\, \mathrm{d}A_0 - \int_{\mathcal{B}^e_0} N^i_{\bar{\varphi}}\, \boldsymbol{b}_0\, \mathrm{d}V_0 = \boldsymbol{0} \\ \mathbf{R}^J_{\tilde{\Phi}} &= \mathop{\mathbf{A}}_{e=1}^{n_{el}} \int_{\mathcal{B}^e_t} \nabla_{\boldsymbol{x}} N^j_{\tilde{\Phi}} \cdot \boldsymbol{\pi}\; \mathrm{d}V_t - \int_{\partial\mathcal{B}^{te}_t} N^j_{\tilde{\Phi}}\, \boldsymbol{T}_t\, \mathrm{d}A_t - \int_{\mathcal{B}^e_t} N^j_{\tilde{\Phi}}\, \boldsymbol{B}_t\, \mathrm{d}V_t = \boldsymbol{0} \end{aligned} \tag{15}$$

with the understanding that the operator $\overset{n_{el}}{\underset{e=1}{\mathbf{A}}}$ denotes the assembly over all $e = 1, \ldots, n_{el}$ element contributions at the $i, j = 1, \ldots, n_{en}$ element nodes to the global node point residuals at all $I, J = 1, \ldots, n_{np}$ global node points. Details concerning the solution of this highly nonlinear system of equations and related numerical examples are illustrated in [Kuhl et al., 2003; Askes et al., 2003].

5. DISCUSSION

The central idea of the present contribution was the derivation of an Arbitrary Lagrangian Eulerian formulation which is embedded in a consistent variational framework. Being essentially characterised through a spatial and a material mapping, the formulation is inherently related to the mechanics on the spatial and the material manifold. The governing equations follow from the evaluation of the corresponding ALE Dirichlet principle based on the total variation of the overall potential energy. By reformulating this total variation with respect to fixed reference coordinates as the sum of the variation with respect to the spatial coordinates at fixed material positions plus the variation with respect to the material coordinates at fixed spatial position, we obtained the Euler–Lagrange equations of the ALE formulation, i.e. the spatial and the material motion version of the balance of linear momentum. Their solution renders the spatial and the material configuration which minimise the overall potential energy. Discrete material forces can thus be applied to detect numerical inhomogeneities induced by the finite element discretisation. An additional release of energy can be observed when moving nodes of the finite element mesh in the direction opposite to the material force acting on it. An optimal mesh thus corresponds to vanishing discrete material node point forces. In this sense, the proposed remeshing strategy is objective with respect to energy minimisation.

References

Askes, H. (2000). *Advanced spatial discretisation strategies for localised failure.* PhD thesis, Technische Universiteit Delft, Delft, Nederlands.

Askes, H., Kuhl, E., and Steinmann, P. (2003). An ALE formulation based on spatial and material settings of continuum mechanics. Part 2: Classification and applications. *submitted for publication.*

Braun, M. (1997). Configurational forces induced by finite–element discretization. *Proc. Estonian Acad. Sci. Phys. Math.*, 46:24–31.

Donea, J. (1980). Finite element analysis of transient dynamic fluid–structure interaction. In Donéa, J., editor, *Advanced Structural Dynamics*, pages 255–290. Applied Science Publishers.

Gurtin, M. E. (2000). *Configurational Forces as Basic Concepts of Continuum Physics.* Springer Verlag.

Hughes, T. J. R., Liu, W. K., and Zimmermann, T. K. (1981). Lagrangian–Eulerain finite element formulation for viscous flows. *Comp. Meth. Appl. Mech. Eng.*, 29:329–349.

Kienzler, R. and Herrmann, G. (2000). *Mechanics in Material Space with Applications to Defect and Fracture Mechanics.* Springer Verlag.

Kuhl, E., Askes, H., and Steinmann, P. (2003). An ALE formulation based on spatial and material settings of continuum mechanics. Part 1: Generic hyperelastic formulation. *submitted for publication.*

Kuhl, E. and Steinmann, P. (2003). On spatial and material settings of thermo–hyperelastodynamics for open systems. *Acta Mechanica*, 160:179–217.

Maugin, G. A. (1993). *Material Inhomogenities in Elasticity.* Chapman & Hall.

Mueller, R. and Maugin, G. A. (2002). On material forces and finite element discretizations. *Comp. Mech.*, 29:52–60.

Steinmann, P. (2002a). On spatial and material settings of hyperelastodynamics. *Acta Mechanica*, 156:193–218.

Steinmann, P. (2002b). On spatial and material settings of thermo–hyperelastodynamics. *J. Elasticity*, 66:109–157.

Thoutireddy, P. (2003). *Variational Arbitrary Lagrangian–Eulerian method.* PhD thesis, California Institute of Technology, Passadena, California.

V

DISLOCATIONS & PEACH-KOEHLER-FORCES

Chapter 13

SELF-DRIVEN CONTINUOUS DISLOCATIONS AND GROWTH

Marcelo Epstein
Department of Mechanical and Manufacturing Engineering
The University of Calgary, Calgary, Alberta T2N 1N4, Canada
epstein@enme.ucalgary.ca

1. INTRODUCTION

That the Eshelby stress is, at least in part, at the very root of the motion of material defects and other inhomogeneities is a thermodynamic truth that few people doubt, although different people arrive at this conclusion in somewhat different ways [11, 12, 15, 17]. As a matter of record, I will briefly revue one of these avenues, which will serve also to establish my notation and conceptual framework. This done, I will proceed to suggest an evolution law that, at least in principle, is independent of the Eshelby stress. This thermodynamic impasse is not resolved in my mind, but I suspect that the Clausius-Duhem inequality may not include a complete description of the phenomenon at hand. I am not taking a strong position on this issue and I am willing to accept criticism, even of the destructive kind. This is so because I am not deeply committed to this exercise, but I just consider it an exercise worth exploring. For the purpose of the resolution of the thermodynamic issue, one may suspend judgement at this point and simply imagine the existence of some extraneous entropy sink that fixes matters (for example: a distributed array of micro-actuators with an unlimited external power to effect the desired evolution, just as a distributed array of refrigerators might push the heat flux in the wrong direction).

2. DISTRIBUTED INHOMOGENEITIES AND THEIR EVOLUTION

Inhomogeneities (dislocations, defects, etc.) and their time evolution (plasticity, growth, etc.) dwell in the body manifold. For simplicity of the exposition, we assume that a *material archetype*, that is a material "point" with a given constitutive response, has been established as a model. The geometric nature of this archetype depends on the type of body under consideration: If the material is *simple*, then the archetype is a three-dimensional Euclidean vector space (or, more pictorially, an "infinitesimal parallelepiped"); if, on the other hand, the body has some internal structure (à la Cosserat), then the archetype must be suitably supplemented with a frame of higher order; if non-local response is in the cards, the archetype might be an entire neighbourhood of $\mathbb{R}^3$, and so on. I have spent a considerable part of my life researching these topics [2, 3, 4, 5, 6, 7, 8, 10], a pursuit that can lead to truly fascinating vistas in differential geometry and its amazing physical interpretations. It is important to notice that what determines the nature of the archetype is not its actual constitutive response but rather the kinematic nature of its configuration space. Focusing our attention on local simple materials, the archetype (as mentioned above) is just a Euclidean vector space representing an infinitesimal die, whose homogeneous deformation into infinitesimal parallelepipeds is all that the material needs to know from the purely mechanical point of view in order to decide the value of its constitutive functionals. For this case, the necessary theory reduces to Noll's [18] and Wang's [19] formulation (1967) and, further back in time, to the ideas of Kondo and his collaborators [16].

That a body is *materially uniform* means that its response at all points is that of the basic archetypal behaviour. In other words, there exists a (smooth) field of *material implants* $P(X)$ of the archetype such that the (scalar, say) constitutive functional ψ is related to the archetypal response $\bar{\psi}$ by composition as follows:

$$\psi(F;X) = J_P^{-1}\bar{\psi}(FP(X))\ . \tag{1}$$

Here, F denotes the deformation gradient at the point X in a fixed reference configuration of the body, and J denotes the determinant of its matrix subscript. The appearance of the determinant is due to the fact that we interpret ψ as a density (of free energy, say) per unit volume in the reference configuration. The material implant $P(X)$ is, technically, a non-singular linear map between the archetype and the tangent space at the point X of the body in the reference configuration.

Is the materially uniform body also *homogeneous*? This would be the case if the field of inverse implants $P^{-1}(X)$ is integrable (namely, it derives from a displacement field). This can be verified analytically (at least locally) by checking the equality of certain partial derivatives. From the geometric point of view, on the other hand, the integrability (homogeneity) condition has a striking interpretation. Indeed, the implants $P(X)$ may be regarded as defining a *distant parallelism* in the body manifold by, for example, considering the local image at each point X of a fixed basis in the archetype. Two vectors at different points of the body are *materially parallel* if they have the same respective components in these local bases. The homogeneity condition can then be seen as the vanishing of the *torsion tensor* of the connection associated with this distant parallelism. If the torsion vanishes, one can find a change of reference configuration that renders the local bases parallel and identical to each other.

But what can be said if the P's are not unique? In fact, they aren't in general. It can be shown that the degree of freedom in the choice of material implants is governed by the *symmetry group* of the archetypal response function $\bar{\psi}$. More specifically, if G is a symmetry of the archetype and $P(X)$ is an implant, then so is the composition $P(X)G$. If the symmetry group happens to be continuous (as in the case of a fully or transversely isotropic solid) then the material connection described above is not unique and its torsion becomes meaningless as a measure of local inhomogeneity. In the particularly important case of a fully isotropic archetype (the symmetry group being the whole orthogonal group), it can be shown that the Riemannian (metric) connection induced by the metric tensor

$$g = (PP^T)^{-1} \tag{2}$$

is a true measure of the inhomogeneity. More specifically, if the curvature tensor $\mathcal{R}$ of this metric vanishes, the body is homogeneous.[1]

3. MATERIAL EVOLUTION

Material evolution can be defined as a change of the inhomogeneity pattern in time, namely, the field of implants is now regarded as:

$$P = P(X, t) . \tag{3}$$

This restricted kind of change in the material response of a body implies that the basic constitution of the material is not changing, but only the way in which the invariable material archetype is accommodated at

[1]For the sake of clarity of the exposition, we are not emphasizing here the subtle distinction between global and local homogeneity.

the different points. The physical interpretation of this kind of evolution varies with the context. Thus, for example, if the determinant of $P(X,t)$ is not constant in time, we may interpret this evolution as a process of volumetric growth or resorption. With a constant determinant, on the other hand, we may obtain plasticity and other such anelastic phenomena.

Having alluded above to the relation between thermodynamics, evolution and the Eshelby tensor, we consider now briefly the implementation of the chain rule in calculating, say, the material time derivative $\dot{\psi}$ of the constitutive functional, such as needed for the free energy density in the Clausius-Duhem inequality. By virtue of the basic Equation (1), a rather straightforward calculation shows that the participation of the implants P in the residual inequality is given by the following dissipative term:

$$tr(bL_P)\ , \tag{4}$$

where b is the Eshelby stress tensor and $L_P = \dot{P}P^{-1}$ is the inhomogeneity "velocity" gradient. The attractive features of this way of revealing the role of the Eshelby stress are: (i) the Eshelby stress appears naturally in the thermodynamic context by a mere use of the chain rule of differentiation; (ii) the derivation is close to the spirit of Eshelby's idea that the Eshelby stress is a change of energy brought about by a "movement" of a defect. Here, indeed, the Eshelby stress is essentially the partial derivative $\partial\psi/\partial P$; and (iii) this derivation clearly shows that the Eshelby stress is the driving force behind the motion of inhomogeneities.

Let us now consider an evolution constitutive law of the first order, such as:

$$\dot{P} = f(P, b, \text{other arguments})\ . \tag{5}$$

We say that the evolution is *intrinsic* or *self driven* if neither b nor any measure of stress or strain appear in the list of arguments, but only properties of the inhomogeneity pattern itself. This would mean that the inhomogenity moves just by virtue of its being there, perhaps in its effort to relax itself. If necessary, we may imagine, as mentioned above, that there exists a control micro-mechanism in charge of measuring the defect density and taking action accordingly. In previous work [9], we have considered the case of an intrinsic evolution of a triclinic solid body in which stresses remain identically zero throughout the evolution. Even in its simplicity this treatment afforded a qualitative description of such phenomena as dislocation pile-ups and annealing.[2] In this paper, we will therefore consider the more difficult case of a fully isotropic body.

We start by considering the following generic intrinsic evolution law:

$$\dot{P} = f(P, \mathcal{R}, \nabla\mathcal{R}; X) \,, \tag{6}$$

where ∇ denotes the gradient operator in the reference configuration. Upon a change of reference configuration, whose gradient at X we denote by $H(X)$, the arguments of this evolution law change in the following way:

$$P \longrightarrow HP \,, \tag{7}$$

$$\dot{P} \longrightarrow H\dot{P} \,, \tag{8}$$

$$\mathcal{R}^I_{JKL} \longrightarrow \mathcal{R}^I_{JKL} H^A_I H^{-J}_B H^{-K}_C H^{-L}_D \,, \tag{9}$$

$$\mathcal{R}^I_{JKL,M} \longrightarrow \mathcal{R}^I_{JKL,M} H^A_I H^{-J}_B H^{-K}_C H^{-L}_D H^{-M}_E \,, \tag{10}$$

where we have used indicial notation wherever extra clarity was necessary. In particular, choosing instantaneously $H = P^{-1}$, and invoking uniformity, we see that the evolution is characterized by a function *pulled back to the archetype* of the form:

$$\bar{L}_P = P^{-1}\dot{P} = f(\bar{\mathcal{R}}, \nabla \bar{calR}) \,, \tag{11}$$

where an overbar denotes quantities pulled back to the archetype.

If we should now effect an arbitrary orthogonal transformation of the archetype and invoke its full isotropy, we would obtain the most general reduced form of the evolution law in terms of the invariants of the curvature tensor $\mathcal{R}$. In particular, we will concentrate our attention on an evolution law of the following form:

$$\bar{L}_P = f(\bar{\kappa}, \nabla\bar{\kappa})I \,, \tag{12}$$

where κ is the scalar curvature and I denotes the identity tensor. Notice that the skew-symmetric part of $\bar{L}_P$ is unimportant in this case since it would imply an evolution within the symmetry group (cf. the *principle of actual evolution* in [13]).

[2]Moreover, because of the identical vanishing of the stress, the thermodynamic question did not arise.

4. EXAMPLE

4.1. GENERALITIES

To investigate the possibility of travelling waves of dislocations we now specialize the results of the previous sections by restricting the implant maps to the following particular form:

$$[\,P\,] = \begin{bmatrix} p(X,t) & 0 & 0 \\ 0 & 1/p(X,t) & 0 \\ 0 & 0 & 1 \end{bmatrix} \tag{13}$$

The curvature invariant κ in this particular case can be obtained by the simple formula:

$$\kappa = (p^2)'' \tag{14}$$

where we use primes to indicate partial X-derivatives. The general evolution law, therefore, can be stated as:

$$\left(\frac{\dot{p}}{p}\right)^2 = f((p^2)'', ((p^2)'''p)^2) \tag{15}$$

where f is an arbitrary function, and where a superimposed dot is used to denote partial time derivatives.

4.2. INHOMOGENEITY FADING

Excluding the gradient-of-curvature term, the evolution equation reduces to:

$$\left(\frac{\dot{p}}{p}\right)^2 = f((p^2)'')\,, \tag{16}$$

and further reducing the dependence to be linear we obtain:

$$\left(\frac{\dot{p}}{p}\right)^2 = C(p^2)''\,, \tag{17}$$

where C is a material constant. Denoting $u = p^2$, we may write this equation as:

$$\dot{u} = 2Cuu''\,, \tag{18}$$

which is a nonlinear heat equation. In fact, linearizing for small implants (namely u close to 1) we obtain the classical heat equation. This shows that for this case the inhomogeneity will tend to disappear provided C is positive. As mentioned in the introduction, this positivity does not emerge from the un-doctored Clausius-Duhem inequality, since b has been excluded a priori from the list of arguments.

4.3. HYPOCYCLOIDAL TRAVELLING WAVES

Having observed that without the inclusion of the gradient of the curvature the evolution law is purely diffusive (and cannot, therefore, sustain travelling waves), we investigate now the case in which the evolution law includes just the gradient of the curvature. In particular, we consider the linear relation:

$$\left(\frac{\dot{p}}{p}\right)^2 = K^2((p^2)'''p)^2 \tag{19}$$

where K is a positive material constant. A travelling wave is, by definition, a solution of this PDE of the form:

$$p(X,t) = p(X - vt) = p(\xi) \tag{20}$$

where v is the speed of the travelling wave. With some abuse of notation, we will keep using primes to denote ordinary derivatives with respect to the new variable ξ. Introducing the assumed form of the solution into Equation (19), we obtain the following ODE to be satisfied by $p(\xi)$:

$$v^2\left(\frac{p'}{p}\right)^2 = K^2((p^2)'''p)^2 \tag{21}$$

The fact that the speed v appears squared implies that for every solution of the form $p(X - vt)$ the function $p(X + vt)$ is also a solution. We consider now the negative square root of Equation(21), namely:

$$\frac{p'}{p} = -\frac{K}{v}(p^2)'''p \tag{22}$$

We consider the whole of space-time $\mathbb{R}^3 \times \mathbb{R}$ as the domain and we seek bounded solutions. We start by noting that Equation (22) can be rewritten as:

$$\frac{u'}{2u^{3/2}} = -\frac{K}{v}u''' \tag{23}$$

where $u = p^2$. This equation is easily seen to have the first integral:

$$u^{-1/2} = \frac{K}{v}u'' + C_1 \tag{24}$$

where C_1 is a constant of integration. If we regard this equation, for the sake of analogy, as the equation of "motion" of a particle, we conclude that the "mass" is $\frac{K}{v}$ and the "force field" is $u^{-1/2} - C_1$. The "potential energy" then is $-2u^{1/2} + C_1 u$. This analogy permits us to see immediately (see, e.g., [1]) that, provided C_1 is positive, there will be continuous periodic (and, therefore, bounded) solutions. The "total energy" of any such solution is given by the first integral:

$$\frac{K}{2v}(u')^2 - 2u^{1/2} + C_1 u = C_2 \tag{25}$$

From the particular shape of our potential energy, we conclude that the total energy C_2 must be bounded as:

$$\frac{-1}{C_1} < C_2 < 0 \tag{26}$$

For any given value of C_2 within these bounds, the solution will oscillate between the values corresponding to the vanishing of the "kinetic energy", namely:

$$\frac{1 - \sqrt{1 + C_1 C_2}}{C_1} < u^{1/2} < \frac{1 + \sqrt{1 + C_1 C_2}}{C_1} \tag{27}$$

Notice that the integration constants C_1 and C_2 are related to the "initial" conditions u_0, u_0' and u_0'' (for some value ξ_0 of ξ) by:

$$C_1 = u_0^{-1/2} - \frac{K}{v}u_0'' \tag{28}$$

and

$$C_2 = -u_0^{1/2} - \frac{K}{v}u_0 u_0'' + \frac{K}{2v}(u_0')^2 \tag{29}$$

From Equation(25) we obtain:

$$\frac{du}{\sqrt{C_2 - C_1 u + 2\sqrt{u}}} = \sqrt{\frac{2v}{K}} d\xi \tag{30}$$

which can be integrated by standard methods to:

$$\sqrt{\frac{2v}{K}}\xi = \frac{2}{C_1^{3/2}}\left[Arcsin\left(\frac{C_1\sqrt{u}-1}{\sqrt{1+C_1C_2}}\right)\right.$$
$$\left. - \sqrt{1+C_1C_2-(C_1\sqrt{u}-1)^2}\right] + C_3 \qquad (31)$$

where C_3 is an immaterial constant of integration fixing the "initial" value ξ_0.

To further clarify the shape of this solution, we notice that Equation (30) can be rewritten as the implicit formula:

$$C_1\sqrt{u}-1 = \sqrt{1+C_1C_2}sin\left[\frac{C_1^{3/2}}{2}(\sqrt{\frac{2v}{K}}\xi - C_3)\right.$$
$$\left. + \sqrt{1+C_1C_2-(C_1\sqrt{u}-1)^2}\right] \qquad (32)$$

which clearly shows the periodicity of the solution. The period is easily seen to be given by $T = 4\pi\sqrt{\frac{K}{2vC_1^3}}$, while the bounds agree with those anticipated in Equation (27).

The implicit form of the solution given in Equation (31) can be written more conveniently in parametric form. Introducing the parameter θ, as suggested by Equation (31), as:

$$sin\theta = \frac{C_1\sqrt{u}-1}{\sqrt{1+C_1C_2}} \qquad (33)$$

it follows that

$$\theta = \frac{C_1^{3/2}}{2}(\sqrt{\frac{2v}{K}}\xi - C_3) + \sqrt{1+C_1C_2-(C_1\sqrt{u}-1)^2} \qquad (34)$$

which, by virtue of (33), can be expressed as:

$$\theta = \frac{C_1^{3/2}}{2}(\sqrt{\frac{2v}{K}}\xi - C_3) + \sqrt{1+C_1C_2}\ cos\theta \qquad (35)$$

Collecting the preceding results and restoring the notation $p = \sqrt{u}$, we arrive at the following parametric equations for the travelling wave shape:

$$\frac{C_1^{3/2}}{2}(\sqrt{\frac{2v}{K}}\xi - C_3) = \theta - \sqrt{1+C_1C_2}\ cos\theta \tag{36}$$

and

$$C_1p = 1 + \sqrt{1+C_1C_2}sin\theta \tag{37}$$

These equations show that, in properly chosen scales, the graph of the travelling wave has the shape of a hypocycloid. We note that there is no a-priori limitation on the size of these travelling waves. The treatment just presented shows that there are definite relations between the size, shape and speed of travel. Specifically, given the maximum $p_{max} > 1$, the minimum $p_{min} < 1$ and the period T, we obtain:

$$\frac{1}{C_1} = \frac{p_{min} + p_{max}}{2} \tag{38}$$

$$\frac{1}{C_2} = -\frac{1}{2}\left(\frac{1}{p_{min}} + \frac{1}{p_{max}}\right) \tag{39}$$

$$v = \frac{\pi^2 K}{T^2}(p_{min} + p_{max})^3 \tag{40}$$

Thus, except for the immaterial shift C_3, the exact shape and speed of propagation of the travelling wave are completely determined by its maximum and minimum values and by the period. The larger the average and the shorter the period of the wave shape, the faster the speed of propagation.

Acknowledgment This work has been partially supported by the Natural Sciences and Engineering Research Council of Canada.

References

[1] Arnold, V., *Mathematical methods of Classical Mechanics*, Springer, 1978

[2] Elzanowski, M., Epstein, M. and Sniatycki, J., (1990), G-structures and material homogeneity, *J. Elasticity*, **23**, 167-180.

[3] Epstein, M., (1992), Eshelby-like tensors in thermoelasticity, *Nonlinear thermomechanical processes in continua*, (W. Muschik and G.A. Maugin, eds.), TUB-Dokumentation, Berlin, **61**, 147-159.

[4] Epstein, M., (1999), On the anelastic evolution of second-grade materials, *Extracta Mathematicae*, **14**, 157-161.

[5] Epstein, M., (1999), Towards a complete second-order evolution law, *Math. Mech. Solids*, **4**, 251-266.

[6] Epstein, M. and Bucataru, I., (in press), Continuous distributions of dislocations in bodies with microstructure, *J. Elasticity.*

[7] Epstein, M. and de León, M., (1996), Homogeneity conditions for generalized Cosserat media, *J. Elasticity*, **43**, 189-201.

[8] Epstein, M. and de León, M., (1998), Geometrical theory of uniform Cosserat media, *J. Geom. Physics*, **26**, 127-170.

[9] Epstein, M. and Elzanowski, M., (in press), A model of the evolution of a two-dimensional defective structure, *J. Elasticity*.

[10] Epstein, M., Elzanowski, M. and Sniatycki, J., (1987), Locality and uniformity in global elasticity,", in *Lecture Notes in Mathematical Physics*, **1139**, Springer Verlag, 300-310.

[11] Epstein, M. and Maugin, G.A., (1990), Sur le tenseur de moment matériel d'Eshelby en élasticité non linéaire, *C. R. Acad. Sci. Paris*, **310**/II, 675-768.

[12] Epstein, M. and Maugin, G.A., (1990), The energy-momentum tensor and material uniformity in finite elasticity, *Acta Mechanica*, **83**, 127-133.

[13] Epstein, M. and Maugin, G.A., (1996), On the geometrical material structure of anelasticity, *Acta Mechanica*, **115**, 119-134.

[14] Eshelby, J.D., (1951), The force on an elastic singularity, *Phil. Trans. Roy. Soc. London*, **A244**, 87-112.

[15] Gurtin, M.E., (2000), *Configurational forces as basic concepts of continuum physics*, Springer-Verlag.

[16] Kondo, K., (1955), Geometry of elastic deformation and incompatibility, *Memoirs of the unifying study of the basic problems in engineering science by means of geometry*, Tokyo Gakujutsu Benken Fukyu-Kai.

[17] Maugin, G.A., (1993), *Material inhomogeneities in elasticity*, Chapman and Hall.

[18] Noll, W., (1967), Materially uniform simple bodies with inhomogeneities, *Arch. Rational Mech. Anal.*, **27**, 1-32.

[19] Wang, C.C., (1967), On the geometric structure of simple bodies: a mathematical foundation for the theory of continuous distributions of dislocations, *Arch. Rational Mech. Anal.*, **27**, 33-94.

Chapter 14

ROLE OF THE NON-RIEMANNIAN PLASTIC CONNECTION IN FINITE ELASTOPLASTICITY WITH CONTINUOUS DISTRIBUTION OF DISLOCATIONS

Sanda Cleja-Tigoiu
Faculty of Mathematics and Informatics, str. Academiei 14, Bucharest, Romania
tigoiu@math.math.unibuc.ro

Abstract In the proposed model, the plastic distortion and the plastic connection with a non-zero Cartan's torsion and a non-metric structure, characterize the irreversible behaviour of the crystalline materials, via the appropriate evolution equations. The material behaves like an elastic material element with respect to a second order reference configuration, attached to plastic material structure.

Keywords: Plastic, elastic, connection, dislocations, torsion, couple stresses

Introduction

In this paper we propose an elasto-plastic model for crystalline materials with continuous distribution of dislocations, following the assumptions elaborated in [4]. The material geometric structure is characterized for any local motion at time t by the plastic second order pair, which generates the plastic distortion $\mathbf{F}^p$ and the plastic connection $\overset{(p)}{\Gamma}$. The plastic connection with non-zero Cartan's torsion has not a metric structure (see [12], [8]) unlike the models developed in [4], [5], where $\overset{(p)}{\Gamma}$ defines a distant parallelism.

The differential geometry concepts, see [11], [7], [10], [8], [9], [13], describe the defects of crystalline materials, which, through by their generation and motion produce plastic deformations and involve changes of the internal mechanical structure during the deformation process. Namely, Cartan torsion describes dislocations,and the non-metric connection is related to the point defects, vacancy, self interstitial, shear fault, etc. The appropriate energetic framework, that accounts for microstructural behaviour of crystalline materials with continuously distributed dislocations, is developed for generalized elasto-plastic materials in [9], for single crystal plasticity in [6], and for the consideration of the influence of the dislocation density on the hardening behaviour in [13]. The energetical selection, of the stress variables and of the appropriate second order deformation measures, is proved in [4] within the framework of elasto-plastic non-Cosserat body. The force vector and the couple vector, acting on the material surface element in the actual configuration, are generated by the non-symmetric tensors, $\mathbf{T}$, Cauchy stress tensor and $\mathbf{M}$, couple stresses.

In our model, the body behaves like an elastic material element (generalizing the description proposed in [14]), but with respect to the second-order reference configuration attached to the pair $(\mathbf{F}^p, \overset{(p)}{\mathbf{\Gamma}})$, i.e. a non-holonomic configuration. The evolution of the irreversible variables is produced by the driving forces defined by Eshelby's stress tensor, see [1], and by the appropriate stress momentum (i.e. a third order tensor generated by the couple stresses) pulled back to the plastically deformed configuration with torsion. In [3], [2] the evolution equations preserve their form under arbitrary changes of reference configurations

Unlike the models proposed in [4], [5] , [9] involving only the torsion, and the model for the description of volumetric growth in uniform bodies, based on an appropriate (motion) symmetric connection (see [2]), our model is constructed with a full non-Riemannian plastic connection.

1. GENERAL PLASTIC CONNECTION

For any motion $\chi : \mathcal{N}_X \times R \longrightarrow \mathcal{E}$ there exists a second order pair of gradients defined on the neighbourhood $\mathcal{N}_X$ of the fixed material point

$$\begin{aligned} \mathbf{F}(\mathbf{X},t) &= \nabla(\chi(\cdot,t)\circ k^{-1})\,|_{\mathbf{X}} \equiv \mathbf{F}_k(\mathbf{X},t), \\ \mathbf{G}(\mathbf{X},t) &= \nabla^2(\chi(\cdot,t)\circ k^{-1})\,|_{\mathbf{X}} \equiv \mathbf{G}_k(\mathbf{X},t), \quad \mathbf{X} = k(X,t), \end{aligned} \tag{1}$$

both of them being related to a fixed global reference configuration k. $\mathbf{F}$ the first deformation gradient is an invertible second order field, $\mathbf{G}$ (the second order deformation gradient) is a third order tensor, which satisfies the symmetry condition (or the first integrability condition, since $\chi(\cdot,t)$ and k are global configurations), for all $\mathbf{Z} \in k(\mathcal{N}_X)$

$$(\mathbf{G}(\mathbf{Z})\mathbf{u})\mathbf{v} - (\mathbf{G}(\mathbf{Z})\mathbf{v})\mathbf{u} = 0, \quad \forall \quad \mathbf{u},\mathbf{v} \in \mathcal{V}_k. \tag{2}$$

A change of reference configurations from k to $\bar{k}$ may be characterized by the pair $\mathbf{H}, \mathbf{K}$

$$\mathbf{H}(\mathbf{X}) = \nabla(\bar{k} \circ k^{-1}) \mid_{\mathbf{X}}, \quad \mathbf{K}(\mathbf{X}) = \nabla^2(\bar{k} \circ k^{-1}) \mid_{\mathbf{X}} \tag{3}$$

which satisfies the symmetry condition (2). Let us consider the motion χ referred to these configurations. The composition of the pairs of the second order gradients is defined by the following rule (see [14], [10])

$$(\mathbf{F}_{\bar{k}}, \mathbf{G}_{\bar{k}}) \circ (\mathbf{H}, \mathbf{K}) = (\mathbf{F}_{\bar{k}}\mathbf{H}, \mathbf{F}_{\bar{k}}\mathbf{K} + \mathbf{G}_{\bar{k}}[\mathbf{H}, \mathbf{H}]), \tag{4}$$

where the following notation was introduced

$$(\mathbf{G}_{\bar{k}}[\mathbf{H}, \mathbf{H}]\mathbf{u})\mathbf{v} = (\mathbf{G}_{\bar{k}}(\mathbf{H}\mathbf{u}))\mathbf{H}\mathbf{v} \quad \forall \mathbf{u}, \mathbf{v} \in \mathcal{V}_k. \tag{5}$$

The left hand side in (4) is just the pair $(\mathbf{F}_k, \mathbf{G}_k)$, defined in (1).

The products in (4) have the components forms:

$$(\mathbf{FH})^i_{AB} = F^i_C H^C_{AB}, \quad (\mathbf{G}[\mathbf{H}, \mathbf{H}])^i_{AB} = M^i_{CD} H^C_A H^D_B. \tag{6}$$

We introduce now the assumption concerning the existence of the local configuration with torsion, or of the non-holonomic configuration $\mathcal{K}$.

A1. For any motion $\chi : \mathcal{N}_X \times R \longrightarrow \mathcal{E}$ there exist a second order pair $\mathbf{H} : \mathcal{V}_k \longrightarrow \mathcal{V}_\mathcal{K}$ linear invertible tensor field, called plastic distortion, $\mathbf{N} : \mathcal{V}_k \longrightarrow \mathcal{V}_k \times \mathcal{V}_\mathcal{K}$ linear, such that the non-symmetry condition holds

$$(\mathbf{N}(\mathbf{X})\mathbf{u})\mathbf{v} - (\mathbf{N}(\mathbf{X})\mathbf{v})\mathbf{u} \neq 0. \tag{7}$$

As a consequence of **A1.** there exists a configuration $\mathcal{K}(\cdot, t)$, called local configuration with torsion, or non-holonomic configuration, such that

$$\nabla(\mathcal{K}(\cdot, t) \circ k^{-1}) \mid_{\mathbf{X}} = \mathbf{H}(\mathbf{X}, t), \text{ but } \nabla^2(\mathcal{K}(\cdot, t) \circ k^{-1}) \mid_{\mathbf{X}} \neq \mathbf{N}(\mathbf{X}, t). \tag{8}$$

Proposition 1. There exists $(\mathbf{F}_{\mathcal{K}(t)}, \mathbf{G}_{\mathcal{K}(t)})$ a second order pair, with respect to the non-holonomic configuration, defined via the relationships:

$$(\mathbf{F}, \mathbf{G}) = (\mathbf{F}_\mathcal{K}, \mathbf{G}_\mathcal{K}) \circ (\mathbf{H}, \mathbf{N}) \tag{9}$$

or equivalently by

$$\begin{gathered} \mathbf{F}_\mathcal{K} = \mathbf{F}\mathbf{H}^{-1}, \quad \mathbf{F}^p \equiv \mathbf{H}, \quad \mathbf{F}^e = \mathbf{F}_\mathcal{K} \\ \mathbf{G}_\mathcal{K} = \mathbf{G}[(\mathbf{F}^p)^{-1}, (\mathbf{F}^p)^{-1}] - \mathbf{F}(\mathbf{F}^p)^{-1}\mathbf{N}[(\mathbf{F})^{-1}, (\mathbf{F}^p)^{-1}]. \end{gathered} \tag{10}$$

We defined the elastic distortion by the formula $(10)_1$ and consequently the multiplicative decomposition of the deformation gradient into its components (elastic and plastic distortion) follows.

The elastic and plastic connections relative to the non-holonomic configuration and to the actual configuration can be introduced:

$$\overset{(e)}{\Gamma}_{\mathcal{K}}= (\mathbf{F}^e)^{-1}\mathbf{G}_{\mathcal{K}}, \quad \overset{(e)}{\Gamma}_{\chi}= -\mathbf{G}_{\mathcal{K}}[(\mathbf{F}^e)^{-1}, \mathbf{F}^e)^{-1}], \tag{11}$$

as well as the plastic connection relative to the initial configuration and to the non-holonomic configuration:

$$\overset{(p)}{\Gamma}_{k}= (\mathbf{F}^p)^{-1}\mathbf{N}, \quad \overset{(p)}{\Gamma}_{\mathcal{K}}= -\mathbf{N}[(\mathbf{F}^p)^{-1}, (\mathbf{F}^p)^{-1}], \tag{12}$$

in terms of the second order gradients $\mathbf{G}_{\mathcal{K}}$ and $\mathbf{N}$, being also dependent on the elastic and plastic distortion.

Proposition 2. The connections are related by

$$\Gamma_k = \overset{(p)}{\Gamma}_k + (\mathbf{F}^p)^{-1} \overset{(e)}{\Gamma}_{\mathcal{K}} [\mathbf{F}^p, \mathbf{F}^p], \quad \Gamma_k = \overset{(p)}{\Gamma}_k - (\mathbf{F})^{-1} \overset{(e)}{\Gamma}_{\chi} [\mathbf{F}, \mathbf{F}], \tag{13}$$

as a consequence of the formula $(10)_2$, where the motion connection relative to the initial configuration k is defined as

$$\Gamma_k = \mathbf{F}^{-1}\mathbf{G} \tag{14}$$

The formulae (13) are similar to these derived in [4] and [13].

Unlike the motion connection, the elastic and plastic connections have non-zero torsion, moreover $\overset{(e)}{\Gamma}_{\mathcal{K}}$ and $\overset{(p)}{\Gamma}_{\mathcal{K}}$ have the same torsion

$$\begin{aligned}(\mathbf{S}_{\mathcal{K}}\tilde{\mathbf{u}})\tilde{\mathbf{v}} &\equiv (\mathbf{F}^e)^{-1}(\mathbf{G}_{\mathcal{K}}\tilde{\mathbf{u}})\tilde{\mathbf{v}} - (\mathbf{G}_{\mathcal{K}}\tilde{\mathbf{v}})\tilde{\mathbf{u}} = \\ &= -(\mathbf{N}[(\mathbf{F}^p)^{-1}, (\mathbf{F}^p)^{-1}]\tilde{\mathbf{u}})\tilde{\mathbf{v}} + (\mathbf{N}[(\mathbf{F}^p)^{-1}, (\mathbf{F}^p)^{-1}]\tilde{\mathbf{v}})\tilde{\mathbf{u}} \equiv \\ &\equiv (\overset{(p)}{\Gamma}_{\mathcal{K}}\, \tilde{\mathbf{u}})\tilde{\mathbf{v}} - (\overset{(p)}{\Gamma}_{\mathcal{K}}\, \tilde{\mathbf{v}})\tilde{\mathbf{u}} \quad \forall \tilde{\mathbf{v}}, \tilde{\mathbf{u}} \in \mathcal{V}_{\mathcal{K}}.\end{aligned} \tag{15}$$

A2. We assume the existence of non-vanishing third order field $\overset{(p)}{\mathbf{Q}}$

$$\overset{(p)}{\mathbf{Q}}\, \mathbf{u} = (\mathbf{N}\mathbf{u})^T\mathbf{F}^p + (\mathbf{F}^p)^T\mathbf{N}\mathbf{u} - (\nabla_k \mathbf{C}^p)\mathbf{u}, \quad \mathbf{C}^p = (\mathbf{F}^p)^T\mathbf{F}^p, \tag{16}$$

as the non-metricity of the (material) space or equivalently expressed

$$\overset{(p)}{\mathbf{Q}}\, \mathbf{u} = (\overset{(p)}{\Gamma}_k\, \mathbf{u})^T\mathbf{C}^p + \mathbf{C}^p(\overset{(p)}{\Gamma}_k\, \mathbf{u}) - (\nabla_k \mathbf{C}^p)\mathbf{u}, \quad \overset{(p)}{\mathbf{Q}} \neq 0. \tag{17}$$

Theorem of the decomposition of the plastic connection ([12])

1. The plastic connection $\overset{(p)}{\Gamma}$ is expressed by $\mathbf{C}^p$, the contortion $\mathbf{W}^p$ and the measure of the non-metricity $\overset{(p)}{\mathbf{Q}}$ via the formulae

$$\overset{(p)}{\Gamma} = \gamma^p + \mathbf{W}^p + \{\overset{(p)}{\mathbf{Q}}\}, \quad \mathbf{C}^p = (\mathbf{F}^p)^T\mathbf{F}^p. \tag{18}$$

2. The torsion $\mathbf{S}$ and the contortion $\mathbf{W}^p$ are defined by

$$(\mathbf{S}\mathbf{u})\mathbf{v} = (\overset{(p)}{\mathbf{\Gamma}}\,\mathbf{u})\mathbf{v} - (\overset{(p)}{\mathbf{\Gamma}}\,\mathbf{v})\mathbf{u}, \quad \forall\, \mathbf{u},\mathbf{v} \in \mathcal{V}_k,$$
$$(\mathbf{W}^p\mathbf{u})\mathbf{v} = \frac{1}{2}((\mathbf{S}\mathbf{u})\mathbf{v} - (\mathbf{S}\mathbf{u})^T\mathbf{v} - (\mathbf{S}\mathbf{v})^T\mathbf{u}). \tag{19}$$

3. γ^p is Riemann connection attached to $\mathbf{C}^p$:

$$(\gamma^p\mathbf{u})\mathbf{v}\cdot\mathbf{w} \equiv \frac{1}{2}(\mathbf{C}^p)^{-1}\mathbf{w}\cdot[((\nabla_k\mathbf{C}^p)\mathbf{v})\mathbf{u} + (\nabla_k\mathbf{C}^p\mathbf{u})\mathbf{v}] -$$
$$-\frac{1}{2}\nabla_k\mathbf{C}^p((\mathbf{C}^p)^{-1}\mathbf{w})\mathbf{u}\cdot\mathbf{v}, \quad \mathbf{C}^p = (\mathbf{F}^p)^T\mathbf{F}^p. \tag{20}$$

4. The third order tensor $\{\overset{(p)}{\mathbf{Q}}\}$ from (18) is related to $\overset{(p)}{\mathbf{Q}}$, see (17), by

$$(\{\overset{(p)}{\mathbf{Q}}\}\mathbf{u})\mathbf{v}\cdot\mathbf{z} = \frac{1}{2}((\overset{(p)}{\mathbf{Q}}\,\mathbf{u})\mathbf{v}\cdot\mathbf{z} + (\overset{(p)}{\mathbf{Q}}\,\mathbf{v})\mathbf{u}\cdot\mathbf{z} - (\overset{(p)}{\mathbf{Q}}\,\mathbf{z})\mathbf{u}\cdot\mathbf{v}). \tag{21}$$

Remark. $\{\overset{(p)}{\mathbf{Q}}\}$ and $\mathbf{W}^p$ describe different things since they are generated by the third order tensor fields with different meanings and symmetries, [12],

$$(\{\overset{(p)}{\mathbf{Q}}\}\mathbf{u})\mathbf{v} = (\{\overset{(p)}{\mathbf{Q}}\}\mathbf{v})\mathbf{u}, \quad (\mathbf{W}^p\mathbf{u})^T = -\mathbf{W}^p\mathbf{u}, \quad \forall\, \mathbf{u},\mathbf{v} \in \mathcal{V}_k. \tag{22}$$

The relationship between the torsions $\mathbf{S}$ and $\mathbf{S}_{\mathcal{K}}$ is derived starting from the adopted definitions (12), (15) and (19).

The decomposition theorem plays a fundamental role in discussing the compatibility between the plastic distortion and plastic connections, when we take into account the Riemann-Cristoffel curvature tensor.

2. COUPLE STRESSES AND ELASTIC MATERIAL ELEMENT

The local balance equations in the actual configuration are given, [4],

$$\rho\mathbf{a} = \operatorname{div}\mathbf{T}^T + \rho\mathbf{b}_f$$
$$0 = 2\,\overset{\times}{\mathbf{T}} + \operatorname{div}\mathbf{M}^T + \rho\mathbf{b}_m, \quad \text{where} \quad \mathbf{u}\times\overset{\times}{\mathbf{T}} = \mathbf{T}^a\mathbf{u}, \quad \forall\,\mathbf{u}\in\mathcal{V}_\chi, \tag{23}$$

where $\mathbf{b}_f$, $\mathbf{b}_m$ are densities of the body forces and body couples, $\mathbf{a}$ denotes the acceleration at the same material point.

Subsequently the appropriate conjugate expressions for the forces and the velocity like quantities can be put into evidence through the elastic and plastic power

$$\frac{1}{2}\frac{\mathbf{\Pi}_{\mathcal{K}}}{\tilde{\rho}}\cdot\dot{\mathbf{C}}^e + \frac{1}{\tilde{\rho}}\boldsymbol{\mu}\cdot\nabla_\chi\mathbf{L}^e, \quad \mathbf{\Sigma}_{\mathcal{K}}\cdot\mathbf{L}^p + \frac{1}{\tilde{\rho}}\boldsymbol{\mu}_{\mathcal{K}}\cdot\nabla_{\mathcal{K}}\mathbf{L}^p, \tag{24}$$

We recall that the the stress momentum $\boldsymbol{\mu}$ is a third order tensor attached to the couple stress second order tensor field $\mathbf{M}$ by

$$\boldsymbol{\mu} = \frac{1}{2}\underline{\epsilon}\mathbf{M}^T, \quad \mu_{ijk} = \frac{1}{2}\epsilon_{ijm}\,(\mathbf{M}^T)_{mk}, \tag{25}$$

with ϵ_{ijm} the permutation symbols.

The rate of elastic distortion $\mathbf{L}^e$ and the rate of the plastic distortion $\mathbf{L}^p$ are written with respect to the actual and the non-holonomic configurations, being related by the kinematic relations:

$$\begin{aligned} &\mathbf{L} = \dot{\mathbf{F}}^e(\mathbf{F}^e)^{-1} + \mathbf{F}^e\mathbf{L}^p(\mathbf{F}^e)^{-1}, \\ &\mathbf{L}^e = \dot{\mathbf{F}}^e(\mathbf{F}^e)^{-1}, \quad \mathbf{L}^p = \dot{\mathbf{F}}^p(\mathbf{F}^p)^{-1}, \quad \mathbf{L} = \nabla_\chi \mathbf{v}. \end{aligned} \tag{26}$$

$\mathbf{v}$ represents the velocity field in the actual configuration.
Piola-Kirchhoff stress tensor $\mathbf{\Pi}_\mathcal{K}$ and Mandel's non-symmetric stress measure (or Eshelby stress tensor, see [1]) $\mathbf{\Sigma}_\mathcal{K}$ are related to the symmetric part of the Cauchy stress tensor $\mathbf{T}^S$, while the pull back to the non-holonomic configuration $\mathcal{K}$ of the stress momentum (third order tensor field) $\boldsymbol{\mu}$, with $\det\mathbf{F}^e = \tilde{\rho}/\rho$, can be also introduced as it follows

$$\begin{aligned} &\mathbf{\Pi}_\mathcal{K} \equiv \mathbf{\Pi} = \det(\mathbf{F}^e)(\mathbf{F}^e)^{-1}\mathbf{T}^s(\mathbf{F}^e)^{-T}, \quad \mathbf{\Sigma}_\mathcal{K} \equiv \mathbf{\Sigma} = \mathbf{C}^e\frac{\mathbf{\Pi}_\mathcal{K}}{\tilde{\rho}}, \\ &\boldsymbol{\mu}_\mathcal{K} = (\det\mathbf{F}^e)(\mathbf{F}^e)^T\boldsymbol{\mu}[(\mathbf{F}^e)^{-T}, (\mathbf{F}^e)^{-T}], \quad \mathbf{C}^e = (\mathbf{F}^e)^T\mathbf{F}^e. \end{aligned} \tag{27}$$

The crystalline body behaves as an elastic material element with respect to the non-holonomic configuration, postulated under the form **A4:**

$$\mathbf{T}^s(\mathbf{X}, t) = \mathbf{f}_{\mathcal{K}(t)}(\mathbf{F}_{\mathcal{K}(t)}, \mathbf{G}_{\mathcal{K}(t)}), \quad \boldsymbol{\mu}(\mathbf{X}, t) = \mathbf{g}_{\mathcal{K}(t)}(\mathbf{F}_{\mathcal{K}(t)}, \mathbf{G}_{\mathcal{K}(t)}). \tag{28}$$

3. CONSTITUTIVE ASSUMPTIONS

Due to the fact that the constitutive equations have to be frame indifferent, we deduce the objectivity restrictions starting from
A5. Objectivity assumption: The pair $(\mathbf{F}^p, \mathbf{N})$, attached to χ, corresponds also to any $\chi^* : \mathcal{N}_X \times \mathbf{R} \longrightarrow \mathcal{E}$ related to the motion χ by:

$$\chi^*(X, t) = \mathbf{x}_0^*(t) + \mathbf{Q}(t)(\mathbf{x} - \mathbf{x}_0), \quad \mathbf{x}_0^*(t), \mathbf{x}_0 \in \mathcal{E}, \quad \mathbf{Q}(t) \in Ort, \tag{29}$$

here $\mathbf{x} = \chi(X, t)$.

Thus if a rigid motion (i.e. a homogeneous deformation) is applied to a pair of second order gradient of the motion then

$$\begin{aligned} &(\mathbf{F}^*(X, t), \mathbf{G}^*(X, t)) = (\mathbf{Q}(t), 0) \circ (\mathbf{F}(\mathbf{X}, t), \mathbf{G}(\mathbf{X}, t)) \quad \text{and} \\ &(\mathbf{F}^*_\mathcal{K}, \mathbf{G}^*_\mathcal{K}) = (\mathbf{Q}\mathbf{F}_\mathcal{K}, \mathbf{Q}\mathbf{G}_\mathcal{K}). \end{aligned} \tag{30}$$

Consequently, the constitutive equations (28) are material frame indifferent if and only if the equivalent constitutive equations can be formulated:

$$\mathbf{\Pi}_{\mathcal{K}} = \overline{f}_{\mathcal{K}}(\mathbf{C}^e, \overset{(e)}{\mathbf{\Gamma}}_{\mathcal{K}}) \text{ or } \mathbf{\Sigma}_{\mathcal{K}} = \overline{h}_{\mathcal{K}}(\mathbf{C}^e, \overset{(e)}{\mathbf{\Gamma}}_{\mathcal{K}}), \ \ \boldsymbol{\mu}_{\mathcal{K}} = \overline{g}_{\mathcal{K}}(\mathbf{C}^e, \overset{(e)}{\mathbf{\Gamma}}_{\mathcal{K}}). \tag{31}$$

A7. The evolution equations for irreversible behaviour with respect to the non-holonomic configuration $\mathcal{K}$ are written under the form

$$\begin{aligned} &\mathbf{L}^p \equiv \dot{\mathbf{F}}^p(\mathbf{F}^p)^{-1} = \mathcal{B}_{\mathcal{K}}(\mathbf{\Sigma}_{\mathcal{K}}, \boldsymbol{\mu}_{\mathcal{K}}, \dot{\mathbf{F}}, \frac{d}{dt}(\nabla_k\mathbf{F}), \mathbf{F}, \nabla_k\mathbf{F}) \\ &\dot{\mathbf{N}} = \mathcal{A}_{\mathcal{K}}(\mathbf{\Sigma}_{\mathcal{K}}, \boldsymbol{\mu}_{\mathcal{K}}, \dot{\mathbf{F}}, \frac{d}{dt}(\nabla_k\mathbf{F}), \mathbf{F}, \nabla_k\mathbf{F}) \end{aligned} \tag{32}$$

related for instance to the yield surface $\mathcal{F}_{\mathcal{K}}(\mathbf{\Sigma}_{\mathcal{K}}, \boldsymbol{\mu}_{\mathcal{K}})$, using the standard procedure in finite plasticity. $(\mathbf{F}^p, \mathbf{N})$ can appears among the variables in the right hand side of (32), for instance as in [3].

Remark. When we pass to the reference configuration, using (27) and (13), we get

$$\mathbf{C}^e = (\mathbf{F}^p)^{-T}\mathbf{C}(\mathbf{F}^p)^{-1}, \ \overset{(e)}{\mathbf{\Gamma}}_{\mathcal{K}} = \mathbf{F}^p(\mathbf{F}^{-1}\nabla_k\mathbf{F} - \overset{(p)}{\mathbf{\Gamma}}_k)[(\mathbf{F}^p)^{-1}, (\mathbf{F}^p)^{-1}]. \tag{33}$$

The irreversible behaviour can be equivalently expressed using $(\mathbf{F}^p, \overset{(p)}{\mathbf{\Gamma}}_k)$, instead of the pair $(\mathbf{F}^p, \mathbf{N})$, since from (12) we have

$$\frac{d}{dt}(\overset{(p)}{\mathbf{\Gamma}}_k) = \dot{\mathbf{N}} - (\mathbf{F}^p)^{-1}\mathbf{L}^p\mathbf{N}. \tag{34}$$

Conclusions. 1. The constitutive functions, written with respect to the non-holonomic configurations, can be explicitly expressed in terms of the pair $(\mathbf{F}^p, \overset{(p)}{\mathbf{\Gamma}}_k)$, for any given pair of time dependent second order deformation gradients $(\mathbf{F}, \nabla_k\mathbf{F})$. By the push forward procedure, the constitutive functions with respect to the actual configuration can be also derived using formulae (28), (27), (31) and (33).

3. The evolution of the irreversible variables, $(\mathbf{F}^p, \overset{(p)}{\mathbf{\Gamma}}_k)$, is prescribed again for any given pair of time dependent second order pair $(\mathbf{F}, \nabla_k\mathbf{F})$, by an appropriate system of differential equations. The system is derived by composing the evolution functions from (32) with (31) and (33).

4. In this general plastic connection case, the compatibility of the irreversible variables $(\mathbf{F}^p, \overset{(p)}{\mathbf{\Gamma}}_k)$ to the specific Riemann curvature structure (via Bianchi's identities) is still an open problem. The appropriate compatibility with vanishing Riemann curvature has been investigated for the proposed models in [4] and [5].

References

[1] S. Cleja-Ţigoiu and G.A. Maugin, Eshelby's stress tensors in finite elastoplasticity. *Acta Mechanica,* **139** (2000) 231–249.

[2] M. Epstein and G. Maugin, Thermomechanics of volumetric growth in uniform bodies. *Int. J. Plast.* **16** (2000) 951- 978.

[3] M. Epstein, On Material Evolution Laws, in M. Maugin, Geometry Continua and Microstructure, Collection Travaux en Cours, **60**. Hermann, Paris (1999) 1–9.

[4] S. Cleja-Ţigoiu, Couple stresses and non-Riemannian plastic connection in finite elasto-plasticity. *ZAMP* **53** (2002) 996–1013.

[5] S. Cleja-Ţigoiu, Small elastic strains in finite elasto-plastic materials with continuously distributed dislocations. Theoretical and Apllied Mechanics. **28-29** (2002) 93–112.

[6] M.E. Gurtin, On the plasticity of single crystal: free energy, microforces, plastic-strain gradients. *J. Mech. Phys. Solids* **48** (2000) 989-1036.

[7] K. Kondo and M. Yuki, On the current viewpoints of non-Riemannian plasticity theory. *In RAAG* II (D), 1958, 202-226.

[8] E. Kröner, The internal mechanical state of soids with defects, *Int.J. Solids Structures* **29** (1992) 1849–1857.

[9] K.C. Le and H. Stumpf, Nonlinear continuum with dislocations. *Int. J. Engng. Sci.* **34** (1996) 339–358.

[10] M. De León and M. Epstein, The Geometry of Uniformity in Second-Grade Elasticity. *Acta Mechanica* **114** (1996) 217–224.

[11] W. Noll, Materially uniform simple bodies with inhomogeneities. *Arch. Rat. Mech. Anal.* **27** (1967) 1–32.

[12] J.A. Schouten, *Ricci-Calculus.* Springer-Verlag 1954.

[13] P. Steinmann, View on multiplicative elastoplasticity and the continuum theory of dislocations. *Int.J.Engng. Sci.* **34** (1996) 1717–1735.

[14] C.C. Wang and I.S. Liu, A note on material symmetry. *Arch. Rat. Mech. Anal.* **74** (1980) 277-294.

Chapter 15

PEACH-KOEHLER FORCES WITHIN THE THEORY OF NONLOCAL ELASTICITY

Markus Lazar

Laboratoire de Modélisation en Mécanique,
Université Pierre et Marie Curie,
4 Place Jussieu, Case 162,
F-75252 Paris Cédex 05, France
lazar@lmm.jussieu.fr

Abstract We consider dislocations in the framework of Eringen's nonlocal elasticity. The fundamental field equations of nonlocal elasticity are presented. Using these equations, the nonlocal force stresses of a straight screw and a straight edge dislocation are given. By the help of these nonlocal stresses, we are able to calculate the interaction forces between dislocations (Peach-Koehler forces). All classical singularities of the Peach-Koehler forces are eliminated. The extremum values of the forces are found near the dislocation line.

Keywords: Nonlocal elasticity, material force, dislocation

1. INTRODUCTION

Traditional methods of classical elasticity break down at small distances from crystal defects and lead to singularities. This is unfortunate since the defect core is a very important region in the theory of defects. Moreover, such singularities are unphysical and an improved model of defects should eliminate them. In addition, classical elasticity is a scale-free continuum theory in which no characteristic length appears. Thus, classical elasticity cannot explain the phenomena near defects and at the atomic scale.

In recent decades, a theory of elastic continuum called nonlocal elasticity has been developed. The concept of nonlocal elasticity was originally proposed by Kröner and Datta [1, 2], Edelen and Eringen [3, 4, 5], Kunin [6] and some others. This theory considers the inner structures of materials and takes into account long-range (nonlocal) interactions.

It is important to note that the nonlocal elasticity may be related to other nonstandard continuum theories like gradient theory [7, 8] and gauge theory of defects [9, 10, 11]. In all these approaches characteristic inner lengths (gradient coefficient or nonlocality parameter), which describe size (or scale) effects, appear. One remarkable feature of solutions in nonlocal elasticity, gradient theory and gauge theory is that the stress singularities which appear in classical elasticity are eliminated. These solutions depend on the characteristic inner length, and they lead to finite stresses. Therefore, they are applicable up to the atomic scale.

In this paper, we investigate the nonlocal force due to dislocations (nonlocal Peach-Koehler force). We consider parallel screw and edge dislocations. The classical singularity of Peach-Koehler force is eliminated and a maximum/minimum is obtained.

2. FUNDAMENTAL FIELD EQUATIONS

The fundamental field equations for an isotropic, homogeneous, nonlocal and infinite extended medium with vanishing body force and static case have been given by the nonlocal theory [3, 4, 5]

$$\begin{aligned} &\partial_j \sigma_{ij} = 0, \\ &\sigma_{ij}(r) = \int_V \alpha(r - r')\, \overset{\circ}{\sigma}_{ij}(r')\, \mathrm{d}v(r'), \\ &\overset{\circ}{\sigma}_{ij} = 2\mu \left(\overset{\circ}{\epsilon}_{ij} + \frac{\nu}{1-2\nu}\, \delta_{ij} \overset{\circ}{\epsilon}_{kk} \right), \end{aligned} \tag{1}$$

where μ, ν are shear modulus and Poisson's ration, respectively. In addition, $\overset{\circ}{\epsilon}_{ij}$ is the classical strain tensor, $\overset{\circ}{\sigma}_{ij}$ and σ_{ij} are the classical and nonlocal stress tensors, respectively. The $\alpha(r)$ is a nonlocal kernel. The field equation in nonlocal elasticity of the stress in an isotropic medium is the following inhomogeneous Helmholtz equation

$$\left(1 - \kappa^{-2}\Delta\right)\sigma_{ij} = \overset{\circ}{\sigma}_{ij}, \tag{2}$$

where $\overset{\circ}{\sigma}_{ij}$ is the stress tensor obtained for the same traction boundary-value problem within the "classical" theory of dislocations. The factor κ^{-1} has the physical dimension of a length and it, therefore, defines an internal characteristic length. If we consider the two-dimensional problem and using Green's function of the two-dimensional Helmholtz equation (2), we may solve the field equation for every component of the stress field (2) by the help of the convolution integral and the two-dimensional Green function

$$\alpha(r-r') = \frac{\kappa^2}{2\pi} K_0\big(\kappa\sqrt{(x-x')^2+(y-y')^2}\big). \tag{3}$$

Here K_n is the modified Bessel function of the second kind and $n = 0, 1, \dots$ denotes the order of this function. Thus,

$$\Big(1-\kappa^{-2}\Delta\Big)\alpha(r) = \delta(r), \tag{4}$$

where $\delta(r) := \delta(x)\delta(y)$ is the two-dimensional Dirac delta function. In this way, we deduce Eringen's so-called nonlocal constitutive relation for a linear homogeneous, isotropic solid with Green's function (3) as the nonlocal kernel. This kernel (3) has its maximum at $r = r'$ and describes the nonlocal interaction. Its two-dimensional volume-integral yields

$$\int_V \alpha(r-r')\,\mathrm{d}v(r) = 1, \tag{5}$$

which is the normalization condition of the nonlocal kernel. In the classical limit ($\kappa^{-1} \to 0$), it becomes the Dirac delta function

$$\lim_{\kappa^{-1}\to 0} \alpha(r-r') = \delta(r-r'). \tag{6}$$

In this limit, Eq. (1) gives the classical expressions. Note that Eringen [4, 5] found the two-dimensional kernel (3) by giving the best match with the Born-Kármán model of the atomic lattice dynamics and the atomistic dispersion curves. He used the choice $e_0 = 0.39$ for the length, $\kappa^{-1} = e_0\, a$, where a is an internal length (e.g. atomic lattice parameter) and e_0 is a material constant.

3. NONLOCAL STRESS FIELDS OF DISLOCATIONS

Let us first review the nonlocal stress fields of screw and edge dislocations in an infinitely extended body. The dislocation lines are along the z-axis. The nonlocal stress components of a straight screw dislocation with the Burgers vector $\mathbf{b} = (0, 0, b_z)$ is given by [4, 5, 7, 10]

$$\sigma_{xz} = -\frac{\mu b_z}{2\pi}\frac{y}{r^2}\Big\{1-\kappa r K_1(\kappa r)\Big\}, \quad \sigma_{yz} = \frac{\mu b_z}{2\pi}\frac{x}{r^2}\Big\{1-\kappa r K_1(\kappa r)\Big\}, \tag{7}$$

where $r = \sqrt{x^2 + y^2}$. The nonlocal stress of a straight edge dislocation with the Burgers vector $\mathbf{b} = (b_x, 0, 0)$ turns out to be [7, 11]

$$\sigma_{xx} = -\frac{\mu b_x}{2\pi(1-\nu)}\frac{y}{r^4}\Big\{(y^2+3x^2) + \frac{4}{\kappa^2 r^2}(y^2-3x^2) - 2y^2\kappa r K_1(\kappa r) - 2(y^2-3x^2)K_2(\kappa r)\Big\},$$

$$\sigma_{yy} = -\frac{\mu b_x}{2\pi(1-\nu)}\frac{y}{r^4}\Big\{(y^2-x^2) - \frac{4}{\kappa^2 r^2}(y^2-3x^2) - 2x^2\kappa r K_1(\kappa r) + 2(y^2-3x^2)K_2(\kappa r)\Big\},$$

$$\sigma_{xy} = \frac{\mu b_x}{2\pi(1-\nu)}\frac{x}{r^4}\Big\{(x^2-y^2) - \frac{4}{\kappa^2 r^2}(x^2-3y^2) - 2y^2\kappa r K_1(\kappa r) + 2(x^2-3y^2)K_2(\kappa r)\Big\},$$

$$\sigma_{zz} = -\frac{\mu b_x \nu}{\pi(1-\nu)}\frac{y}{r^2}\Big\{1 - \kappa r K_1(\kappa r)\Big\}. \tag{8}$$

It is obvious that there is no singularity in (7) and (8). For example, when r tends to zero, the stresses $\sigma_{ij} \to 0$. It also can be found that the far-field expression ($r > 12\kappa^{-1}$) of Eqs. (7) and (8) return to the stresses in classical elasticity. Of course, the stress fields (7) and (8) fulfill Eq. (2) and correspond to the nonlocal kernel (3).

4. PEACH-KOEHLER FORCES DUE TO DISLOCATIONS

The force between dislocations, according to Peach-Koehler formula in nonlocal elasticity [12], is given by

$$F_k = \varepsilon_{ijk}\sigma_{in}b'_n\xi_j, \tag{9}$$

where b'_n is the component of Burgers vector of the 2nd dislocation at the position r and ξ_j is the direction of the dislocation. Obviously, Eq. (9) is quite similar in the form as the classical expression of the Peach-Koehler force. Only the classical stress is replaced by the nonlocal one. Eq. (9) is particularly important for the interaction between dislocations.

4.1. PARALLEL SCREW DISLOCATIONS

We begin our considerations with the simple case of two parallel screw dislocations. For a screw dislocation with $\xi_z = 1$ we have in Cartesian

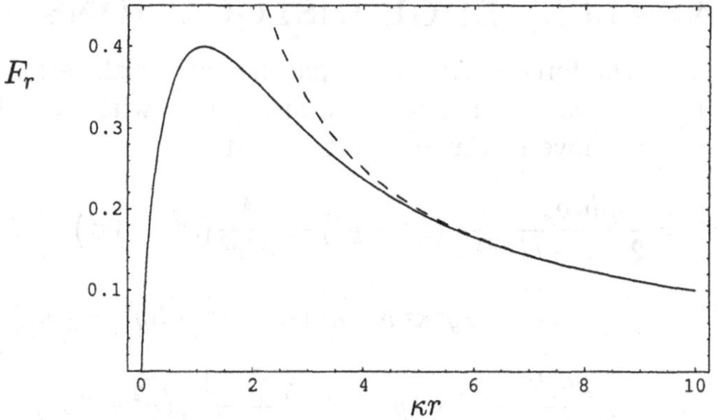

Figure 15.1. The Peach-Koehler force between screw dislocations is given in units of $\mu\kappa b_z b'_z/[2\pi]$. Nonlocal elasticity (solid) and classical elasticity (dashed).

coordinates

$$F_x = \sigma_{yz} b'_z = \frac{\mu b_z b'_z}{2\pi} \frac{x}{r^2} \Big\{1 - \kappa r K_1(\kappa r)\Big\},$$
$$F_y = -\sigma_{xz} b'_z = \frac{\mu b_z b'_z}{2\pi} \frac{y}{r^2} \Big\{1 - \kappa r K_1(\kappa r)\Big\}, \tag{10}$$

and in cylindrical coordinates

$$F_r = F_x \cos\varphi + F_y \sin\varphi = \frac{\mu b_z b'_z}{2\pi r} \Big\{1 - \kappa r K_1(\kappa r)\Big\},$$
$$F_\varphi = F_y \cos\varphi - F_x \sin\varphi = 0. \tag{11}$$

Thus, the force between two screw dislocations is also a radial force in the nonlocal case. The force expression (11) has some interesting features. A maximum of F_r can be found from Eq. (11) as

$$|F_r|_{\max} \simeq 0.399\kappa \frac{\mu b_z b'_z}{2\pi}, \quad \text{at} \quad r \simeq 1.114\kappa^{-1}. \tag{12}$$

When the nonlocal atomistic effect is neglected, $\kappa^{-1} \to 0$, Eq. (11) gives the classical result

$$F_r^{cl} = \frac{\mu b_z b'_z}{2\pi r}. \tag{13}$$

To compare the classical force with the nonlocal one, the graphs from Eqs. (11) and (13) are plotted in Fig. 15.1. It can be seen that near the dislocation line the nonlocal result is quite different from the classical one. Unlike the classical expression, which diverges as $r \to 0$ and gives an infinite force, it is zero at $r = 0$. For $r > 6\kappa^{-1}$ the classical and the nonlocal expressions coincide.

4.2. PARALLEL EDGE DISLOCATIONS

We analyze now the force between two parallel edge dislocations with (anti)parallel Burgers vector. For an edge dislocation with $\xi_z = 1$ with Burgers vector b'_x we have in Cartesian coordinates

$$F_x = \sigma_{yx} b'_x = \frac{\mu b_x b'_x}{2\pi(1-\nu)} \frac{x}{r^4} \Big\{ (x^2 - y^2) - \frac{4}{\kappa^2 r^2}(x^2 - 3y^2) \qquad (14)$$

$$- 2y^2 \kappa r K_1(\kappa r) + 2(x^2 - 3y^2) K_2(\kappa r) \Big\},$$

$$F_y = -\sigma_{xx} b'_x = \frac{\mu b_x b'_x}{2\pi(1-\nu)} \frac{y}{r^4} \Big\{ (y^2 + 3x^2) + \frac{4}{\kappa^2 r^2}(y^2 - 3x^2)$$

$$- 2y^2 \kappa r K_1(\kappa r) - 2(y^2 - 3x^2) K_2(\kappa r) \Big\}.$$

F_x is the driving force for conservative motion (gliding) and F_y is the climb force. The glide force has a maximum/minimum in the slip plane (zx-plane) of

$$|F_x(x,0)| \simeq 0.260\kappa \frac{\mu b_x b'_x}{2\pi(1-\nu)}, \quad \text{at} \quad |x| \simeq 1.494\kappa^{-1}. \qquad (15)$$

The maximum/minimum of the climb force is found as

$$|F_y(0,y)| \simeq 0.546\kappa \frac{\mu b_x b'_x}{2\pi(1-\nu)}, \quad \text{at} \quad |y| \simeq 0.996\kappa^{-1}. \qquad (16)$$

It can be seen that the maximum of the climb force is greater than the maximum of the glide force (see Fig. 15.2). The glide force F_x is zero at $x = 0$ ($\varphi = \frac{1}{2}\pi$). This corresponds to one equilibrium configuration of the two edge dislocations. In classical elasticity the glide force is also zero at the position $x = y$ ($\varphi = \frac{1}{4}\pi$). But in nonlocal elasticity we obtain from Eq. (14) the following expression for the glide force

$$F_x(\varphi = \pi/4) = \frac{\mu b_x b'_x \sqrt{2}}{4\pi(1-\nu)} \frac{1}{r} \Big\{ \frac{4}{\kappa^2 r^2} - \kappa r K_1(\kappa r) - 2K_2(\kappa r) \Big\}. \qquad (17)$$

Its maximum is

$$|F_x(\varphi = \pi/4)|_{\max} \simeq 0.151\kappa \frac{\mu b_x b'_x \sqrt{2}}{4\pi(1-\nu)}, \quad \text{at} \quad r \simeq 0.788\kappa^{-1}. \qquad (18)$$

The $F_x(\varphi = \pi/4)$ gives only a valuable contribution in the region $r \lesssim 12\kappa^{-1}$. Therefore, only for $r > 12\kappa^{-1}$ the position $\varphi = \frac{1}{4}\pi$ is an equilibrium configuration. The climb force F_y is zero at $y = 0$.

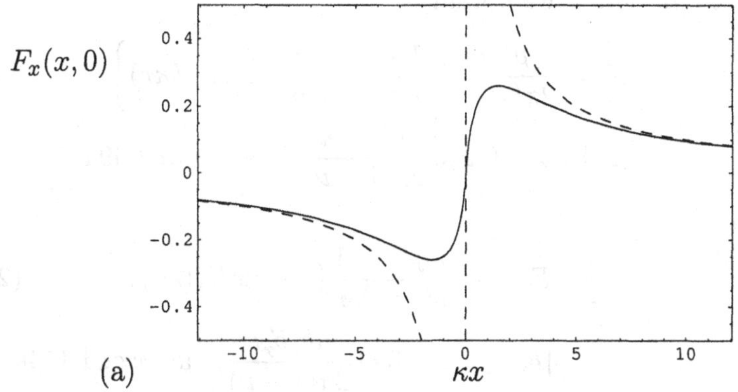

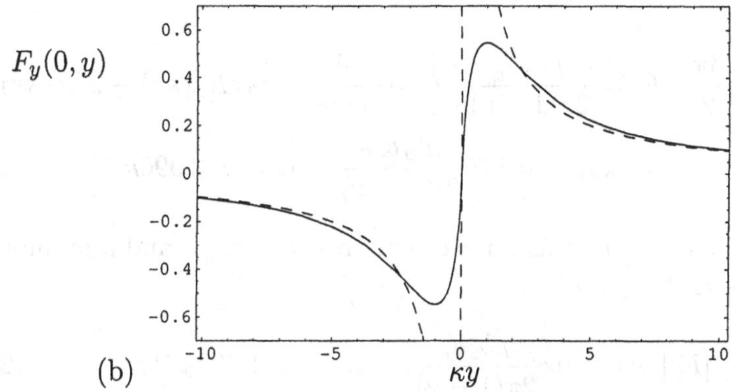

Figure 15.2. The glide and climb force components near the dislocation line: (a) $F_x(x,0)$ and (b) $F_y(0,y)$ are given in units of $\mu b_x b'_x \kappa/[2\pi(1-\nu)]$. The dashed curves represent the classical force components.

From Eq. (14) the force can be given in cylindrical coordinates as

$$F_r = \frac{\mu b_x b'_x}{2\pi(1-\nu)}\,\frac{1}{r}\bigg\{\Big(1-\kappa r K_1(\kappa r)\Big) - \cos 2\varphi\bigg(\frac{4}{\kappa^2 r^2}-\kappa r K_1(\kappa r)-2K_2(\kappa r)\bigg)\bigg\},$$

$$F_\varphi = \frac{\mu b_x b'_x}{2\pi(1-\nu)}\,\frac{\sin 2\varphi}{r}\bigg\{1-\frac{4}{\kappa^2 r^2}+2K_2(\kappa r)\bigg\}. \qquad (19)$$

The force between edge dislocations is not a central force because a tangential component F_φ exists. Both the components F_r and F_φ depend on r and φ. The dependence of φ in F_r is a new feature of the nonlocal result (19) not present in the classical elasticity. Unlike the "classical" case, the force F_r is only zero at $r=0$ and F_φ is zero at $r=0$ and $\varphi = 0, \frac{1}{2}\pi, \pi, \frac{3}{2}\pi, 2\pi$. The force F_r in (19) has an interesting dependence of the angle φ. In detail, we obtain

$$\text{(i) } \varphi = 0, \pi: \quad F_r = \frac{\mu b_x b'_x}{2\pi(1-\nu)} \frac{1}{r} \left\{ 1 - \frac{4}{\kappa^2 r^2} + 2K_2(\kappa r) \right\}, \tag{20}$$

$$|F_r|_{\max} \simeq 0.260 \frac{\mu b_x b'_x \kappa}{2\pi(1-\nu)}, \quad \text{at} \quad r \simeq 1.494\kappa^{-1},$$

$$\text{(ii) } \varphi = \frac{\pi}{4}, \frac{3\pi}{4}, \frac{5\pi}{4}, \frac{7\pi}{4}: F_r = \frac{\mu b_x b'_x}{2\pi(1-\nu)} \frac{1}{r} \{1 - \kappa r K_1(\kappa r)\}, \tag{21}$$

$$|F_r|_{\max} \simeq 0.399 \frac{\mu b_x b'_x \kappa}{2\pi(1-\nu)}, \quad \text{at } r \simeq 1.114\kappa^{-1},$$

$$\text{(iii) } \varphi = \frac{\pi}{2}, \frac{3\pi}{2}: F_r = \frac{\mu b_x b'_x}{2\pi(1-\nu)} \frac{1}{r} \left\{ 1 + \frac{4}{\kappa^2 r^2} - 2\kappa r K_1(\kappa r) - 2K_2(\kappa r) \right\},$$

$$|F_r|_{\max} \simeq 0.547 \frac{\mu b_x b'_x \kappa}{2\pi(1-\nu)}, \quad \text{at } r \simeq 0.996\kappa^{-1}. \tag{22}$$

On the other hand, F_φ has a maximum at $\varphi = \frac{1}{4}\pi, \frac{5}{4}\pi$ and a minimum at $\varphi = \frac{3}{4}\pi, \frac{7}{4}\pi$. They are

$$|F_\varphi| \simeq 0.260\kappa \frac{\mu b_x b'_x}{2\pi(1-\nu)}, \quad \text{at} \quad r \simeq 1.494\kappa^{-1}. \tag{23}$$

The classical result for the force reads in cylindrical coordinates

$$F_r^{cl} = \frac{\mu b_x b'_x}{2\pi(1-\nu)} \frac{1}{r}, \quad F_\varphi^{cl} = \frac{\mu b_x b'_x}{2\pi(1-\nu)} \frac{\sin 2\varphi}{r}. \tag{24}$$

To compare the nonlocal result with the classical one, diagrams of Eqs. (19) and (24) are drawn in Fig. 15.3. The force calculated in nonlocal elasticity is different from the classical one near the dislocation line at $r = 0$. It is finite and has no singularity in contrast to the classical result. For $r > 12\kappa^{-1}$ the classical and the nonlocal expressions coincide.

5. CONCLUSION

The nonlocal theory of elasticity has been used to calculate the Peach-Koehler force due to screw and edge dislocations in an infinitely extended body. The Peach-Koehler force calculated in classical elasticity is infinite near the dislocation core. The reason is that the classical elasticity is invalid in dealing with problems of micro-mechanics. The nonlocal elasticity gives expressions for the Peach-Koehler force which are physically more reasonable. The forces due to dislocations have no singularity.

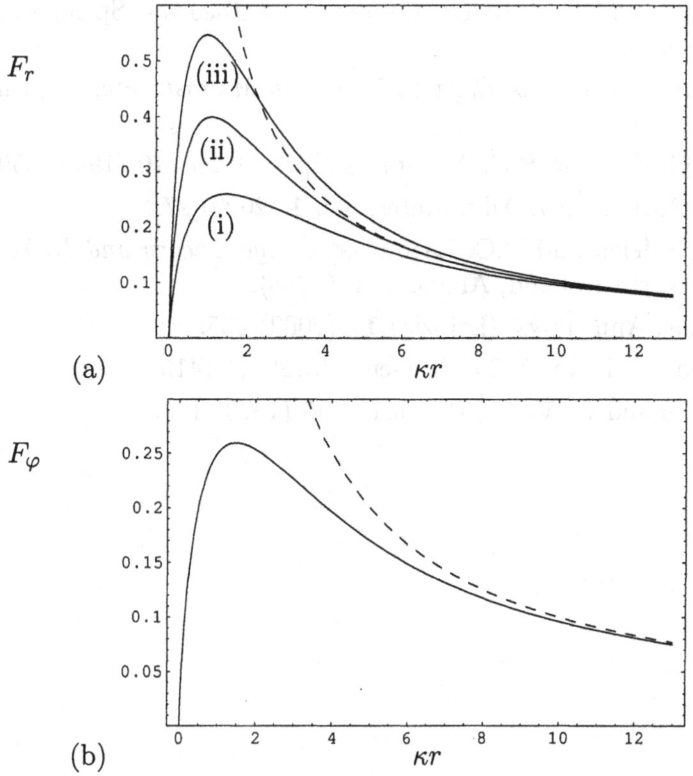

Figure 15.3. The force near the dislocation line: (a) F_r and (b) F_φ with $\varphi = \frac{1}{4}\pi$, $\frac{5}{4}\pi$ are given in units of $\mu b_x b'_x \kappa/[2\pi(1-\nu)]$. The dashed curves represent the classical force components.

They are zero at $r = 0$ and have maxima or minima in the dislocation core. The finite values of the force due to dislocations may be used to analyze the interaction of dislocations from the micro-mechanical point of view.

Acknowledgments

This work was supported by the European Network TMR 98-0229.

References

[1] E. Kröner and B.K. Datta, Z. Phys. **196** (1966) 203.

[2] E. Kröner, Int. J. Solids Struct. **3** (1967) 731.

[3] A.C. Eringen and D.G.B. Edelen, Int. J. Engng. Sci. **10** (1972) 233.

[4] A.C. Eringen, J. Appl. Phys. **54** (1983) 4703.

[5] A.C. Eringen, *Nonlocal Continuum Field Theories*, Springer, New York (2002).

[6] I.A. Kunin, *Theory of Elastic Media with Microstructure*, Springer, Berlin (1986).

[7] M.Yu. Gutkin and E.C. Aifantis, Scripta Mater. **40** (1999) 559.

[8] M.Yu. Gutkin, Rev. Adv. Mater. Sci. **1** (2000) 27.

[9] D.G.B. Edelen and D.C. Lagoudas, *Gauge Theory and Defects in Solids*, North-Holland, Amsterdam (1988).

[10] M. Lazar, Ann. Phys. (Leipzig) **11** (2002) 635.

[11] M. Lazar, J. Phys. A: Math. Gen. **36** (2003) 1415.

[12] I. Kovács and G. Vörös, Physica B **96** (1979) 111.

VI

MULTIPHYSICS & MICROSTRUCTURE

Chapter 16

ON THE MATERIAL ENERGY-MOMENTUM TENSOR IN ELECTROSTATICS AND MAGNETOSTATICS

Carmine Trimarco
University of Pisa, Dipartimento di Matematica Applicata 'U.Dini'
Via Bonanno 25/B. I-56126 Pisa, Italy
trimarco@dma.unipi.it

Abstract The notion of *material* force arises in a natural way in materials with 'defects' or in inhomogeneous materials. This notion, which is intimately related with that of *material* energy-momentum tensor, has been widely and successfully employed in crack-propagation problems, in phase-transition problems and others. In elasticity, there is general agreement on the form of the *material* energy-momentum tensor, also known as the *configurational* stress tensor. In dielectrics, slightly different forms are introduced in the literature for this tensor. One can also find in the literature different expressions for the *material* energy-momentum tensor that is inherent to magnetic materials. Attention is focused here on these differences, which apparently stem from different standpoints and approaches. On the one hand, one can note that the discrepancies can be readily composed. On the other hand, each of the various forms of the aforementioned tensor seems to be useful, though in the appropriate context. Some of them need to be properly interpreted; others are apparently pertinent to specific boundary value problems.

1. INTRODUCTION

The variation of the global energy of a bounded elastic body with respect to a re-location of a material point in its reference configuration produces a mechanical effect on this material point. The effect, which is evident if this point represents an inhomogeneity or a defect of the material, can be viewed as a force, the *material* force [1-8]. This force is governed by an energy-momentum tensor, the name stemming from analogous tensors that typically appears in the more general context of field theories. In electromagnetic materials, at least two energy-momentum tensors, which differ from one another, may be introduced. One of them corresponds to the classical energy-momentum tensor. As this tensor accounts for the traction, it has to be viewed as a Cauchy-like stress tensor. The other one represents the material energy-momentum (or energy-stress tensor), an Eshelby-like tensor, which is intimately related to the aforementioned material force [3,8,9]. Attention is focused hereafter on the form of this tensor in dielectrics or in magnetic materials. Due to slightly different approaches, slightly different forms for the material tensor have been proposed in dielectrics by different authors [10-12]. An analogous dichotomy may emerge in magnetic materials [13,14]. In the framework proposed here, the apparent discrepancy of the various forms for this tensor can be readily composed, if the *material* electromagnetic fields are introduced. These fields satisfy the Maxwell equations in the referential frame of the body [3,7-9]. However, one can note that each of the aforementioned forms addresses boundary conditions that involve different physical quantities. In this respect, a discussion on the boundary conditions along a crack-line of a dielectric, in the presence of electric or magnetic fields, will be helpful, as it will address the question as to the appropriate choice of the material tensor. Due to the presence of electric (or magnetic) fields, the crack-problem is more involved than the classical mechanical one [11,14-16]. The difficulties can be shown in evidence if the crack-line is replaced by a crack-region of finite width, which separates part of the dielectric from a vacuum. As is known, the electric field **E** does not vanish in a vacuum and suffers a jump across the boundary of the crack-region. So does the Maxwell stress. As a result, an electrostatic traction arises across this boundary, contrary to the standard mechanical problem, in which the crack-line is free of traction. All these additional effects have to be taken into account, with the purpose of recovering the model of the single crack-line for a dielectric. In section 2, two limit cases of given external electric fields are briefly discussed for a rigid dielectric. Then, in section 3, elastic dielectrics and magnetic materials are introduced in the general context of finite deformations. In this framework, the total electric and magnetic stresses and the material tensors are eventually written in terms of the electric and magnetic quantities. A specific elastic dielectric, whose constitutive response is a very simple one, is considered in section 4. The total Cauchy-stress that is associated with this dielectric turns out to coincide with that of a rigid dielectric and the related Eshelby energy-momentum tensor turns out to vanish identically. A similar treatment is proposed for specific magnetic materials. The treatment for these materials is based on the magnetic induction **B** instead of the magnetic field **H**, despite an apparent analogy with the dielectric case.

2. PRELIMINARY REMARKS ON CRACK-PROBLEMS IN DIELECTRICS

Hereafter, we refer to the classical problem of a crack-line in a dielectric. With reference to the figures 1 and 2, a crack-region of finite width separates the body from a vacuum.

In the limit as the width tends to zero, the two parallel lines *r* and *s* overlap and reduce to a single crack-line, whereas the contour γ that smoothly connects *r* and *s* reduces to a point, which will represent the tip of the crack. In the purely mechanical problem, the crack-line is currently assumed to be a traction-free part of the boundary. This condition still holds true at the boundary of the crack-region.

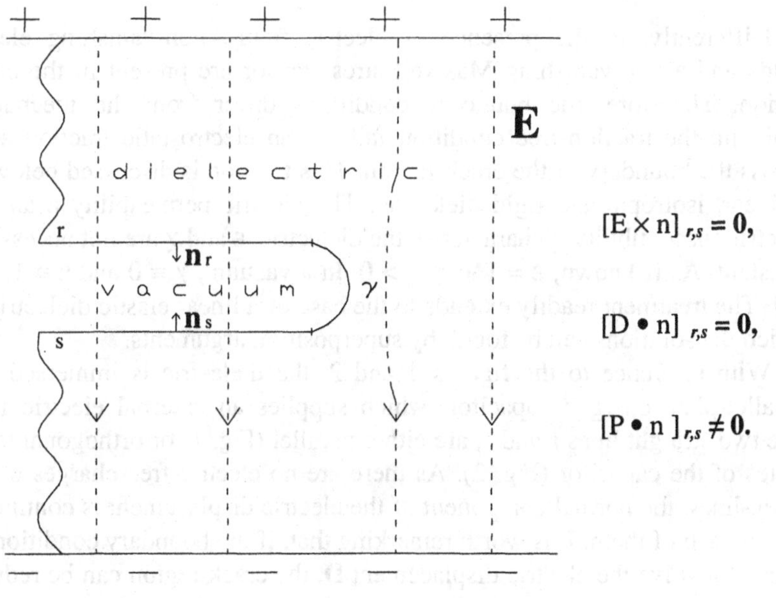

Fig. 1

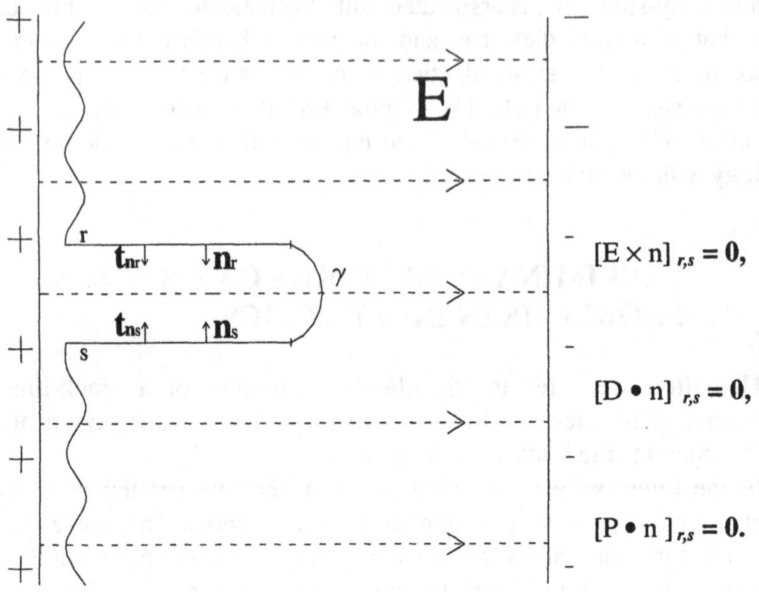

Fig. 2

Differently, in the presence of electric fields, non-vanishing electric fields and a non-vanishing Maxwell stress tensor are present in the crack-region. Therefore, the boundary conditions differ from the mechanical problem: the traction-free condition fails as an electrostatic traction arises across the boundary of the crack-region. This traction is discussed below for a linear, isotropic and rigid dielectric. The electric permeability ε and the electric susceptibility χ characterise the dielectric. ε and χ are not necessarily constant. As is known, $\varepsilon = 1 + \chi$, $\chi > 0$. In a vacuum, $\chi \equiv 0$ and $\varepsilon \equiv 1$, [17-19]. The treatment readily extends to the case of a linear elastic dielectric, in which the solutions can be found by superposition arguments.

With reference to the figures 1 and 2, the dielectric is immersed in a parallel-plate charged capacitor, which supplies an external electric field. The two straight lines *r* and *s*, are either parallel (Fig. 1) or orthogonal to the plates of the capacitor (Fig. 2). As there are no electric free-charges on the crack-lines, the normal component of the electric displacement is continuous across each of them. It is worth remarking that, if the boundary conditions of interest involve the electric displacement **D**, the crack-region can be reduced to a single crack-line, without prejudice for the consistency of the jump of the normal components of **D**. The same reasoning does not apply for the polarisation vector **P**, whose normal component is discontinuous across *r* and *s*. This remark is important in the general nonlinear context. In the linear framework, the continuity of the normal component of **D** across *r* and *s*, is expressed by the equation

$$\varepsilon 0\ (E+-\varepsilon E-)\,|\ r,s \bullet n\ r,s = 0, \tag{2.1}$$

or, equivalently,

$$-\ \varepsilon_0 (\chi \mathbf{E})^- \bullet \mathbf{n}\ |_{r,s} = \varepsilon_0 (\mathbf{E}^+ - \mathbf{E}^-)\ |_{r,s} \bullet \mathbf{n}_{r,s}, \tag{2.2}$$

as $\mathbf{D} = \varepsilon_0 \varepsilon \mathbf{E}$ and $\mathbf{P} = \varepsilon_0 \chi \mathbf{E}$. ε_o represents the electric permeability of a vacuum. Superscripts + and – denote the limits of **E**, as the crack-lines *r* or *s* are approached from the exterior and from the interior domain, respectively. The traction across *r* and *s* is due to the jump of the electrostatic tensor, which corresponds to the Maxwell tensor $\mathbf{T}_M$ in the crack region and to the tensor $(\varepsilon\, \mathbf{T}_M)$ in the dielectric [17-19]. Specifically,

$$\mathbf{t}_n\ |_{r,s} = (\mathbf{T}_M{}^+ - (\varepsilon\, \mathbf{T}_M)^-)\ |_{r,s}\, \mathbf{n}_{r,s}. \tag{2.3}$$

where

$$\mathbf{T}_M = \varepsilon_o\, (\mathbf{E} \otimes \mathbf{E} - \tfrac{1}{2} E^2 \mathbf{I}) \tag{2.4}$$

Note that the following equation holds true in the dielectric:

$$\mathrm{div}\,(\varepsilon\, \mathbf{T}_M) = -\tfrac{1}{2}\, \varepsilon_o\, E^2\, \mathrm{grad}\, \chi. \tag{2.5}$$

For the sake of simplicity, we will consider the traction across *r* and *s*, in the case the external electric field to be directed either along $\mathbf{n}_{r,s}$ or orthogonal to $\mathbf{n}_{r,s}$. With reference to figures 1 and 2, one can note that, in both cases, the resulting electric field, which is solution of the Maxwell equations, is closely aligned with the external field. In addition, the traction is directed in both cases along the respective outward unit normal, far from the end-points. As a result, the mechanical effect of the electric field is that of opposing the opening of the crack. The explicit formula for the electrostatic traction is straightforward if the electric field is orthogonal to $\mathbf{n}_{r,s}$ (Fig. 2). In this case, as **E** is continuous across *r* and across *s*, the jump of the traction is only due to the different values of ε (or of χ) in the dielectric and in a vacuum, respectively.

$$\mathbf{t_n}\big|_{r,s} = -\tfrac{1}{2}\,\varepsilon_o\,[\chi]\,\mathrm{E}^2\big|_{r,s}\,\mathbf{n}_{r,s} \equiv \tfrac{1}{2}\,\varepsilon_o\,\chi\,\mathrm{E}^2\big|_{r,s}\,\mathbf{n}_{r,s} \tag{2.6}$$

If the dielectric is homogeneous, $\chi_r = \chi_s$ and the resultant traction $\mathbf{t_n} \equiv (\mathbf{t_n}|_r + \mathbf{t_n}|_s)$ vanishes in the limit as the distance between *r* and *s* tends to zero. This result provides the means for evaluating through the standard procedure the J-integral and the material force acting at the tip of the crack. This force, in turn, appeals to the Eshelby material tensor, in its appropriate form that is consistent with the boundary conditions. These forms will be presented and discussed in the subsequent section. Analogous arguments can be expounded for magnetised (non-ferromagnetic) materials.

3. STRESS TENSORS AND MATERIAL ENERGY-MOMENTUM TENSORS

Introduce a reference and a current configuration V and $\mathcal{V}$, respectively, for the dielectric of interest. V, $\mathcal{V} \subset E_3$. E_3 represents the 3-dimensional Euclidean space. Assume, as usual that the deformation is a smooth mapping $\boldsymbol{k}$, such that $\boldsymbol{k}$ (V) = $\mathcal{V}$. **F** denotes the deformation gradient and J its determinant, which is strictly positive as usual, [20]. The following *material* fields are introduced in V:

$$\begin{aligned}&\boldsymbol{D} = \mathrm{J}\mathbf{F}^{-1}\,\mathbf{D};\quad \boldsymbol{B} = \mathrm{J}\mathbf{F}^{-1}\,\mathbf{B};\quad \boldsymbol{P} = \mathrm{J}\mathbf{F}^{-1}\,\mathbf{P};\\ &\boldsymbol{E} = \mathbf{F}^{\mathrm{T}}\,\mathbf{E};\boldsymbol{H} = \mathbf{F}^{\mathrm{T}}\,\mathbf{H};\boldsymbol{M} = \mathbf{F}^{\mathrm{T}}\,\mathbf{M};\end{aligned} \tag{3.1}$$

M represents the magnetization. In the absence of free charges and free currents, the Maxwell equations of electrostatic and magnetostatic for these material fields read,

$$\begin{aligned}&\mathrm{div}_R\,\boldsymbol{D} = 0,\\ &\mathrm{div}_R\,\boldsymbol{B} = 0,\\ &\mathrm{curl}_R\,\boldsymbol{E} = 0,\\ &\mathrm{curl}_R\,\boldsymbol{H} = 0, \text{ in V.}\end{aligned} \tag{3.2}$$

Details can be found in [3,7-9]. Based on a Lagrangian density *L* that depends on the aforementioned material fields, the total stress tensor in the Cauchy form is

$$\mathbf{T} = -\,\mathrm{J}^{-1}(\partial L/\partial \mathrm{F})\mathbf{F}^{\mathrm{T}}. \tag{3.3}$$

Two Lagrangian densities L^e and L^m are reported here below:

$$L^e(\mathbf{F}, \boldsymbol{E}, \boldsymbol{P}, \mathbf{X}) = ½\, \varepsilon_o\, \mathrm{J}\, \boldsymbol{E} \bullet \mathbf{C}^{-1}\boldsymbol{E} + \boldsymbol{P}\bullet\boldsymbol{E} - \hat{\mathrm{W}}^e(\mathbf{F}, \mathbf{F}\boldsymbol{P}, \mathbf{X}), \tag{3.4}$$

$$L^m(\mathbf{F}, \mathsf{B}, \boldsymbol{M}, \mathbf{X}) = -½\,(\mu_o\,\mathrm{J})^{-1}\, \mathsf{B}\bullet\mathsf{C}\mathsf{B} + \boldsymbol{M}\bullet\mathsf{B} - \hat{\mathrm{W}}^m(\mathbf{F}, \mathbf{J}\mathbf{F}^{-T}\boldsymbol{M}, \mathbf{X}), \tag{3.5}$$

where $\mathbf{C} = \mathbf{F}^T\mathbf{F}$. L^e is related to electrostatics, whereas L^m to magnetostatics. With reference to the formulas (3.3) and (3.4), one can write the explicit form for the electrostatic stress tensor

$$\mathbf{T}^e = \mathbf{T}_M + (\partial\hat{\mathrm{W}}^e/\partial\mathbf{F}\boldsymbol{P}) \otimes \mathbf{P} + \mathrm{J}^{-1}(\partial\,\hat{\mathrm{W}}^e/\partial\mathbf{F})_{FP}\,\mathbf{F}^T, \tag{3.6}$$

The term $(\partial\hat{\mathrm{W}}^e/\partial\mathbf{F}\boldsymbol{P})$ in the formula (3.6) can be replaced by $\mathbf{E}$, as the following equation, which stems from the variational procedure, holds true [8]:

$$(\partial\hat{\mathrm{W}}^e/\partial\mathbf{F}\boldsymbol{P}) = \mathbf{F}^{-T}\boldsymbol{E} \equiv \mathbf{E}\,. \tag{3.7}$$

Based on the formulas (3.3) and (3.5), the magnetostatic stress tensor reads

$$\mathbf{T}^m = \mathbf{T}_M{}^m + \{(\partial\,\hat{\mathrm{W}}^m/\partial\,(\mathbf{J}\mathbf{F}^{-T}\boldsymbol{M}))\bullet\mathbf{M}\}\,\mathbf{I} - \mathbf{M}\otimes(\partial\,\hat{\mathrm{W}}^m/\partial\,\mathbf{J}\mathbf{F}^{-T}\boldsymbol{M}) +$$

$$+\,\mathrm{J}^{-1}(\partial\,\hat{\mathrm{W}}^m/\partial\mathbf{F})_{JF\text{-}TM}\,\mathbf{J}\mathbf{F}^T\,, \tag{3.8}$$

where

$$\mathbf{T}_M{}^m = (\mu_o)^{-1}(\mathbf{B}\otimes\mathbf{B} - ½\,\mathbf{B}^2\,\mathbf{I}) \tag{3.9}$$

represents the Maxwell stress tensor in magnetostatics, which is defined also in a vacuum. μ_o is the magnetic permeability of a vacuum. The formula (3.8) leads to the following final form:

$$\mathbf{T}^m = \mathbf{H}\otimes\mathbf{B} - ½\,\mu_o\,(\mathrm{H}^2 - \mathrm{M}^2)\,\mathbf{I} + \mathrm{J}^{-1}\{(\partial\hat{\mathrm{W}}^m/\partial\mathbf{F})_{JF\text{-}TM}\}\mathbf{F}^T\,, \tag{3.10}$$

by taking into account that $\mathbf{H} = (\mu_o)^{-1}\mathbf{B} - \mathbf{M}$ and that $\mathbf{B} = \partial\hat{\mathrm{W}}^m/\partial\,(\mathbf{J}\mathbf{F}^{-T}\boldsymbol{M})$. The latter equality holds true in the absence of magnetisation gradients.

The material energy-momentum tensor for a dielectric is given by the formula [7,8]

$$\mathbf{b}^e = (\hat{\mathrm{W}}^e - \boldsymbol{E}\bullet\boldsymbol{P})\,\mathbf{I} - \mathbf{F}^T(\partial\hat{\mathrm{W}}^e/\partial\mathbf{F}) + \boldsymbol{E}\otimes\boldsymbol{P}. \tag{3.11}$$

Differentiation of $\hat{\mathrm{W}}^e$ with respect to $\mathbf{F}$ provides the following explicit expression:

$$(\partial\hat{W}^e/\partial\mathbf{F}) = \mathbf{F}^T(\partial\hat{W}^e/\partial\mathbf{F})_{FP} + (\partial\hat{W}^e/\partial\mathbf{F}\boldsymbol{P})(\partial\mathbf{F}\boldsymbol{P}/\partial\mathbf{F}) =$$

$$= \mathbf{F}^T(\partial\hat{W}^e/\partial\mathbf{F})_{FP} + (\partial\hat{W}^e/\partial\mathbf{F}\boldsymbol{P}) \otimes \boldsymbol{P}. \tag{3.12}$$

Hence, by taking into account the formula (3.7), the formula (3.11) also reads

$$\mathbf{b}^e = (\hat{W}^e - \boldsymbol{E}\bullet\boldsymbol{P})\,\mathbf{I} - \mathbf{F}^T(\partial\hat{W}^e/\partial\mathbf{F})_{FP}\,. \tag{3.13}$$

$\mathbf{b}^e$ is evidently defined only inside the dielectric. An alternate equivalent form is the following [8-11]:

$$\boldsymbol{b}^e = L^e\,\mathbf{I} - \mathbf{F}^T(\partial L^e/\partial\mathbf{F}) + \boldsymbol{E}\otimes\boldsymbol{D}. \tag{3.14}$$

One can note that $\boldsymbol{b}^e$ is apparently defined in a vacuum. However, this occurrence is only apparent. In fact, $\boldsymbol{b}^e$ turns out to be equivalent to $\mathbf{b}^e$ once some redundant terms are cancelled out [12]. Notice also that the all the aforementioned energy-momentum tensors are divergence free in V for homogeneous materials. Across ∂V, the boundary of V, the following equalities hold true:

$$[\mathbf{b}^e]\,\mathbf{N} = -\{(\hat{W}^e - \boldsymbol{E}\bullet\boldsymbol{P})\,\mathbf{N} - \mathbf{F}^T(\partial\hat{W}^e/\partial\mathbf{F})_{FP}\,\mathbf{N}\}^-, \tag{3.15}$$

$$[\mathbf{b}^e]\,\mathbf{N} = -\{(\hat{W}^e - \boldsymbol{E}\bullet\boldsymbol{P})\,\mathbf{N} - \mathbf{F}^T(\partial\hat{W}^e/\partial\mathbf{F})\,\mathbf{N} + [\boldsymbol{E}]\,(\boldsymbol{P}\bullet\mathbf{N})\}^-, \tag{3.16}$$

$$[\boldsymbol{b}^e]\,\mathbf{N} = [L^e]\,\mathbf{N} - [\mathbf{F}^T(\partial L^e/\partial\mathbf{F})]\,\mathbf{N} + [\boldsymbol{E}]\,(\boldsymbol{D}\bullet\mathbf{N}), \tag{3.17}$$

N being the outward unit normal to ∂V.

One may note that across the boundary $\boldsymbol{b}^e$ involves the normal component of the material electric displacement, which is continuous. Differently, $\mathbf{b}^e$ involves the polarisation, whose normal component is discontinuous across the boundary, according to the equalities (3.15) and (3.16). This remark suggests that in crack-problems, $\boldsymbol{b}^e$ would be more appropriate, having in mind the remark expounded in section 2. With reference to the previous section, $\boldsymbol{b}^e$ would be consistent with both models: the two parallel crack-lines and the single crack-line. One can proceed in an analogous fashion for magnetic materials. We only report here below the final expression for $\mathbf{b}^m$, which represents the counterpart of the formulas (3.11) and (3.13)

$$\mathbf{b}^m = \hat{W}^m\mathbf{I} - \boldsymbol{M}\otimes\mathbf{B} - \mathbf{F}^T(\partial\hat{W}^m/\partial\mathbf{F}) =$$

$$= (\hat{W}^m - \boldsymbol{M}\bullet\mathbf{B})\,\mathbf{I} - \mathbf{F}^T(\partial\hat{W}^m/\partial\mathbf{F})_{(J\mathbf{F}^{-T}\boldsymbol{M})}. \tag{3.18}$$

Formulae (3.11)–(3.18) are useful for evaluating the material (or configurational) force at the tip of the crack in dielectrics and magnetic materials. They also account for the material force at the interface of two thermodynamical phases that possibly co-exist in these materials.

4. AN INTERESTING EXAMPLE

***i)* A dielectric material.**

In the following we will consider a dielectric, which may undergo finite deformations. We assume that this dielectric is homogeneous and isotropic in its current configuration and that the relationship $\mathbf{P} = \varepsilon_o\chi\mathbf{E}$, χ = constant, holds true in this configuration. Also, assume that the energy density per unit volume of the current configuration is

$$w^d(\mathbf{P}) = \tfrac{1}{2}(\varepsilon_o\chi)^{-1}P^2. \tag{4.1}$$

The energy density per unit volume of the reference configuration, in terms of the material fields, reads

$$\hat{W}^d(\mathbf{F},\boldsymbol{P}) = \tfrac{1}{2}(\varepsilon_o\chi)^{-1} J P^2 = \tfrac{1}{2}(\varepsilon_o\chi J)^{-1}\boldsymbol{P}\bullet\mathbf{C}\boldsymbol{P} \tag{4.2}$$

With reference to the formulas (3.6), (3.7) and (4.2), one is able to write the electrostatic stress tensor $\mathbf{T}^e$ for this dielectric as

$$\mathbf{T}^e = \mathbf{T}_M + \mathbf{E}\otimes\mathbf{P} - \tfrac{1}{2}\varepsilon_o\chi E^2\mathbf{I} \equiv \varepsilon\mathbf{T}_M. \tag{4.3}$$

It is noticeable that the expression (4.3) for the stress tensor coincides with that for a rigid dielectric, such as introduced in section 2.

With reference to the formula (3.13), the material energy-momentum tensor for this dielectric can be written as

$$\mathbf{b}^e = \{\tfrac{1}{2}(\varepsilon_o\chi J)^{-1}\boldsymbol{P}\bullet\mathbf{C}\boldsymbol{P} - \boldsymbol{E}\bullet\boldsymbol{P} + \tfrac{1}{2}(\varepsilon_o\chi J)^{-1}\boldsymbol{P}\bullet\mathbf{C}\boldsymbol{P})\}\,\mathbf{I}. \tag{4.4}$$

It is easy to check that this tensor turns out to vanish identically, if the formula (3.7) is taken into account. The formulae (4.3) and (4.4) express a surprising and unexpected result. However, should χ depend on **F**, additional *electrostictive* terms would appear in the expression of the stress tensor. In the case of fluids, χ would depend on J and the well-known formula for the stress would be recovered straightforwardly [17-19].

***ii)* A paramagnetic material.**

One can assume that an elastic magnetic material is characterised by the classical constitutive relationship in its current configuration.

$$\mathbf{H} = (\mu_o\mu)^{-1}\mathbf{B}, \tag{4.5}$$

where μ represents the magnetic permittivity of the material. The related energy density in its current configuration is w^m, given by

$$w^m(\mathbf{M}) = \tfrac{1}{2}(\mu_o\mu)(\chi^m)^{-1}\,\mathrm{M}^2, \tag{4.6}$$

where χ^m denotes the magnetic susceptibility. It is worth recalling that $\mu = 1 + \chi^m$ and that $\chi^m > 0$ for a paramagnetic material. **M** has to be understood here as depending on **F** and on ***M***. With reference to all these quantities and to the formula (3.9), the formula (3.10) reduces to [19,21]

$$\mathbf{T}^m = \mu^{-1}\,\mathbf{T}_M{}^m \equiv \mu\,\mu_o\,(\mathbf{H}\otimes\mathbf{H} - \tfrac{1}{2}\mathrm{H}^2\,\mathbf{I}). \tag{4.7}$$

The corresponding material tensor $\mathbf{b}^m$ identically vanishes, as in the case of the dielectric previously discussed.

5. FINAL REMARK

As is known, the mechanical stress is completely undetermined in a rigid body, although procedures can be developed for recovering a specific expression for the stress even in this case. In the presence of electric or magnetic fields, a specific form for the stress tensor is available for a rigid body. This form stems from the electromagnetic theory. It is worth remarking that the stress tensor, which is associated with elastic dielectrics (or magnetic materials), turns out to coincide with that of a rigid body, provided that these materials are homogeneous and isotropic in their current configuration. An additional unexpected result is that the material tensor that is associated with these bodies turns out to vanish identically.

6. REFERENCES

1. D. Eshelby, Force on an elastic singularity, Phil. Trans. Roy. Soc. Lond., A244, (1951), 87-112.
2. J.D. Eshelby, The elastic energy-momentum tensor, J. of Elasticity, 5, (1975), 321-335.
3. G.A. Maugin, Material Inhomogeneities in Elasticity, Series 'Applied Mathematics and Mathematical Computation', Chapman and Hall, London, Vol. 3, 1993.

4. R. Kienzler and G. Herrmann, Mechanics in Material Space, Springer-Verlag, Berlin, 2000.
5. M.E. Gurtin, Configurational Forces as Basic Concepts of Continuum Physics, Springer-Verlag, New York, 2000.
6. G.A. Maugin and C.Trimarco, On material and physical forces in liquid crystals, Int. J. Engng. Sci., 33, (1995), 1663-1678.
7. G.A. Maugin and C.Trimarco, Elements of field theory in inhomogeneous and defective materials, CISM courses and lectures n. 427 on 'Configurational Mechanics of Materials', Springer-Verlag, Wien, (2001), 55-128.
8. C. Trimarco and G.A. Maugin, Material mechanics of electromagnetic bodies, CISM courses and lectures n. 427 on 'Configurational Mechanics of Materials', Springer-Verlag, Wien, (2001), 129-172.
9. C. Trimarco, A Lagrangian approach to electromagnetic bodies, Technische Mechanik, 22 (3), (2002), 175-180.
10. Y.E. Pack and G. Herrmann, Conservation laws and the material momentum tensor for the elastic dielectric, Int. J. Engng. Sci., 24, (1986), 1365-1374.
11. Y.E. Pack and G. Herrmann, Crack extension force in a dielectric medium, Int. J. Engng. Sci., 8, (1986), 1375-1388.
12. G.A. Maugin and M. Epstein, The electroelastic energy-momentum tensor, Proc. Roy. Soc. Lond., A433, (1991), 299-312.
13. R.D. James, Configurational forces in magnetism with application to the dynamics of a small-scale ferromagnetic shape memory cantilever, Continuum Mech. Thermodynamics, 14, (2002), 55-86.
14. M. Sabir, M. and G.A. Maugin, On the fracture of paramagnets and soft ferromagnets, Int. J. Non-Linear Mechanics, 31, (1996), 425-440.
15. Z. Suo, C.M. Kuo, D.M. Barnett, and J.R. Willis, Fracture mechanics for piezoelectric ceramics, J. Mech. Phys. Solids, 40, (1992), 739-765.
16. C. Dascalu, and G.A. Maugin, Energy-release rates and path-independent integrals in electrostatic crack propagation, Int. J. Engng. Sci., 32, (1994), 755-765.
17. R. Becker, Electromagnetic Fields and Interactions, Vol.1, Blackie & Sons, London, 1964.
18. J.D. Jackson, Classical Electrodynamics, J. Wiley & Sons, New York, 1962.
19. J.A. Stratton, Electromagnetic Theory, McGraw-Hill, New York, 1941.
20. C.A. Truesdell, and R.A. Toupin, The Classical Theory of Fields, Handbuch der Physik, Bd.III/1, S. Flügge (ed.), Springer-Verlag, Berlin, 1960.
21. W.F. Brown, Magnetoelastic Interactions, Series 'Springer Tracts in Natural Philosophy, Vol. 9, Springer-Verlag, Berlin, 1966.

Chapter 17

CONTINUUM THERMODYNAMIC AND VARIATIONAL MODELS FOR CONTINUA WITH MICROSTRUCTURE AND MATERIAL INHOMOGENEITY

Bob Svendsen
Chair of Mechanics
University of Dortmund
D-44221 Dortmund, Germany
bob.svendsen@udo.edu

Abstract

The purpose of this work is the continuum thermodynamic and variational formulation of spatial and configurational models for a class of elastic, viscous continua with microstructure and material inhomogeneity. As in the standard case, the variational formulation of such models relies on the basic results of the direct continuum thermodynamic formulation obtained in the context of the total energy balance and dissipation principle. On this basis, spatial field relations in the bulk and across a singular surface, as well as the corresponding boundary conditions, are derived for the positional and microstructure degrees-of-freedom as the rate stationarity conditions of a corresponding rate-functional. Finally, variation of the incremental form of the rate functional with respect to the reference configuration yields the configurational balance and field relations for the materials in question. For simplicity, attention is restricted here to isothermal and quasi-static conditions.

Keywords: Energy balance, dissipation principle, variational formulation, microstructure, configuration fields, multiple lengthscales

Introduction

The behaviour of many materials of engineering interest (*e.g.*, metals, alloys, granular materials, composites, liquid crystals, polycrystals) is often influenced by an existing or emergent microstructure (*e.g.*, phases in multiphase materials, phase transitions, voids, microcracks, dislocation substructures, texture). In general, the components of such a microstructure have different material properties, resulting in a macroscopic material behaviour which is materially inhomogeneous. Attempts to incorporate this fact into the continuum modeling of such materials have lead to a number of approaches to and viewpoints on the issue, depending in part on the nature of the microstructure and corresponding inhomogeneity in question (*e.g.*, Noll, 1967; Capriz, 1989; Maugin, 1993; Gurtin, 2000). Beyond heterogeneous material properties, processes associated with the microstructure which are represented in the model by continuum fields (*e.g.*, damage and order parameter fields, director field) also contribute to configurational fields and processes. Such fields represent additional continuum degrees-of-freedom for which corresponding field relations must be formulated. Contigent on the premise that the corresponding processes contribute to energy flux and energy supply in the material, field relations for such degrees-of-freedom result from the Euclidean frame-indifference of the total rate-of-work (*e.g.*, Capriz, 1989), or more generally from that of the total energy balance (*e.g.*, Capriz and Virga, 1994; Svendsen, 1999; Svendsen, 2001). Once thermodynamically-consistent field relations and reduced constitutive relations have been obtained, one is in a position to formulate and solve initial-boundary-value problems. In the context of elastic material behaviour and thermodynamic equilibrium, such initial-boundary-value problems are often formulated variationally (*e.g.*, for elastic phase transitions: Ball and James, 1987; for elastic liquid crystals: Virga, 1994; for configurational fields in elastic materials: Podio-Guidugli, 2001; see also Šilhavý, 1997, Chs. 13-21). Recently, it has been shown (*e.g.*, Ortiz and Repetto, 1999; Miehe, 2002; Carstensen et al., 2003) that direct variational methods for elastic materials can be carried over to the inelastic case with the help of a so-called incremental variational formulation. The purpose of this short work is the application of this incremental approach to the variational formulation of spatial balance relations as well as of configurational field and balance relations for a material inhomogeneous inelastic continuum containing a singular surface and microstructure. For simplicity, the formulation in this work is restricted to isothermal and quasi-static conditions.

1. BASIC CONTINUUM THERMODYNAMIC FORMULATION

Let E represent 3-dimensional Euclidean point space with translation vector space V and $B \subset E$ an arbitrary reference configuration of some material body containing a stationary coherent singular surface S. The time-dependent deformation or motion of the material body with respect to B and E during some time interval $I \subset \mathbb{R}$ takes as usual the form $\boldsymbol{\xi} : I \times B \rightarrow E \mid (t, \boldsymbol{r}) \mapsto \boldsymbol{x}_t = \boldsymbol{\xi}(t, \boldsymbol{r})$. Since S is coherent, $\boldsymbol{\xi}$ is continuous and piecewise continuously differentiable, implying that the jump $[\![\boldsymbol{\xi}]\!] := \boldsymbol{\xi}_+ - \boldsymbol{\xi}_-$ of $\boldsymbol{\xi}$ at S vanishes, *i.e.*, $[\![\boldsymbol{\xi}]\!] = \mathbf{0}$ holds. Basic kinematic quantities of interest include the material velocity $\dot{\boldsymbol{\xi}}(t, \boldsymbol{r}) \in V$ and the deformation gradient $\boldsymbol{F}(t, \boldsymbol{r}) := \nabla\boldsymbol{\xi}(t, \boldsymbol{r}) \in \mathrm{Lin}^+(V, V)$. Since S is stationary and $\boldsymbol{\xi}$ is continuous and piecewise continuously differentiable, the Hadamard lemma (*e.g.*, Šilhavý, 1997, Prop. 2.1.6) implies that $[\![\dot{\boldsymbol{\xi}}]\!] = \mathbf{0}$ and that $[\![\boldsymbol{F}]\!]$ is rank-one convex.

To account for the effect of microstructural processes and/or properties on the macroscopic behaviour, the standard continuum degrees of freedom as represented by $\boldsymbol{\xi}$ are complemented here by additional microstructural ones, idealized here in the form of a time-dependent field on B, *e.g.*, the Cosserat rotation field, or the director field for uniaxial liquid crystals. ¿From the mathematical point of view, a formulation sufficiently general to encompass such standard models is obtained when this field is assumed to take values in a submanifold[1] $\mathfrak{S}$ of some finite-dimensional inner product space W. To simplify the formulation, however, it is useful to work with the inclusion $\varsigma : I \times B \rightarrow W \mid (t, \boldsymbol{r}) \mapsto s = \varsigma(t, \boldsymbol{r})$ of the structure field into W. For simplicity, attention is restricted here to the case that S is coherent with respect to ς as well, *i.e.*, $[\![\varsigma]\!] = \mathbf{0}$. The Hadamard lemma then yields $[\![\dot{\varsigma}]\!] = \mathbf{0}$ and $[\![\nabla\varsigma]\!]$ rank-one convex.

Assuming now that processes associated with the evolution of $x := (\boldsymbol{\xi}, \varsigma)$ result in mechanical work being done in the system, the approach to the formulation of balance relations for materials with microstructure being pursued here is based on the total energy balance[2]

$$\mathcal{E} = \overline{\int_P \psi}^{\,\cdot} + \int_P \delta - \int_{\partial P} f \cdot \dot{x} - \int_P s \cdot \dot{x} = 0 \tag{1}$$

for any part $P \subset B$ (*e.g.*, Svendsen, 2004). Here, ψ represents the free energy density, δ the dissipation-rate density, F the generalized momentum flux density of normal to ∂B, and s the generalized momentum supply-rate density. Assuming in addition that the material with microstructure in question is materially inhomogeneous and behaves viscoelastically, the invariance of the total energy balance with respect to Euclidean observer (*e.g.*, Capriz and Virga, 1994; Šilhavý, 1997, Ch. 6; Svendsen, 1999; Svendsen, 2001) together combined with the exploitation of the dissipation principle (*e.g.*, Šilhavý, 1997, Ch.

[1] For example, in the case of uniaxial nematic liquid crystals, $\mathfrak{S}$ could be the unit sphere S^2, a smooth compact submanifold of three-dimensional Euclidean vector space $V \cong W$. In this case, we have $\boldsymbol{\iota}(\boldsymbol{e}) = \boldsymbol{e}$ for all $\boldsymbol{e} \in S^2$, and $\boldsymbol{\pi}(\boldsymbol{a}) = \boldsymbol{a}/|\boldsymbol{a}|$ for all non-zero $\boldsymbol{a} \in V$.

[2] The volume dv and surface da measures are left out of the corresponding integral notations in this work for simplicity.

9; Svendsen, 1999; Svendsen, 2001) results in the field relations

$$\begin{aligned} \mathbf{0} &= \operatorname{div}(\partial_{\nabla\dot{x}} r_{\mathrm{I}}) - \partial_{\dot{x}} r_{\mathrm{I}} + s && \text{on } B \setminus S\,, \\ \mathbf{0} &= [\![\partial_{\nabla\dot{x}} r_{\mathrm{I}}]\!]\boldsymbol{n} && \text{on } S\,, \end{aligned} \tag{2}$$

(*e.g.*, Svendsen, 2004) and constitutive restriction $F = \partial_{\nabla\dot{x}} r_{\mathrm{I}}$ in terms of the rate potential $r_{\mathrm{I}} := \dot{\psi} + d_{\mathrm{I}}$. This potential is determined by the reduced forms[3] $\psi = \psi(x, \nabla x, \boldsymbol{r})$ and $d_{\mathrm{I}} = d_{\mathrm{I}}(x, \nabla x, \dot{x}, \nabla\dot{x}, \boldsymbol{r})$ at any[4] $\boldsymbol{r} \in B$. The dissipation potential determines the residual constitutive form $\delta = \partial_{\dot{x}} d_{\mathrm{I}} \cdot \dot{x} + \partial_{\nabla\dot{x}} d_{\mathrm{I}} \cdot \nabla\dot{x} \geq d_{\mathrm{I}}$ of the dissipation-rate density δ, with equality holding in the rate-independent special case. Note that d_{I} is convex and minimal in its rate (*i.e.*, non-equilibrium) arguments $\dot{x}$, $\nabla\dot{\boldsymbol{\xi}}$ and $\nabla\dot{x}$, as well as non-negative.

2. RATE-BASED AND INCREMENTAL VARIATIONAL FORMULATIONS

For concreteness, the variational formulation to follow presumes a loading enviroment for the material under consideration of the displacement-traction[5] type generalized to the current setting, *i.e.*, applying to x. As usual, the boundary ∂B of B is then divided into generalized displacement ∂B_x and generalized traction ∂B_f parts such that $\partial B = \partial B_x \cup \partial B_f$ and $\emptyset = \partial B_x \cap \partial B_f$ hold. By definition, f is compatible with $f = \partial_{\nabla\dot{x}} r_{\mathrm{I}}\, \boldsymbol{n}$ on ∂B_f, and vanishes on ∂B_x. Likewise, x is prescribed on ∂B_x. On this basis, consider the class of loading environments characterized by the non-conservative form

$$\begin{aligned} -\int_B s \cdot \dot{x} - \int_{\partial B} f \cdot \dot{x} &= \overline{\int_B w_s}^{\,\boldsymbol{\cdot}} + \int_B (\partial_{\dot{x}} d_s \cdot \dot{x} + \partial_{\nabla\dot{x}} d_s \cdot \nabla\dot{x}) \\ &+ \overline{\int_{\partial B_f} w_f}^{\,\boldsymbol{\cdot}} + \int_{\partial B_f} \partial_{\dot{x}} d_f \cdot \dot{x} \end{aligned} \tag{3}$$

for the generalized power of external forces or rate of external work, holding for all kinematically-admissible x. Here, $w_s = w_s(x, \nabla x)$ and $w_f = w_f(x)$ represent direct generalizations of the bulk and surface potential energy densities for a conservative loading environment (*e.g.*, Šilhavý, 1997, §13.3) to the case of microstructure, while $d_s = d_s(x, \nabla x, \dot{x}, \nabla\dot{x})$ and $d_f = d_f(x, \dot{x})$ represent corresponding dissipation potentials accounting for the effects of friction and other non-conservative loading processes in the bulk and on the boundary,

[3] Material frame-indifference leads to a further reduction in the forms of ψ and d_{I} not accounted for here.
[4] For notational simplicity, we neglect the dependence of the constitutive relations on $\boldsymbol{r}$ in the notation until it becomes relevant.
[5] Other such environments, *e.g.*, unilateral or bilateral contact (*e.g.*, Šilhavý, 1997, §13.3), can also be generalized to the current context and approach take here.

respectively. In terms of the rate potentials $r_s := \dot{w}_s + d_s$ and $r_f := \dot{w}_f + d_f$ associated with the supply-rate and boundary-flux contributions, respectively, to the rate of external work, where $\delta_q f := \partial_q f - \operatorname{div}(\partial_{\nabla q} f)$ represents the variational derivative (*e.g.*, Abraham et al., 1988, Supplement 2.4C; Šilhavý, 1997, §13.3), (3) together with (2) yields

$$\begin{aligned} \mathbf{0} &= \delta_{\dot{x}} r && \text{on } B \setminus S , \\ \mathbf{0} &= \partial_{\nabla\dot{x}} r\, \boldsymbol{n} + \partial_{\dot{x}} r_f && \text{on } \partial B_f , \\ \mathbf{0} &= [\![\partial_{\nabla\dot{x}} r]\!] \boldsymbol{n} && \text{on } S , \end{aligned} \tag{4}$$

where $r := r_{\mathrm{I}} + r_{\mathrm{S}}$. In particular, $(4)_1$ implies that r is a null Lagrangian in the rates $\dot{x}$ (*e.g.*, Šilhavý, 1997, §13.6). As it turns out, these last relations represent the (rate) stationarity conditions of the (rate) functional[6]

$$R(x, \dot{x}) := \int_B r(x, \nabla x, \dot{x}, \nabla \dot{x}) + \int_{\partial B_f} r_f(x, \dot{x}) \tag{5}$$

with respect to B. Note that this functional is bounded from above; indeed, in the context of (3), the energy balance (1) and result $\delta \geq d_{\mathrm{I}}$ imply $0 \geq R$, in particular via the convexity of r and r_f in their rate arguments. Again, equality holds in the rate-independent case. The vanishing of the variation of R with respect to admissible variations $\delta\dot{x}$ in the rates holding x fixed[7] implies (4). In addition, one can show that the stability of rate stationary points of R is determined by the Hessian matrix of $d := d_{\mathrm{I}} + d_s$ with respect to its rates together with $\partial_{\dot{x}}(\partial_{\dot{x}} d_f)$. Since d and d_f are by definition convex in the rates, this matrix is positive-definite, R is minimal in the rates, and states satisfying (4) are stable in the rates.

Consider next the incremental form of the above rate-based variational formulation. Time-integration of R over the time interval $[t_s, t_e] \subset I$, rearrangement and forward-Euler approximation of the time-averages over $[t_s, t_e]$ yields the functional

$$I_{e,s}(x_e) = \int_B \varphi_{e,s}(x_e, \nabla x_e) + \int_{\partial B_f} \varphi_{fe,s}(x_e) \quad , \tag{6}$$

where

$$\begin{aligned} \varphi_{e,s}(x_e, \nabla x_e) &= \psi(x_e, \nabla x_e) + w_s(x_e, \nabla x_e) \\ &+ t_{e,s}\, d(x_s, \nabla x_s, x_{e,s}/t_{e,s}, \nabla x_{e,s}/t_{e,s}) , \\ \varphi_{fe,s}(x_e) &= w_f(x_e) + \Delta t\, d_f(x_s, x_{e,s}/t_{e,s}) , \end{aligned} \tag{7}$$

---

[6]This is a functional on the tangent bundle $T\mathcal{X}$ of the (infinite-dimensional) manifold $\mathcal{X}$ of all admissible states x.

[7]In the tangent-bundle context, this represents the so-called fibre derivative of R on $T_x\mathcal{X}$ (*e.g.*, Abraham et al., 1988, Supplement 8.1B).

represent the volume and surface density, respectively, of $I_{e,s}(\chi_e)$, $t_{e,s} := t_e - t_s$, $\chi_{e,s} := \chi_e - \chi_s$ and likewise for $\nabla\chi$. Requiring the variation of $I_{e,s}(\chi_e)$ with respect to χ_e to vanish for all admissible $\delta\chi_e$ results in the incremental form

$$\begin{aligned} \mathbf{0} &= \delta_{\chi_e}\varphi_{e,s} && \text{on } B \setminus S\,, \\ \mathbf{0} &= \partial_{\nabla\chi_e}\varphi_{e,s}\,\boldsymbol{n} + \partial_{\chi_e}\varphi_{fe,s} && \text{on } \partial B_f\,, \\ \mathbf{0} &= [\![\partial_{\nabla\chi_e}\varphi_{e,s}]\!]\boldsymbol{n} && \text{on } S\,, \end{aligned} \tag{8}$$

of the system (4). Analogous to r being a null Lagrangian in $\dot{\chi}$ on the basis of $(4)_1$, note that $(8)_1$ implies that $\varphi_{e,s}$ is a null Lagrangian in χ_e (*e.g.*, Šilhavý, 1997, §13.6). Like for the canonical free energy, one can show that $I_{e,s}(\chi_e)$ is a (monotonically) non-increasing function (*e.g.*, Svendsen, 2004), and so a possible Liapunov function for the processes of interest (*e.g.*, Šilhavý, 1997, Ch. 15).

3. INCREMENTAL VARIATIONAL FORMULATION OF CONFIGURATIONAL FIELDS AND RELATIONS

The results of the variational formulation from the last section were obtained by varying the fields holding the reference configuration B of the material under consideration fixed. As is well-known (*e.g.*, Šilhavý, 1997, §14.5; Podio-Guidugli, 2001), variations of the reference configuration at fixed fields yield in the elastic context variational forms of configurational fields and balance relations. The purpose of this section is to derive these in the current setting with respect to the incremental functional $I_{e,s}(\chi_e)$ from (6). To do this, we reintroduce the dependence of the constitutive relations on $\boldsymbol{r} \in B$ and consider a smooth variation of B as represented by a one-parameter family $\boldsymbol{\lambda}_\tau : B \to B_\tau \mid \boldsymbol{r} \mapsto \boldsymbol{r}_\tau = \boldsymbol{\lambda}_\tau(\boldsymbol{r})$ of transformations of B which leave ∂B *fixed.* By definition, $\boldsymbol{r}_0 = \boldsymbol{\lambda}_0(\boldsymbol{r}) = \boldsymbol{r}$ and $\nabla\boldsymbol{\lambda}_0 = \mathbf{1}$. This one-parameter family induces the corresponding parameterized form[8]

$$\bar{I}(\tau) := \int_{\boldsymbol{\lambda}_\tau[B]} \varphi(\chi_\tau, \nabla\chi_\tau, \boldsymbol{r}_\tau) + \int_{\partial B_f} \varphi_f(\chi, \boldsymbol{r}) \tag{9}$$

of $I_{e,s}$, where $\chi_\tau := \chi \circ \boldsymbol{\lambda}_\tau^{-1}$. Pulling (9) back to B then yields

$$\bar{I}(\tau) = \int_B \varphi(\chi, \nabla\chi_\tau \circ \boldsymbol{\lambda}_\tau, \boldsymbol{\lambda}_\tau(\boldsymbol{r}))\, \det(\nabla\boldsymbol{\lambda}_\tau) + \int_{\partial B_f} \varphi_f(\chi, \boldsymbol{r}) \quad , \tag{10}$$

[8] Dropping the e and s subscripts for the moment.

and so the result

$$\partial_\tau \bar{I}|_{\tau=0} = \int_B \partial_{\nabla x}\varphi \cdot \partial_\tau(\nabla x_\tau \circ \boldsymbol{\lambda}_\tau)|_{\tau=0} + \partial_{\boldsymbol{r}}\varphi \cdot \boldsymbol{v} + \varphi\,\mathbf{1} \cdot \nabla \boldsymbol{v} \tag{11}$$

for its variation with respect to τ, where $\boldsymbol{v} := (\partial_\tau \boldsymbol{\lambda}_\tau)|_{\tau=0}$. Since $\boldsymbol{\lambda}_\tau$ leaves the boundary ∂B of B fixed by definition, note that $\boldsymbol{v}$ vanishes on ∂B. Now, from the result $\mathbf{0} = \partial_\tau(\nabla x) = \partial_\tau(\nabla x_\tau \circ \boldsymbol{\lambda}_\tau)\,(\nabla \boldsymbol{\lambda}_\tau) + (\nabla x_\tau \circ \boldsymbol{\lambda}_\tau)\,\partial_\tau(\nabla \boldsymbol{\lambda}_\tau)$, we obtain $\partial_\tau(\nabla x_\tau \circ \boldsymbol{\lambda}_\tau)|_{\tau=0} = -(\nabla x)\,(\nabla \boldsymbol{v})|_{\tau=0}$. Substituting this into (11), it reduces to

$$\begin{aligned}
\partial_\tau \bar{I}|_{\tau=0} &= \int_B \partial_{\boldsymbol{r}}\varphi \cdot \boldsymbol{v} + \{\varphi\mathbf{1} - (\nabla x)^{\mathrm{T}}(\partial_{\nabla x}\varphi)\} \cdot \nabla \boldsymbol{v} \\
&= \int_B \{\partial_{\boldsymbol{r}}\varphi - \operatorname{div}[\varphi\mathbf{1} - (\nabla x)^{\mathrm{T}}(\partial_{\nabla x}\varphi)]\} \cdot \boldsymbol{v} \\
&\quad - \int_S [\![\varphi\mathbf{1} - (\nabla x)^{\mathrm{T}}(\partial_{\nabla x}\varphi)]\!]\boldsymbol{n} \cdot \boldsymbol{v}\,,
\end{aligned} \tag{12}$$

again since $\boldsymbol{v}$ vanishes on ∂B. Consequently, the requirement that $I_{e,s}$ be independent of (compatible) change of reference configuration, *i.e.*, that $\partial_\tau \bar{I}_{e,s}|_{\tau=0}$ vanish for all variations $\boldsymbol{v}$ leaving ∂B fixed, then implies

$$\begin{aligned}
\mathbf{0} &= \operatorname{div} \boldsymbol{E}_{e,s} - \partial_{\boldsymbol{r}}\varphi_{e,s} && \text{on } B \setminus S\,, \\
\mathbf{0} &= [\![\boldsymbol{E}_{e,s}]\!]\boldsymbol{n} && \text{on } S\,.
\end{aligned} \tag{13}$$

Here,

$$\begin{aligned}
\boldsymbol{E}_{e,s} &:= \varphi_{e,s}\,\mathbf{1} - (\nabla x_e)^{\mathrm{T}}(\partial_{\nabla x_e}\varphi_{e,s}) \\
&= \varphi_{e,s}\,\mathbf{1} - (\nabla \boldsymbol{\xi}_e)^{\mathrm{T}}(\partial_{\nabla \boldsymbol{\xi}_e}\varphi_{e,s}) - (\nabla \varsigma_e)^{\mathrm{T}}(\partial_{\nabla \varsigma_e}\varphi_{e,s})
\end{aligned} \tag{14}$$

represents the generalized total Eshelby or configurational stress tensor in the context of the incremental formulation.

The generalized form (14) for the configurational stress is formally analogous to that for the elastic case (*e.g.*, in the standard context: Maugin, 1993; in the microstructural context: Svendsen, 2001) in which $\boldsymbol{E}$ is determined by the free energy density ψ. Indeed, the role of ψ in the elastic case is played in the current inelastic context by $\varphi_{e,s}$. Finally, note that, if the material behaviour is homogeneous, then $\varphi_{e,s}$ is translationally invariant, *i.e.*, $\varphi_{e,s}(x_e, \nabla x_e, \boldsymbol{r} + \boldsymbol{a}) = \varphi_{e,s}(x_e, \nabla x_e, \boldsymbol{r})$ holds for all $\boldsymbol{a} \in V$. In this case, $\partial_{\boldsymbol{r}}\varphi_{e,s}$ vanishes, and $(13)_1$ reduces to $\mathbf{0} = \operatorname{div} \boldsymbol{E}_{e,s}$. In this case, the total Eshelby stress tensor is analogous to a null divergence (*e.g.*, Olver, 1986; Šilhavý, 1997, §13.6).

References

Abraham, R., Marsden, J. E., Ratiu, T., *Manifolds, Tensor Analysis, and Applications*, Applied Mathematical Sciences 75, 1988.

Ball, J. M., James, R. D., Fine phase mixtures as minimizers of energy, *Arch. Rat. Mech. Anal.* **100**, 13–52, 1987.

Carstensen, C., Hackl, K., Mielke, A., Nonconvex potentials and microstructures in finite-strain plasticity, *Proc. Roy. Soc. A* **458**, 299–317, 2003.

Noll, W., Material uniform inhomogeneous material bodies, *Arch. Rat. Mech. Anal.* **27**, 1–32, 1967.

Capriz, G., Continua with microstructure, *Springer Tracts in Natural Philosophy* **37**, 1989.

Capriz, G., Virga, E., On singular surfaces in the dynamics of continua with microstructure, *Quart. Appl. Math.* **52**, 509–517, 1994.

Gurtin, M. E., *Configurational forces as basic concepts of continuum physics*, Springer Series on Applied Mathematical Sciences **137**, Springer-Verlag, 2000.

Maugin, G., *Material Inhomogeneities in Elasticity*, Chapmann Hall, 1993.

Miehe, C., Strain-driven homogenization of inelastic microstructures and composites based on an incremental variational formulation, *Int. J. Numer. Meth. Engng.* **55**, 1285–1322, 2002.

Olver, P., *Applications of Lie Groups to Differential Equations*, Springer Verlag, 1986.

Ortiz, M., Repetto, E. A., Nonconvex energy minimization and dislocation structures in ductile single crystals, *J. Mech. Phys. Solids* **47**, 397–462, 1999.

Podio-Guidugli, P., Configurational balances via variational arguments, *Interfaces and Free Boundaries* **3**, 223-232, 2001.

Šilhavý, M., *The Mechanics and Thermodynamics of Continuous Media*, Springer Verlag, 1997.

Svendsen, B., On the thermodynamics of isotropic thermoelastic materials with scalar internal degrees of freedom, *Cont. Mech. Thermodyn.* **11**, 247–262, 1999.

Svendsen, B., On the formulation of balance relations and configurational fields for continua with microstructure and point defects via invariance, *Int. J. Solids Struct.* **38**, 1183–1200, 2001a.

Svendsen, B., On the thermodynamic- and variational-based formulation of models for inelastic continua with internal lengthscales, *Comp. Meth. Appl. Mech. Engrg.*, in press, 2004.

Virga, E., *Variational theories for liquid crystals*, Chapman & Hall, 1994.

Chapter 18

A CRYSTAL STRUCTURE-BASED EIGENTRANSFORMATION AND ITS WORK-CONJUGATE MATERIAL STRESS

Chien H. Wu
Department of Civil and Materials Engineering (MC 246), University of Illinois at Chicago, 842 West Taylor Street, Chicago, Illinois 60607-7023, USA
cwu@uic.edu

Abstract In the abstract of his 1970 paper, Eshelby stated: "*The force on a dislocation or point defect, as understood in solid-state physics, and the crack extension force of fracture mechanics are examples of quantities which measure the rate at which the total energy of a physical system varies as some kind of departure from uniformity within it changes its configuration.*" He then went on to demonstrate that the elastic energy-momentum tensor proves to be a useful tool in calculating such forces. The 'forces' turn out to be the appropriate traction vectors associated with the energy-momentum (stress) tensor. It is therefore natural and perhaps even fundamental to look for the 'strain tensor' that can be paired with the 'stress tensor' to form work. The 'strain rate' would then be that some kind of departure from uniformity within a physical system. In this paper, we examine the configurational changes brought about by atomic diffusion in a nonuniform alloy crystal. The transformation from a reference, single-parameter simple cubic cell to a six-parameter alloy crystal cell, called the eigentransformation, is identified as the needed kinematic tensor.

1. INTRODUCTION

We again quote Eshelby (1970) from the introduction of the same paper: "In solid-state theory, theoretical metallurgy, fracture mechanics, and elsewhere, there are departures from uniformity in a material on various scales which, for want of a better term, we shall call defects. The configuration of the defects can be specified by a number, possibly infinite, of parameters." Here, the term defect is used to describe departures from uniformity in materials, and the geometric configurations of such defects are to be represented by a number of parameters. Toward the end of formulating a continuum theory, however, the possibly infinite number of parameters can be meaningfully converted into a finite number of field variables. In case atomic diffusion is the appropriate physical mechanism, a field of nonuniform molar concentrations becomes the defect of an alloy crystal. What is then the configuration, which is nothing but the time-geometry of the crystal, affected by this defect? In general, the composition of a crystal can vary from the stoichiometric value without altering the type of structure, although the edge lengths and interaxial angles may change slightly (Barrett, 1973). The transformation from a reference cell, which may be conveniently taken as a simple cubic cell, to a triclinic cell becomes the natural kinematic variable for describing the defect.

Eshelby continued in his introduction: "Following the terminology of analytical mechanics and thermodynamics we call the rate of decrease of the total energy of the system with respect to a parameter the generalized force acting on that parameter, or, in simple cases, on the defect itself. ... It is always the total energy which is important, the energy of the system we concentrate our attention on, and which contains the defect, plus the energy of the environment with which it interacts, in our case some mechanical loading device. ... In thermodynamics the matter is handled by introducing enthalpy and Gibbs free energy, quantities which though nominally referring to the system under observation actually relate to the energy of the system plus the energy of its environment." The concept conveyed in these statements, which are directly applicable to defects specified by parameters, can be straightforwardly extended to situations where defects are field variables. What is needed in this transition is an energy density per unit reference volume that is actually a function of the defect field. Such an energy density is constructed in this paper in terms of the molar Gibbs energy for zero stress and the strain energy density per unit stress-free volume. To convert the latter from per stress-free volume to per reference volume, three sets of kinematical variables are needed: an eigentransformation, which describes the defect field, a deformation gradient, which ties the reference configuration to the spatial configuration, and finally an elastic transformation. It should be noted that the mathematic structure of this scheme was first recognized by Epstein and Maugin (1990) as a way to obtain the energy-momentum tensor as the derivative of the energy density with respect to a kinematic tensor that we call eigentransformation. That the eigentransformation can actually be tied to the molar volume of a physical system, so that atomic diffusion and elastic deformation become coupled nonlinear phenomena in isotropic solids, appears to have been first recognized by Wu (2001). This result is now extended to anisotropic solids in this paper. We use the fact that the composition of a crystal can vary from the stoichiometric value without altering the type of structure, although the edge lengths and interaxial angles may change slightly (Barrett, 1973). Thus, the eigentransformation associated with an anisotropic crystal is in general fully populated, instead of showing a mere isotropic expansion.

Eshelby concluded his introduction in the quoted paper with the admission: "The writer should perhaps admit that this tensor has become an obsession with him since he first noticed its connection with the force on a defect, and no doubt it appears at some points in the argument where one could get along without." This writer must admit that finding the work-conjugate tensor of the energy-momentum tensor has also become an obsession with him ever since he seemed to have understood the implication of this so-called materials stress. The result presented in this paper is but one specific case that can only be applied to atomic diffusion in nonuniform crystals. Eigentransformations for other physical systems must be built upon specific physical mechanisms. It appears, however, that once a mechanism is identified and properly formulated, one just could not get along without the knowledge of the associated energy-momentum tensor in studying the evolution of the underlying mechanism.

2. KINEMATICS

Kinematics deals with the time-geometry of motions of bodies. It is not in any way concerned with the causes of motions. Let **X** denote the coordinates of points in a continuum in some reference configuration $V(\mathbf{X})$ that is just a region of the ambient Euclidean space. The coordinate of the place occupied by **X** at time t is given by

$$\mathbf{x} = \mathbf{y}(\mathbf{X}, t), \tag{2.1}$$

and $v(\mathbf{x})$ is the spatial configuration, another region of the Euclidean space, taken up by $V(\mathbf{X})$ at time t. The velocity of the point **X** is denoted by

$$\dot{\mathbf{y}}(\mathbf{X},t) = \frac{\partial}{\partial t}\mathbf{y}(\mathbf{X},t). \tag{2.2}$$

The spatial configuration $v(\mathbf{x})$ is customarily referred to as the **deforming body**, and the above velocity the **motion velocity** when the transformation

$\mathbf{X} \rightarrow \mathbf{x}$ is caused by Newtonian body forces and surface tractions. The transformation may also be the result of **eigentransformations** brought about by such nonelastic strains as thermal expansion, phase transformation, initial strains, plastic and misfit strains (Mura, 1982). For such cases, the above velocity may be referred to as the **configurational velocity**, even though the function **y**(**X**, t) is the so-called total deformation that consists of the generally incompatible eigentransformation and an accompanying elastic transformation. This observation is however not pursued further in this paper. In the most general case, of course, the effects of Newtonian forces and eigentransformations are coupled. The deformation gradient **F** or the Jacobian matrix, and the associated Jacobian J are

$$\mathbf{F} = \nabla \mathbf{y}(\mathbf{X}, t), \quad J = \det \mathbf{F}. \tag{2.3}$$

The transformation of volumes between reference and spatial coordinates is given by

$$dv(\mathbf{x}) = J dV(\mathbf{X}). \tag{2.4}$$

The eigentransformations studied in this paper are the result of nonuniform composition in a $\mathscr{C}$–component system. Let a single-phase mixture of $\mathscr{C}$ components be defined by $\mathscr{C}$ molar concentrations C_i [mol/m^3], so that the total molar concentration (or simply molar density) C and the associated mole fractions $x_i = C_i / C$ satisfy

$$C = \sum_{i=1}^{\mathscr{C}} C_i, \qquad 1 = \sum_{i=1}^{\mathscr{C}} x_i. \tag{2.5}$$

The use of x_i for mole fractions is a historical one that is commonly found in chemical treatment of mixtures and should not cause any confusion with the three components of the spatial coordinate representation **x**. In fact, we will follow the practice of Sandler (1999) and use $\underline{x}$ to represent the first $\mathscr{C}$-1 mole fractions for short, as only $\mathscr{C}$-1 of the mole fractions are independent by the second of (2.5). The $\mathscr{C}$ components are the components of a triclinic system of edge lengths a, b, and c (along the corresponding crystal axes); and interaxial angles α, β and γ. We shall use $p \equiv (a, b, c; \alpha, \beta, \gamma)$ to denote the six crystal lattice parameters for short. It is known that a composition of an alloy can vary from the stoichiometric value without altering the type of structure, although the crystal parameters generally change somewhat with composition. Thus, if a simple cubic cell in the reference configuration *V*(**X**) is filled with the $\mathscr{C}$ components in accordance with the local molar concentrations $C_i(\mathbf{X},t)$, the cell by itself will develop into a stress-free crystal cell, of which the lattice parameters may be determined by the local composition, i.e.,

$$p = p(\underline{x}(\mathbf{X},t)) = (a,b,c;\alpha,\beta,\gamma)(\underline{x}(\mathbf{X},t)). \tag{2.6}$$

Let $\mathbf{e}_I$ (I = 1,2,3) be the fixed unit vectors in **X,** and $\mathbf{a}_\alpha$ (α=1,2,3) the three edge vectors of the stress-free crystal cell. The mapping that takes $\mathbf{e}_I$ to $\mathbf{a}_\alpha$ is a linear transformation that will be denoted by $\mathbf{F}^*$ in the development to follow. One convenient representation for $\mathbf{F}^*$ would be $\mathbf{a}_\alpha = F^*_{\alpha I}\mathbf{e}_I$. While there is not a unique way to define $\mathbf{F}^*$, it suffices to say that $\mathbf{F}^*$ is a function of the lattice parameters, which, in turn, are functions of the mole fractions that are actually functions of position and time in a nonuniform system.

Unless the stress-free crystal cells are identical they cannot be stacked together to form stress-free crystals. To form crystals from nonuniform stress-free cells additional elastic deformation, $\mathbf{F}^e$, must be imposed on all the cells and the final product is a crystal with residual stress, but without the influence of Newtonian body and surface forces. Such a crystalline body is termed a **free body** by Mura (1982). The result of applying $\mathbf{F}^*$ to a differential element d**X** in *V*(**X**) is another differential element, which will be denoted by $d\mathbf{X}^{SF}$, a stress-free element, i.e.,

$$d\mathbf{X}^{SF} = \mathbf{F}^* d\mathbf{X}, \quad dV^{SF} = \det \mathbf{F}^* dV = J^* dV. \tag{2.7}$$

The transformation $\mathbf{F}^*$ is in general incompatible in the sense that the first of the above cannot be integrated to obtain $\mathbf{X}^{SF}$ as single-valued functions of **X**. The solid body, which occupies *V* (**X**), is therefore forced to deform into a new configuration that occupies *v*(**x**) in a spatial reference coordinate system **x.** The associated transformation is the deformation gradient $\mathbf{F} = \partial\mathbf{x}/\partial\mathbf{X}$. Finally, the combination of **F** and $\mathbf{F}^*$ is termed the elastic transformation $\mathbf{F}^e = \mathbf{F}\mathbf{F}^{*-1}$. We have

$$d\mathbf{x} = \mathbf{F} d\mathbf{X}, \quad dv = J dV, \quad J = \det\mathbf{F} \tag{2.8}$$

$$d\mathbf{x} = \mathbf{F}^e d\mathbf{X}^{SF}, \quad dv = J^e dV^{SF}, \quad J^e = \det\mathbf{F}^e = J / J^* \tag{2.9}$$

which, together with (2.7), complete the needed three-frame kinematics illustrated in Fig. 1.

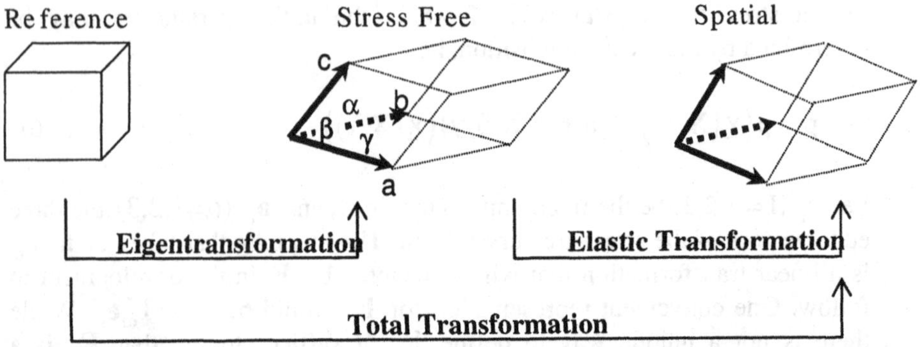

Fig.1. The three-frame kinematics

In terms of the volume elements dV, dV^{SF} and dv, the three sets of densities are:

$$C=\sum_{i=1}^{C} C_i(\mathbf{X},t),\quad C^{SF}=\sum_{i=1}^{C} C_i^{SF}(\mathbf{X},t),\quad c=\sum_{i=1}^{C} c_i(\mathbf{x},t). \qquad (2.10)$$

There are also the identities:

$$J^* = dV^{SF}/dV = C/C^{SF}(\mathbf{X},t) = C_i(\mathbf{X},t)/C_i^{SF}(\mathbf{X},t), \qquad (2.11)$$

$$J = dv/dV = C/c(\mathbf{x},t) = C_i(\mathbf{X},t)/c_i(\mathbf{x},t), \qquad (2.12)$$

$$J^e = dv/dV^{SF} = C^{SF}(\mathbf{X},t)/c(\mathbf{x},t) = C_i^{SF}(\mathbf{X},t)/c_i(\mathbf{x},t), \qquad (2.13)$$

$$x_i = c_i/c = C_i/C = C_i^{SF}/C^{SF} \quad (i = 1 \text{ to } C). \qquad (2.14)$$

We conclude this section by defining the following convenient symbols for a number of inverse quantities:

$$\mathbf{f} \equiv \mathbf{F}^{-1},\ j \equiv 1/J;\ \mathbf{f}^* \equiv \mathbf{F}^{*-1},\ j^* \equiv 1/J^*;\ \mathbf{f}^e \equiv \mathbf{F}^{e-1},\ j^e \equiv 1/J^e \qquad (2.15)$$

3.   THE HELMHOLTZ FREE ENERGY

The desired Helmholtz free energy per unit volume of $V(\mathbf{X})$ is

$$A(\mathbf{F},T,C_1,...,C_\square) = C\underline{A}(\mathbf{F},T,\underline{x}) = C\underline{G}(\mathbf{S},T,\underline{x}) + \mathbf{S}\cdot\mathbf{F}$$
$$\text{and } \mathbf{F} = \mathbf{F}^e\mathbf{F}^* \tag{3.1}$$

where T is the temperature, $\underline{A}(\mathbf{F},T,\underline{x})$ the molar Helmholtz energy, $\underline{G}(\mathbf{S},T,\underline{x})$ the molar Gibbs energy and $\mathbf{S}$ the Piola stress. The following conditions and substitutions are used:

$$\underline{A}(\mathbf{I},T_o,\underline{x}_o) = 0, \tag{3.2}$$

$$\underline{A}(\mathbf{F}^*,T,\underline{x}) = \underline{G}^{SF}(T,\underline{x}) \equiv \underline{G}(\mathbf{S},T,\underline{x})\big|_{S=0}, \tag{3.3}$$

$$\underline{A}(\mathbf{F}^e\mathbf{F}^*,T,\underline{x}) = \underline{A}(\mathbf{F}^*,T,\underline{x}) + \frac{C^{SF}}{C^{SF}}[\underline{A}(\mathbf{F}^e\mathbf{F}^*,T,\underline{x}) - \underline{A}(\mathbf{F}^*,T,\underline{x})]$$
$$= \underline{A}(\mathbf{F}^*,T,\underline{x}) + \frac{1}{C^{SF}}W^{SF}(\mathbf{F}^e,T) \tag{3.4}$$

The condition (3.2) sets the uniform state as a reference, (3.3) follows from the fact that $\mathbf{F}^*$ is the stress-free eigentransformation at temperature T and composition $\underline{x}$, and the definition of W^{SF} indicates that it is just the strain energy density per unit stress-free volume and $W^{SF}(\mathbf{I},T,\underline{x}) = 0$. Using the above properties, we obtain from (3.1)

$$A(\mathbf{F},T,C_1,...,C_\square) = C\underline{A}(\mathbf{F}^e\mathbf{F}^*,T,\underline{x})$$
$$= C[\underline{A}(\mathbf{F}^*,T,\underline{x}) + \frac{1}{C^{SF}}W^{SF}(\mathbf{F}^e,T)] \tag{3.5}$$

or

$$A(\mathbf{F},T,C_1,...,C_\square) = C\underline{G}(0,T,\underline{x}) + \frac{C}{C^{SF}}W^{SF}(\mathbf{F}^e,T,\underline{x})$$
$$= C\underline{G}^{SF}(T,\underline{x}) + J^*W^{SF}(\mathbf{F}^e,T) \tag{3.6}$$

where the dependence of W^{SF} on concentration has been ignored, as it is usually negligible. Before proceeding, we note that

$$J^* = J^*\left(\underline{x}(\mathbf{X},t)\right),\ \ \mathbf{F}^e = \mathbf{F}\left(\mathbf{F}^*\right)^{-1} \ \text{and}\ \mathbf{F}^* = \mathbf{F}^*\left(\underline{x}(\mathbf{X},t)\right). \tag{3.7}$$

It is also noted that the very last term of (3.6) may be interpreted as the strain energy density per unit volume in $V(\mathbf{X})$, i.e.,

$$W(\mathbf{F},t) = J^* W^{SF}(\mathbf{F}^e,T). \tag{3.8}$$

With the last two equations in mind and for isothermal conditions, we have from (3.6)

$$\begin{aligned}\dot{A}(\mathbf{F},T,C_1,\ldots,C_C) &= \frac{\partial A}{\partial \mathbf{F}}\cdot\dot{\mathbf{F}} + \sum_i \frac{\partial A}{\partial C_i}\cdot\dot{C}_i \\ &= \frac{\partial W}{\partial \mathbf{F}}\cdot\dot{\mathbf{F}} + \sum_i \frac{\partial}{\partial C_i}\left[C\underline{G}^{SF}(T,\underline{x}) + J^* W^{SF}(\mathbf{F}^e,T)\right]\cdot\dot{C}_i \\ &= \frac{\partial W}{\partial \mathbf{F}}\cdot\dot{\mathbf{F}} + \sum_i \left\{\bar{G}_i^{SF}(T,\underline{x}) + \frac{\partial}{\partial C_i}\left[J^* W^{SF}(\mathbf{F}^e,T)\right]\right\}\cdot\dot{C}_i \\ &= \mathbf{S}\cdot\dot{\mathbf{F}} + \sum_i \mu_i\cdot\dot{C}_i\end{aligned} \tag{3.9}$$

In the above equation, the Piola stress $\mathbf{S}$ and chemical potential μ_i are given by

$$\mathbf{S} = \frac{\partial A}{\partial \mathbf{F}} = \frac{\partial W}{\partial \mathbf{F}} = J^*\mathbf{S}^e\left(\mathbf{f}^*\right)^T\ ,\ \mathbf{S}^e \equiv \frac{\partial W^{SF}}{\partial \mathbf{F}^e},\ \ \mathbf{f}^* = \left(\mathbf{F}^*\right)^{-1} \tag{3.10}$$

$$\mu_i = \bar{G}^{SF}(T,\underline{x}) + J^*\mathbf{C}^e\cdot\left[\frac{\partial \mathbf{F}^*}{\partial C_i}\left(\mathbf{f}^*\right)^T\right] = \bar{G}^{SF}(T,\underline{x}) + \mathbf{C}\cdot\left[\mathbf{f}^*\frac{\partial \mathbf{F}^*}{\partial C_i}\right] \tag{3.11}$$

where $\mathbf{C}^e$ and $\mathbf{C}$ are the generalized materials or configurational stresses given by

$$\mathbf{C}^e = W^{SF}\mathbf{1} - \left(\mathbf{F}^e\right)^T\mathbf{S}^e,\ \ \mathbf{C} = W\mathbf{1} - (\mathbf{F})^T\mathbf{S}. \tag{3.12}$$

It is now clear that the above tensors are needed in defining the chemical potentials, but the chemical potentials themselves are not tensors (Truskinovskiy, 1983).

4. CONCLUSIONS

The use of a simple cubic lattice as a common reference framework is found to be most convenient in describing nonuniform anisotropic crystal structures via the introduction of an eigentransformation. This practice is expected to be most useful in the treatment of bicrystal interfaces. The known phenomena of finite elastic deformation and atomic diffusion can now be seamlessly merged into a unified theory.

5. ACKNOWLEDGEMENT

The research support of the National Science Foundation, CMS-0010077, is gratefully acknowledged.

6. REFERENCES

Eshelby, J.D. (1970) "Energy relations and the energy-momentum tensor in continuum mechanics," Inelastic behavior of Solids. Eds. Kanninen, M. F., Adler, W. F. Rosenfeld, A. R., and Jaffee, R. I., McGraw-Hill, NY, 77-114.

Barrett, C.S. (1973) "Crystal Structure," Metals Handbook, 8th ed., Vol. 8, Metallography, Structures and Phase Diagrams, American Society for Metals

Mura, T. (1982) Micromechanics of Defects in Solids, Martinus Nijhoff Publishers.

Truskinovskiy, L.M. (1983) "The chemical potential tensor," Geokhimiya, No. 12, 1730-1744.

Epstein, M. and Maugin, G. A. (1990) "The energy momentum tensor and material uniformity in finite elasticity." Acta Mechanica **83**, 127-133.

Sandler, S. I. (1999) Chemical and Engineering Thermodynamics, 3rd edition. John Wiley & Sons, Inc., New York.

Wu, C. H. (2001) "The role of Eshelby stress in composition-generated and stress-assisted diffusion." J. Mech. Phys. Solids **49**, 1771-1794.

VII

FRACTURE & STRUCTURAL OPTIMIZATION

Chapter 19

TEACHING FRACTURE MECHANICS WITHIN THE THEORY OF STRENGTH–OF–MATERIALS

Reinhold Kienzler

Department of Production Engineering, University of Bremen

rkienzler@uni-bremen.de

George Herrmann

Division of Mechanics and Computation, Stanford University

g.herrmann@dplanet.ch

Abstract Some elements of fracture mechanics, such as energy–release rates and stress–intensity factors, might be examined not on the basis of continuum theories, but on the basis of the much simpler theories of strength–of–materials. Thus it becomes possible to teach fracture mechanics in undergraduate courses for engineers.

Keywords: strength–of–materials, stress–intensity factors, energy–release rates, bars/beams/shafts

Introduction

The object of the present paper is to apply the concept of configurational mechanics, i. e., material forces, to the one–dimensional theories of strength–of materials, and to show that the calculation of energy–release rates and stress–intensity factors for structural elements with cracks can be discussed in an undergraduate course for engineers.

We suppose that near to the end of the first year, students in engineering are familiar with the theories of tension–compression of bars, bending of beams and torsion of shafts. They know how to evaluate the strain–energy density per unit of length and the potential of external forces for these elements and how to apply the virtual–work theorem and Castigliano's principle. Parallel to the course on strength–of–materials, the students usually take a course on material science and have been introduced to the concepts of stress–intensity factors K and energy–release rates $\mathcal{G}$. Possibly, they had to use the laboratory equipment to measure those quantities. Then they know that it is much easier to measure $\mathcal{G}$ and not K and calculate, in turn, K from the Irwin relation

$$\mathcal{G} = \frac{K^2}{E^*} \tag{1}$$

with $E^* = E$ for plane stress and $E^* = \frac{E}{1-\nu^2}$ for plane strain, Young's modulus E and Posson's ratio ν. These are the necessary preliminaries for the following. Within the theories of strength–of–materials we introduce the one–dimensional counterpart of the Eshelby tensor b_{ij}, i. e., the material force B , and — by considering discontinues bars/beams/shafts — we derive remarkably simple formulae for the calculation of stress–intensity factors. Here, bars in tension–compression are treated. The results for beams in bending and shafts in torsion are given by analogy considerations. A detailed outline of the theory and the application to arches, plates and shells with cracks may be found in [1].

1. GOVERNING EQUATIONS

As a reminder for the students (and in order to introduce the notation used throughout the paper) we collect the basic equations of the elementary bar theory. We consider an elastic bar of unspecified length l and cross-section A , which may be subjected to end loads N^l, N^r, and distributed applied axial loads $n(x)$, measured per unit length of the bar, as sketched in Figure 1

$$Equilibrium: \quad N' = -n \ , \tag{2}$$

$$Kinematics: \quad \epsilon = u' \ , \tag{3}$$

$$Hooke's\ law: \quad N = EA\epsilon \ , \tag{4}$$

$$Strain-energy-density: \quad W = \frac{1}{2}EA\epsilon^2 = \frac{1}{2}\frac{N^2}{EA} \ , \tag{5}$$

$$Potential\ of\ ext.\ forces: \quad V = -nu \ , \tag{6}$$

$$Total\ energy: \quad \Pi = \int_0^l (W+V)dx \ , \tag{7}$$

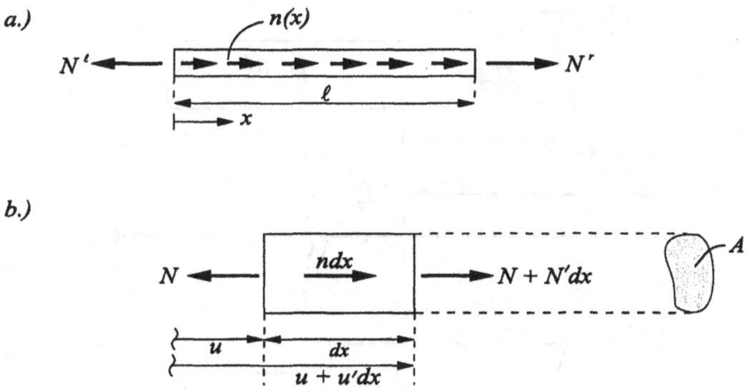

Figure 19.1. a. Bar under tension-compression b. Deformed infinitesimal element (cross-section A)

where N is the resultant internal axial force, ϵ is the axial strain,u is the axial displacement and primes indicate differentiation with respect to the axial coordinate x. The stiffness EA of the bar may be inhomogeneous, because either $E(x)$ or $A(x)$ or both may vary along the $x-$ axis.

The distributed forces n and, in turn, the potential of external forces V might be omitted without loss of clarity, if the available time for teaching in restricted (as is usual).

In a first step, we differentiate the potential energy per unit of length $(W+V)$ with respect to x

$$(W+V)' = \frac{1}{2}EA'u'^2 + EAu'u'' - n'u - nu' . \tag{8}$$

With (3) and (4), EAu' is replaced by N and the resulting product $Nu''$is modified by the product rule resulting in

$$(W+V)' = (Nu')' - (N'+n)u' + \frac{1}{2}EA'u'^2 - n'u . \tag{9}$$

Use of (2) and rearrangement leads to

$$B' = -b \tag{10}$$

with

$$B = W+V - Nu' , \tag{11}$$

$$b = -\frac{1}{2}EA'u'^2 + n'u . \tag{12}$$

Equation (10) is a balance law, which reduces to a conservation law

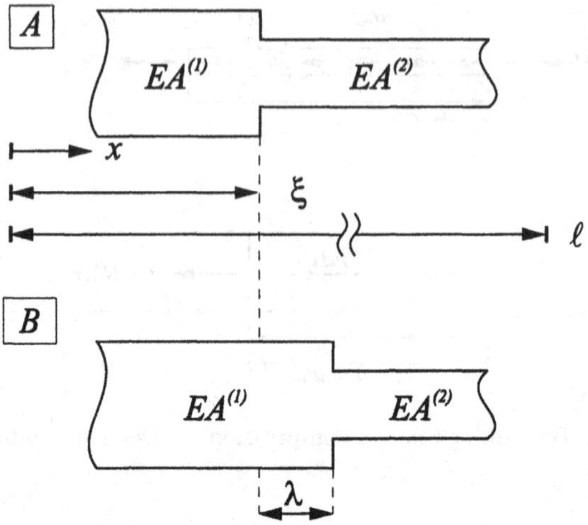

Figure 19.2. Bar with jump in axial stiffness

$$B' = 0 \; , \tag{13}$$

if the bar is homogeneous, both materially, $EA = const$, as well as physically, $n = const.$

The quantity b is, therefore, called inhomogeneity force, and the quantity B is called material force as will be explained in the following.

2. ENERGY–RELEASE RATES AND STRESS–INTENSITY FACTORS

Let us now explore the physical significance of the material force B by considering a bar containing a jump in the axial stiffness EA at an arbitrarily fixed position $x = \xi$ given by

$$EA(x) = \begin{cases} EA^{(1)} = const. \; for \; x < \xi \, , \\ EA^{(2)} = const. \; for \; x > \xi \end{cases} \tag{14}$$

(see Figure 2, state $\boxed{A}$). At the transition point $x = \xi$, we can distinguish between continuous and discontinuous variables. If we assume that the axial load n is smooth at ξ then, obviously, N and u are continuous, while EA and u' are discontinuous. The expression for the material force B , given by (11), might be rearranged by (3)–(6) as

$$B = -\frac{1}{2}\frac{N^2}{EA} - nu \; . \tag{15}$$

Therefore, it follows that the material force B is discontinuous. The jump term $[B]$ in B is easily calculated to be

$$[B] \; = \; B^{+} \; - \; B^{-} \; = \; \frac{1}{2}\, N^{2}\, (\xi)\, [C] \, , \tag{16}$$

where [C] is the jump in the compliance $C \; = \; \frac{1}{EA}$

$$[C] \; = \; \frac{1}{EA^{(2)}} \; - \; \frac{1}{EA^{(1)}} \, . \tag{17}$$

To provide a physical interpretation of [B], a bar (length l) is considered to be composed of two sections, $EA^{(1)}$ (length ξ) and $EA^{(2)}$ (length $l \; - \; \xi$), subjected to end forces N_0. The potential Π^i of the internal forces, i. e., the strain energy of the system, is given by

$$\Pi^i \; = \; \int_0^l W dx \; = \; \frac{1}{2} \int_0^l \frac{N_0^2}{EA} dx \, , \tag{18}$$

and, due to Clapeyron's theorem, the potential of the external forces follows to be

$$\Pi^a \; = \; - \; 2\Pi^i \, . \tag{19}$$

Thus the complete potential energy Π is given by

$$\Pi \; = \; \Pi^i \; + \; \Pi^a \; = \; - \; \Pi^i \, . \tag{20}$$

We wish now to calculate the change of energy when the cross–section $x \; = \; \xi$ is translated by a small amount λ (Figure 2 state $\boxed{B}$). Using (18) and (20), the result is

$$\lambda \; = \; 0 : \qquad \Pi_{\boxed{\mathrm{A}}} \; = \; - \frac{1}{2} \left(\frac{N_0^2 \xi}{EA^{(1)}} \; + \; \frac{N_0^2 (l \; - \; \xi)}{EA^{(2)}} \right) \, ,$$

$$\lambda \; \neq \; 0 : \qquad \Pi_{\boxed{\mathrm{B}}} \; = \; - \frac{1}{2} \left(\frac{N_0^2 (\xi \; + \; \lambda)}{EA^{(1)}} \; + \; \frac{N_0^2 (l \; - \; \xi \; - \; \lambda)}{EA^{(2)}} \right) \, . \tag{21}$$

The change of energy due to this material translation turns out to be

$$\Delta\Pi \; = \; \Pi_{\boxed{\mathrm{B}}} \; - \; \Pi_{\boxed{\mathrm{A}}} \; = \; \frac{1}{2} \lambda \, N_0^2 \, [C] \; = \; - \, \lambda \, [B] \, . \tag{22}$$

The quantity $\lambda[B]$ may be, therefore, interpreted as the work of the material (concentrated) force $[B]$ in the material translation λ. The force acts at the cross–section $x \; = \; \xi$ and points horizontally in the

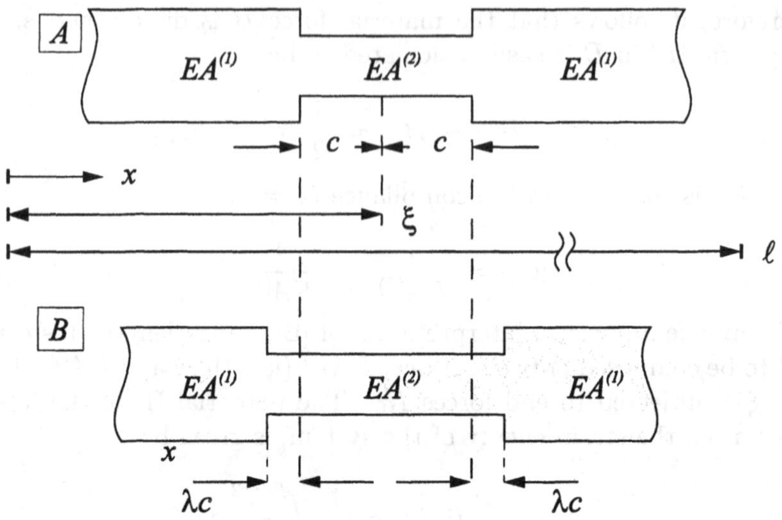

Figure 19.3. Bar with partly reduced stiffness

direction of the stiffer material. The force is likely to lower the amount of Π, by weakening the bar, i.e., by shifting the discontinuity such that the weaker part of the beam increases (removal of mass). With

$$\bar{\mathcal{G}} = - \lim_{\lambda \to 0} \frac{\Pi(\xi + \lambda) - \Pi(\xi)}{\lambda} = - \frac{\partial \Pi}{\partial \lambda} = + [B] , \tag{23}$$

$[B]$ may be identified as the energy–release rate $\bar{\mathcal{G}}$ due to a translation of the cross–section $x = \xi$ in the x – direction. Equations (22) and (23) are also valid for arbitrary loading and boundary conditions. Here, N_0 merely has to be replaced by $N(\xi)$. For the calculation of $[B]$ it is necessary to know only $N(\xi)$ and $[C]$.

Next, we consider a bar (length l) containing at $x = \xi$ a segment of length $2c$ with reduced stiffness (Figure 3 state A).

$$\begin{cases} EA^{(1)} = const.\ for\ 0 \le x < \xi - c\ and\ \xi + c < x \le l , \\ EA^{(2)} < EA^{(1)}\ for\ \xi - c < x < \xi + c . \end{cases} \tag{24}$$

Again, the bar is subjected to pure tension $N_0 = const.$ The change of total energy, now due to a selfsimilar expansion $c \to c(1 + \lambda)$ (Figure 3, state B) is calculated in the same way as above yielding

$$\Delta \Pi = - \lambda c\, N^2\, [C] = 2 \lambda c\, [B] \tag{25}$$

The energy release rate due to the transformation $c \to c(1 + \lambda)$ is

$$\bar{\mathcal{G}} = - \frac{\partial \Pi}{\partial(\lambda c)} = - 2\,[B]\,. \tag{26}$$

In the following, we will assume that the length c in Figure 3 is small in comparison to the length l, such that the bar, with partly reduced stiffnes, might be regarded as a structure with two symmetric edge cracks.

As mentioned in the Introduction, energy–release rates play an essential role in fracture mechanics. The energy–release rate $\mathcal{G}$ due to crack extension $a \to a + \Delta a$ is related in linear elasticity to the stress–intensity factors K by equation (1).

Via Rice's J integral, $J = J_1 = \mathcal{G}$ may be interpreted as the crack–driving force, i. e., a material force acting at the crack tip and pointing in the direction of crack extension.

In Kienzler & Herrmann [1] it was assumed that the energy–release rate $\mathcal{G}$ for crack extension is equal to that for crack widening, i. e., $c \to c(1 + \lambda)$. Thus we postulated

$$\mathcal{G} = \frac{\bar{\mathcal{G}}}{d}\,. \tag{27}$$

For dimensional reasons ($\mathcal{G}$ is defined in plane elasticity whereas $\bar{\mathcal{G}}$ is defined in a bar theory) a measure of thickness d needs to be introduced.

The justification for the postulate (27) is as follows (see Fig. 4). The uniform state of stress in the bar is disturbed by the presence of cracks. The stress trajectories, initially straight, are now "flowing" around the crack, leaving the triangles 012 and 023 stress–free. Guided by St. Venant's principle and engineering experience, the angle of the triangle as being 45 ° is estimated as a good approximation, hence $\beta \approx 1$. If the crack is widened into a band of width Δb in the direction of constant stress, the size of the triangular zones remains the same so that the stress relief zone is changed from area 1231 to area 45784 (Figure 4 b). In comparison, when the crack is extended by Δa, strain energy is released from strips 2683 and 2641 (Figure 4 a). It is seen that the stress relief zone 123876541 for crack extension differs from 12387541 for crack widening only by a triangular area 56725. This triangular area is proportional to Δa^2 and may be neglected in comparison with the strips (proportional to Δa), since Δa is small in comparison to a. In the limit Δa, $\Delta b \to 0$ the energy–release rates $\mathcal{G}$ and $\frac{\bar{\mathcal{G}}}{d}$ are approximately equal and (27) is applicable.

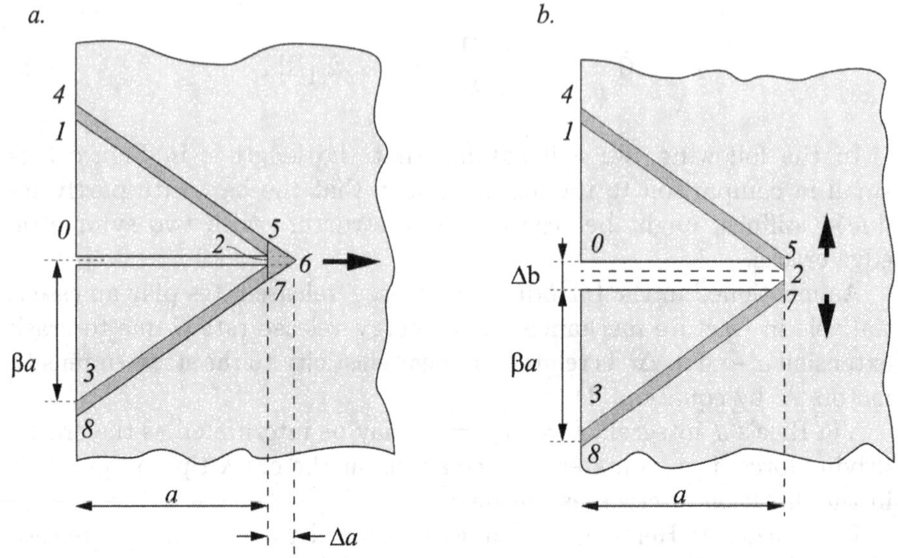

Figure 19.4. Energy–release zones at a crack tip for (a) crack extension and (b) crack widening into a fracture band

Combining (16), (17), (26), (27) and (1) leads to a remarkably simple formula to calculate stress–intensity factors for bars with cracks under tension

$$K = N\sqrt{\frac{1}{d}(\frac{1}{A^{(2)}} - \frac{1}{A^{(1)}})} . \tag{28}$$

In complete analogy, the corresponding formulae for beams and shafts are given by

$$K = M\sqrt{\frac{1}{d}(\frac{1}{I^{(2)}} - \frac{1}{I^{(1)}})} , \tag{29}$$

$$K = T\sqrt{\frac{1}{d^*}(\frac{1}{I_p^{(2)}} - \frac{1}{Ip^{(1)}})} , \tag{30}$$

respectively,with bending Moment M, torque T, moment of inertia I and polar moment of inertia I_p.

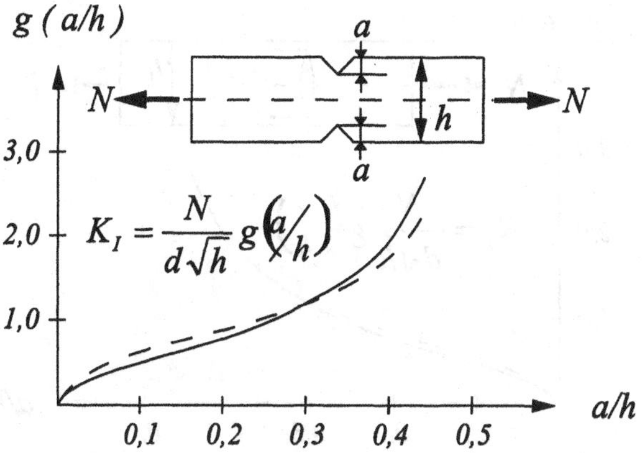

Figure 19.5. Stress–intensity factor vs. dimensionless crack length for a bar under simple tension with symmetrical edge cracks (— (32), - - - [2])

3. EXAMPLES

As a first example, consider a bar of rectangular cross–section $d \star h$ with symmetrical edge cracks of length a under pure tension (Figure 5). The cross–section areas are

$$A^{(1)} = dh \; ,$$

$$A^{(2)} = d\,(h - 2a) \; ,$$

With (28) the stress–intensity factors for the bar under tensile loading are given by

$$K = \frac{N}{d\sqrt{h}} \, g\,(\frac{a}{h}) \tag{31}$$

with

$$g\,(\frac{a}{h}) = \sqrt{\frac{1}{1 - 2\,\frac{a}{h}} - 1} \; . \tag{32}$$

The results are compared graphically with that of Benthem & Koiter [2] in Figure 5. The agreement is quite satisfactory.

As a second example, consider a bar of rectangular cross–section $d \star h$ with a centered crack of length $2\,a$ under pure tension (Figure 6). As in the example above, the cross–sectional areas are

$$A^{(1)} = dh \; ,$$

$$A^{(2)} = d(h - 2a) \; .$$

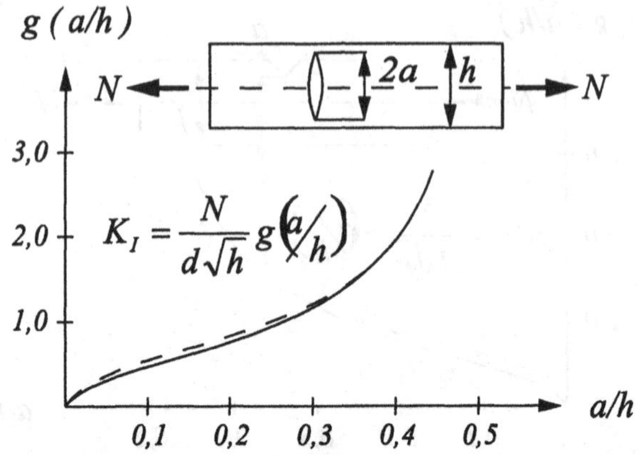

Figure 19.6. Stress–intensity factor vs. dimensionless crack length for a bar under simple tension with center crack (— (32), - - - [2]).

It is, therefore, not possible to distinguish between edge cracks and a center crack for the tension loading. Equation (32) is applicable likewise. The result is compared graphically with that of [2] in Figure 6.

In [1] further applications are given and it may be concluded that the far–reaching and rather unexpected applicability of theories of strength–of–materials to cracked structural elements of great variety confirms the power of these theories which rest mainly on their simplicity and accuracy as compared to theories of elastic continua.

References

[1] Kienzler, R. and Herrmann, G. (2000). *Mechanics in Material Space.* Berlin: Springer.

[2] Benthem, J.P. and Koiter, W.T. (1973). *Asymptotic approximations to crack problems. In: Sih, G. C. (ed.), Mechanics of Fracture I.* Leyden: Noordhoff, 131-178.

Chapter 20

CONFIGURATIONAL THERMOMECHANICS AND CRACK DRIVING FORCES

Cristian Dascalu

Laboratoire Sols-Solides-Structures, Université Joseph Fourier, Institut National Polytechnique, CNRS UMR 5521, BP 53, 38041 Grenoble cedex 9, FRANCE

Cristian.Dascalu@inpg.fr

Vassilios K. Kalpakides

Department of Mathematics, Division of Applied Mathematics and Mechanics, University of Ioannina, GR-45110, Ioannina, GREECE

vkalpak@cc.uoi.gr

Abstract We present a new formulation of the configurational force balance in a constitutive-independent framework of thermomechanics. To this end we use invariance requirements for the configurational working - here defined following the ideas of Green and Naghdi on the basic postulates of continua. This new approach has the essential property of providing an expression of the driving force on cracks in accordance with the well-known formula of the energy release rate in thermoelasticity.

1. INTRODUCTION

Since the introduction of the concept of configurational force, as a "force" acting on a defect by Eshelby, a great number of researchers have presented contributions on the subject which could be called *configurational mechanics* or *mechanics in material space* (e.g. Maugin 1993, 1995; Gurtin 2000, Podio-Guidugli 2001).

Following Gurtin (2000) and co-workers, in this work the concepts of configurational force and traction are postulated and the corresponding balance equations are derived by the use of invariance requirements of the working. The advantage of this procedure consists in the derivation of the basic configurational force balance without invocation of constitutive relations. The idea of invariance of the energy equation under a group of transformations is applied in the material space, introducing for this purpose the notion of configurational working.

The pseudomomentum balance equation in the framework of thermoelasticity and the consequent thermoelastic material forces on inhomogeneities have attracted the attention of many researchers within the last years (Maugin 1993; Dascalu and Maugin 1995; Epstein and Maugin 1995; Maugin 2000, Gurtin 2000; Steinmann 2002).

In Dascalu and Maugin (1995) thermoelastic material forces have been derived through direct manipulations (essentially pull-back transformations on a reference manifold). A special feature of this approach was the use of the "thermal displacement" variable introduced by Green and Naghdi (1991,1993), a scalar field playing a role similar to mechanical displacements, for the macroscopic description of the thermal motion. In the context of material forces acting on crack tips, the use of such a variable appeared to be a key argument for the compatibility with the energy description of the fracture process. Indeed, it was shown that by calculating the crack driving force with the obtained pseudomomentum balance, the classical expression of the thermoelastic energy-release rate is recovered.

As concerns Gurtin's approach to thermoelasticity (e.g. Gurtin 2000), we note that thermal quantities enter into the energy equation without work conjugates; as a consequence of this, they do not affect the invariance of the working, hence they can be introduced in configurational balance equation only through constitutive relations. To our view, one can account of thermal quantities from the very beginning by inserting them into the expression of the working. To obtain this the Green-Naghdi considerations on thermomechanics (Green and Naghdi 1991, 1993) can be invoked, according to which thermal and kinematical quantities are introduced on an equal footing. Here we give a modified version of Gurtin's approach for the derivation of configurational force balance, based on Green and Naghdi's view of thermomechanics. For thermoelastic materials with cracks, we show the compatibility between configurational force and energy descriptions of fracture.

2. PRELIMINARIES

Consider a body $\mathbf{B}$ the elements of which occupy at some arbitrary time t_0 a region B_r of the three dimensional Euclidean space $\mathcal{E}$, called reference configuration of the body. A motion of the body is a smooth mapping χ that, for each $t \in I \subset \Re$ (I an interval of the reals) and for each $\mathbf{X} \in B_r$, assigns a point $\mathbf{y} = \chi(\mathbf{X}, t)$ in $\mathcal{E}$. The set $B_t = \chi(B_r, t)$

is the current configuration of the body at time t. We shall distinguish between the Euclidean space which contains B_r and the corresponding one which contains B_t denoting $\mathcal{E}_{mat}$ the former and $\mathcal{E}_{sp}$ the latter. All the vectors belonging to $\mathcal{E}_{mat}$ are called material vectors while the vectors which belong to $\mathcal{E}_{sp}$ are called spatial vectors.

Consider a *control volume* $P = P(t)$ that migrates through B_r. If U is the normal velocity of its boundary ∂P, then an admissible velocity field for ∂P will be constrained by the relation $\mathbf{u} \cdot \mathbf{n} = U$. The migration of a control volume $P(t)$ can arise from a *time dependent change in reference* $\mathbf{X} = \hat{X}(\overset{*}{\mathbf{X}}, t)$, where $\overset{*}{\mathbf{X}}$ are referred as the *reference labels.* Given a fixed region $\overset{*}{P}$ in the space of reference labels, we take the migrating control volume $P(t) = \hat{X}(\overset{*}{P}, t)$ with velocity $\mathbf{u}(\mathbf{X}, t)$. Then *the motion velocity following* $\partial P(t)$ is defined to be

$$\overset{\circ}{\mathbf{y}}(\mathbf{X}, t) = \frac{\partial \hat{\chi}}{\partial t}(\overset{*}{\mathbf{X}}, t)|_{\overset{*}{\mathbf{X}} = \hat{X}^{-1}(\mathbf{X}, t)},$$

where $\hat{\chi} = \chi \circ \hat{X}$. Note that $\overset{\circ}{\mathbf{y}} = \dot{\mathbf{y}} + \mathbf{F}\,\mathbf{u}$, which is nothing else but the velocity of $\partial \bar{P}(t)$, where $\bar{P}(t) = \chi(P(t), t)$ is the deformed control volume.

Green and Naghdi (1991, 1993) in their exploration of the basic postulates of thermomechanics considered as a primitive quantity the *thermal displacement* $\alpha = \hat{\alpha}(\mathbf{X}, t)$, a scalar quantity the (time) derivative of which provides the temperature

$$\dot{\alpha} = \frac{\partial \hat{\alpha}}{\partial t}(\mathbf{X}, t) = \theta. \tag{1}$$

Their procedure leads to the derivation of equations for momentum and entropy as the basic equations of their theory:

$$\text{Div } \mathbf{T} + \mathbf{f} = \rho \ddot{\mathbf{y}}, \quad \rho \dot{\hat{\eta}} = r + \xi - \text{Div } \mathbf{s}, \quad \xi \geq 0\,; \quad k = \mathbf{s} \cdot \mathbf{n}, \tag{2}$$

where $\hat{\eta}$ is the entropy density per unit mass, ρ is the mass density, ξ and r are the internal and external rate of entropy supply per unit volume, respectively, $-k$ is the internal flux of entropy per unit area, $\mathbf{s}$ the entropy flux vector and $\mathbf{f}$ the body force per unit volume.

3. CONFIGURATIONAL THERMOMECHANICS

The concept of working and its invariance comes from the work of Green and Rivlin (1966), where they proved that the equations of motion and angular momentum can be deduced from the equation of energy by making use of invariance condition under superimposed rigid body motions. Such an invariance condition affects only the mechanical terms that expend power. The other terms also contribute to the power balance but they can not be caught by the invariance requirement because they can not be associated with the kinematical quantities. To insert the thermal agents into the working we invoke the considerations of Green and Naghdi (1991, 1993), where one can find the proper kinematical quantities for this purpose. Thus, apart from the standard forces which expend power over the motion velocity $\dot{\mathbf{y}}$, the entropy flux k, the entropy sources r, ξ and the entropy density $\eta = \rho\hat{\eta}$ will also expend power over the "velocity" of the "thermal motion" $\dot{\alpha}$, that is the temperature. Hence, after these considerations the *total working* for a control volume P can be defined by the relation

$$\mathcal{W}(P) = \int_{\partial P} \mathbf{Tn}\cdot\dot{\mathbf{y}}\, ds + \int_P \mathbf{b}\cdot\dot{\mathbf{y}}\, dv - \int_{\partial P} \mathbf{s}\cdot\mathbf{n}\dot{\alpha}\, ds + \int_P (s+\xi)\dot{\alpha}\, dv, \quad (3)$$

where we have incorporated in body force $\mathbf{b}$ and in entropy source s the inertia and the entropy density, respectively, i.e.,

$$\mathbf{b} = -\rho\ddot{\mathbf{y}} + \mathbf{f}, \quad s = -\rho\dot{\hat{\eta}} + r. \quad (4)$$

Thus, we can reasonably say that the total working in the framework of thermomechanics consists of two parts: the first one (the first two integral terms) is the standard mechanical working and the second part called *heating.*

To check the reliability of the proposed working we examine whether it is able to provide (by invariance requirement in spatial observer changes) the known equations of Green and Naghdi. Hence, we need a new kind of observer changes covering the "motion" described by the mapping α. Taking the view of Green and Naghdi, we consider that $\alpha(\mathbf{X}, t)$ represents a "second motion" of the continuum taking place at a different scale in comparison with the macroscopic motion $\mathbf{y}(\mathbf{X}, t)$. This could be a continuous representation of the lattice vibration, a phenomenon of quantum-mechanical origin. This justifies us to postulate complete independence in the corresponding observers of the two motions, hence to introduce the observer changes of the macroscopic motion:

$$\dot{\mathbf{y}} \to \dot{\mathbf{y}} + \mathbf{w} + \boldsymbol{\omega} \times \mathbf{y}, \quad \dot{\alpha} \to \dot{\alpha}. \quad (5)$$

and observer changes of the "thermal motion":

$$\dot{\mathbf{y}} \to \dot{\mathbf{y}}, \quad \dot{\alpha} \to \dot{\alpha} + \beta, \quad (6)$$

where $\mathbf{w}$, $\boldsymbol{\omega}$ are arbitrary spatial vectors and β is an arbitrary scalar.

By requiring that $\mathcal{W}(P)$ be invariant under changes in macroscopic observers, one deduces (Kalpakides and Dascalu 2002) the balance laws for momentum and angular momentum, while the requirement of invariance under changes of observers of the "thermal motion" provides the entropy balance law.

Next, we proceed to the configurational framework. In addition to usual material fields (Gurtin 2000): configurational stress $\mathbf{C}$, internal body force $\mathbf{g}$ and external body force $\mathbf{e}$, we introduce the *configurational heating* Q, a scalar field aiming at capture of configurational changes related with thermal phenomena. The flow of heat and entropy into $P(t)$ related to material transfer through the boundary ∂P will be

$$\int_{\partial P} QU\, ds, \quad \int_{\partial P} \frac{Q}{\theta} U\, ds, \tag{7}$$

respectively. As concerns the other thermal quantities, the entropy flux k expends power over $\overset{\circ}{\alpha}$ which is given by the relation

$$\overset{\circ}{\alpha} = \dot{\alpha} + \boldsymbol{\beta} \cdot \mathbf{u}, \quad \boldsymbol{\beta} = \nabla \alpha \tag{8}$$

namely, the "velocity" of thermal displacement following the boundary ∂P. The work-conjugate of entropy sources s and ξ is the quantity $\dot{\alpha}$.

Consequently, we define the configurational working for a migrating control volume $P(t)$ as

$$\begin{aligned} \mathcal{W}(P(t)) = \int_{\partial P(t)} \mathbf{Cn} \cdot \mathbf{u}\, ds + \int_{\partial P(t)} \mathbf{Tn} \cdot \overset{\circ}{\mathbf{y}}\, ds + \int_{P(t)} \mathbf{b} \cdot \dot{\mathbf{y}}\, dv \\ + \int_{\partial P(t)} Q\mathbf{n} \cdot \mathbf{u}\, ds - \int_{\partial P(t)} \mathbf{s} \cdot \mathbf{n} \overset{\circ}{\alpha}\, ds + \int_{P(t)} (s + \xi) \dot{\alpha}\, dv. \end{aligned} \tag{9}$$

Following the arguments of Gurtin (2000), we first require invariance of $\mathcal{W}(P(t))$ under changes in material observer. This leads (Kalpakides and Dascalu 2002) to the local form of configurational force balance

$$\mathrm{Div}(\mathbf{C} + Q\mathbf{I}) + \mathbf{g} + \mathbf{e} = \mathbf{0}, \quad \text{for all } \mathbf{X} \in B_r. \tag{10}$$

We then impose that $\mathcal{W}(P)$ be independent of the choice of tangential component of the velocity field $\mathbf{u}$, because such an arbitrary choice leaves unaffected the prescribed normal velocity U of the boundary ∂P. In this way we deduce the following expression for $\mathbf{C}$

$$\mathbf{C} = (\pi - Q)\mathbf{I} - \mathbf{F}^T\mathbf{T} + \boldsymbol{\beta} \otimes \mathbf{s}, \tag{11}$$

for some scalar π.

Next, by invariance under time-dependent changes in reference configurations (Kalpakides and Dascalu 2002), we get the expressions of body forces

$$\mathbf{e} = -\mathbf{F}^T\mathbf{b} - \boldsymbol{\beta}(s+\xi), \quad \mathbf{g} = -\nabla\pi + \mathbf{T} : \nabla\mathbf{F} - \nabla\boldsymbol{\beta}\cdot\mathbf{s}. \qquad (12)$$

Invariance under changes in instantaneously coinciding migrating volumes (Gurtin 2000, p. 42) here applied to the energy and entropy balances leads to

$$\pi = \epsilon \qquad \text{and} \qquad \frac{Q}{\dot{\alpha}} = \eta. \qquad (13)$$

where ϵ is the internal energy density. Defining the free energy function as $\Psi \equiv \epsilon - \theta\eta$ we get $\Psi = \epsilon - Q = \pi - Q$.

By introducing the *Eshelby stress tensor*, the *pseudomomentum* and the *material body force*, respectively, as follows

$$\mathcal{C} := -((K-\Psi)\mathbf{I} + \mathbf{F}^T\mathbf{T} - \boldsymbol{\beta}\otimes\mathbf{s}), \quad \mathcal{P} := -\rho\mathbf{F}^T\dot{\mathbf{y}} - \eta\boldsymbol{\beta},$$

$$\mathbf{f}^{th} := -\nabla\Psi + \frac{1}{2}\nabla\rho\dot{\mathbf{y}}^2 - \nabla\theta\eta + \mathbf{T} : \nabla\mathbf{F} + \nabla\boldsymbol{\beta}\cdot\mathbf{s} - \mathbf{F}^T\mathbf{f} - \boldsymbol{\beta}(r+\xi)$$

we obtain the local balance

$$\frac{d\mathcal{P}}{dt} = \text{Div}\,\mathcal{C} + \mathbf{f}^{th} \qquad (14)$$

In the particular case of classical thermoelasticity one shows that the material body force $\mathbf{f}^{th}$ becomes

$$\mathbf{f}^{th} = \theta\nabla\left(\frac{\boldsymbol{\beta}}{\theta}\right)\mathbf{s} = \nabla\left(\frac{\boldsymbol{\beta}}{\theta}\right)\mathbf{q}, \qquad (15)$$

where $\mathbf{q}$ is the heat flux vector fulfilling the standard relation $\mathbf{s} = \mathbf{q}/\theta$.

4. APPLICATION TO FRACTURE

Consider a two-dimensional homogeneous thermoelastic body B_r containing a smooth crack $C(t)$. We suppose that the crack faces are traction free and thermally isolated, i.e.

$$\mathbf{T}^{\pm}\mathbf{n} = \mathbf{0}\ ,\ \mathbf{q}^{\pm}\cdot\mathbf{n} = 0, \qquad (16)$$

where $\mathbf{n}$ is a unit normal vector on the crack curve.

The global balance of energy for the fractured body can be deduced in the form (Kostov and Nikitin 1970; Gurtin 1979; Bui, Erlacher and Nguyen 1980):

$$\frac{d}{dt}\int_{B_r}(\epsilon + K)\ dv + \mathcal{G} = \int_{\partial B_r}\mathbf{Tn}\cdot\dot{\mathbf{y}}\ da - \int_{\partial B_r}\mathbf{q}\cdot\mathbf{n}\ da \qquad (17)$$

with ∂B_r being the exterior boundary of B_r. Here, $\mathcal{G}$ is the rate of energy released by the body during fracture:

$$\mathcal{G} = \lim_{\Gamma \to 0} \int_\Gamma [(\epsilon + K)(\mathbf{u} \cdot \mathbf{n}) + \mathbf{Tn} \cdot \dot{\mathbf{y}} - \mathbf{q} \cdot \mathbf{n}] \, ds \tag{18}$$

with Γ a closed contour encircling the crack tip and shrinking to it and $\mathbf{u}$ the crack tip velocity.

Motivated by the crack-tip behavior of the solutions for linear materials (e.g. Atkinson and Craster 1992) we assume that, as approaching the crack tip, one has:

$$\dot{\alpha} \sim -\boldsymbol{\beta} \cdot \mathbf{u}\,; \quad \dot{\mathbf{y}} \sim -\mathbf{F}\,\mathbf{u}. \tag{19}$$

The $\sim$ symbol means that the two fields have the same order of singularity near the tip of the crack and their difference has a vanishing contribution to the limit in (18). Note that the first assumption in (19) is a common one in the fracture theory (see for instance Bui, Erlacher and Nguyen (1980); Gurtin (2000)).

From eq. (19) one deduces that (Kalpakides and Dascalu 2002) :

$$(\epsilon + K)\mathbf{u} + \mathbf{T}\dot{\mathbf{y}} - \mathbf{q} \sim [(\Psi - K)\mathbf{I} - \mathbf{F}^T\mathbf{T} + \boldsymbol{\beta} \otimes \frac{\mathbf{q}}{\theta}]\mathbf{u} \\ -[(\rho \mathbf{F}^T \dot{\mathbf{y}} + \eta \boldsymbol{\beta})] \cdot \mathbf{u}. \tag{20}$$

Using eq. (20) in the formula (18) of the energy-release rate, we can express $\mathcal{G}$ as:

$$\mathcal{G} = \mathcal{F} \cdot \mathbf{u} \tag{21}$$

with the driving force vector $\mathcal{F}$ given by

$$\mathcal{F} = \lim_{\Gamma \to 0} \int_\Gamma [(\Psi - K)\mathbf{I} - \mathbf{F}^T\mathbf{T} + \boldsymbol{\beta} \otimes \frac{\mathbf{q}}{\theta}]\mathbf{n} - [(\rho \mathbf{F}^T \dot{\mathbf{y}} + \eta \boldsymbol{\beta})] \cdot \mathbf{n} \; ds. \tag{22}$$

The vectorial quantity $\mathcal{F}$ should be used in a crack propagation criterion. This material-force expression was derived by direct manipulations on the energy-release rate.

On the other hand, by considering the global form of the material momentum balance (14–15), for the fractured body, through a classical procedure (Gurtin 1979; Bui, Erlacher and Nguyen 1980) one obtains:

$$\frac{d}{dt} \int_{B_r} \mathcal{P} \, dv + \mathcal{F} = \int_{\partial B_r} \mathcal{C} \cdot \mathbf{n} \, da + \int_{\partial B_r} \mathbf{f}^{th} \, dv, \tag{23}$$

where $\mathcal{F}$ is the same as in eq. (22).

In conclusion, by making use of the thermal variables introduced by Green and Naghdi, we have constructed a balance of configurational forces in thermomechanics which is consistent with the Griffith's approach to fracture.

References

Atkinson, C. and Craster, R.V. 1992 Some fracture problems in fully coupled dynamic thermoelasticity, *J. Mech. Phys. Solids*, **40**, 1415–1432.

Bui, H.D., Erlacher, A. and Nguyen, Q.S. 1980 Propagation de fissure en thermoélasticité dynamique, *J. Mécanique*, **19**, 697–725.

Dascalu C. and Maugin G.A. 1995 The thermoelastic material-momentum equation, *J. Elasticity*, **39**, 201–212.

Epstein, M. and Maugin G.A. 1995 Thermoelastic material forces: definitions and geometric aspects, *C.R. Acad.Sci. Paris*, **II-320**, 63–68.

Green, A.E. and Naghdi, P.M. 1991 A re-examination of the basic postulates of thermomechanics, *Proc. R. Soc. Lond. A*, **432**, 171–194.

Green, A.E. and Naghdi, P.M. 1993 Thermoelasticity without energy dissipation, *J. Elasticity*, **31**, 189–208.

Green, A.E. and Rivlin, R.S. 1964 On Cauchy's equations of motion, *Z. Angew. Math. Phys.*, **15**, 290–292.

Gurtin, M.E. 1979 Thermodynamics and the Griffith criterion for brittle fracture, *Int. J. Solids Structures*, **15**, 553–560.

Gurtin, M.E. and P. Podio-Guidugli 1996 Configurational forces and the basic laws for crack propagation, *J. Mech. Phys. Solids*. **44**, 905–927.

Gurtin, M.E. 2000 *Configurational Forces as Basic Concepts of Continuum Physics*, Applied Mathematical Sciences, **37**, New York, Springer.

Kalpakides V.K. and Dascalu, C. 2002 On the configurational force balance in thermomechanics, *Proc. R. Soc. Lond. A*, **458**, 3023–3039.

Kostrov, B.V. and Nikitin, L.V. 1970 Some general problems of mechanics of brittle fracture, *Arch. Mech.*, **22**, 749–776.

Maugin, G.A. 1993 *Material Inhomogeneities in Elasticity*, London, Chapman and Hall.

Maugin, G.A. 1995 Material Forces: Concepts and applications, *Appl. Mech. Rev.*, **48**, 213–245.

Maugin, G.A. 2000 On the universality of the thermodynamics of forces driving singular sets, *Arch. Appl. Mech.* , **70**, 31–45.

Podio-Guidugli, P. 2001 Configurational balances via variational arguments, *Interfaces and Free Boundaries*, **3**, 223–232.

Steinmann, P., 2002 On spatial and material settings of thermo-hyperelastodynamics, *J. Elasticity*, **66**, 109–157.

Chapter 21

STRUCTURAL OPTIMIZATION BY MATERIAL FORCES

Manfred Braun

Universität Duisburg-Essen

Institut für Mechatronik und Systemdynamik

m.braun@uni-duisburg.de

Abstract The overall stiffness of a truss is optimized by choosing the nodal coordinates of the undeformed truss such that the strain energy of the loaded truss attains a minimum. The derivatives of the strain energy with respect to the nodal coordinates are interpreted as material forces acting on the nodes of the undeformed truss in "design space".

Keywords: truss, optimization, configurational forces

Introduction

According to the principle of virtual work the deformation of an elastic truss under a given load minimizes the total potential energy. The strain energy of the deformed truss still depends on the applied load and on the data of the truss. If the topology, the elastic properties of the members, and the applied load are prescribed, the strain energy of the loaded structure is a function of the nodal coordinates of the undeformed truss. By moving the nodes in an appropriate manner, the strain energy can be lowered and even minimized, thus enhancing the rigidity of the truss.

The derivatives of strain energy with respect to the nodal coordinates can be interpreted in terms of configurational or material forces, which keep the design of the truss in its given shape. Releasing these forces allows the truss to adopt a shape with minimal strain energy, which means that the truss becomes more rigid. In the optimization process some nodes have to be fixed due to constructional reasons. For instance, the locations of supports and applied loads should not be changed. These restrictions can be interpreted as supports fixing the nodes within the "design space" in the same way as usual supports restrict the nodal displacements in the physical space. The movable nodes, however, can be chosen such that they are free of configurational forces.

1. A SIMPLE EXAMPLE

Consider a primitive truss consisting of only two members, hinged at fixed supports on top of each other and connected at a common free node (Figure 21.1). Both members have the same tensile stiffness EA, the horizontal member has length a, the other one is inclined at an angle φ.

The strain energy of the truss, expressed in terms of the displacements u_x, u_y of the free node, is the quadratic form

$$\Pi = \frac{EA}{2a} \begin{bmatrix} u_x & u_y \end{bmatrix} \begin{bmatrix} 1+\cos^3\varphi & -\sin\varphi\cos^2\varphi \\ -\sin\varphi\cos^2\varphi & \sin^2\varphi\cos\varphi \end{bmatrix} \begin{bmatrix} u_x \\ u_y \end{bmatrix}. \quad (1)$$

According to the principle of virtual work, the partial derivatives of the strain energy with respect to the displacements yield the corresponding nodal forces. If the free node is loaded by a vertical force P, it undergoes the displacements

$$u_x = -\frac{Pa}{EA}\cot\varphi \quad \text{and} \quad u_y = -\frac{Pa}{EA} \cdot \frac{1+\cos^3\varphi}{\sin^2\varphi\cos\varphi} \quad (2)$$

in horizontal and vertical directions, respectively.

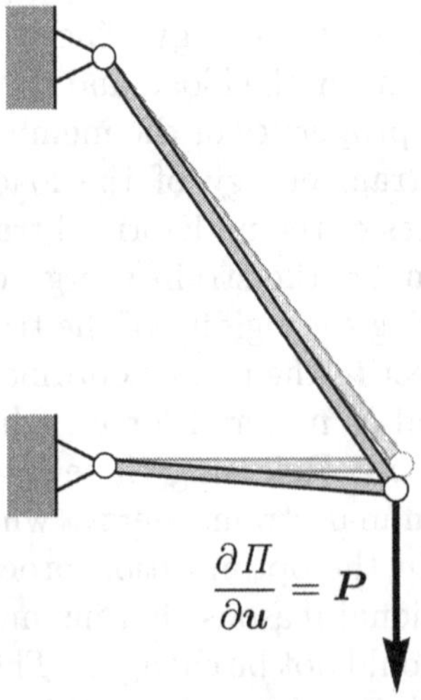

Figure 21.1. Truss in physical space

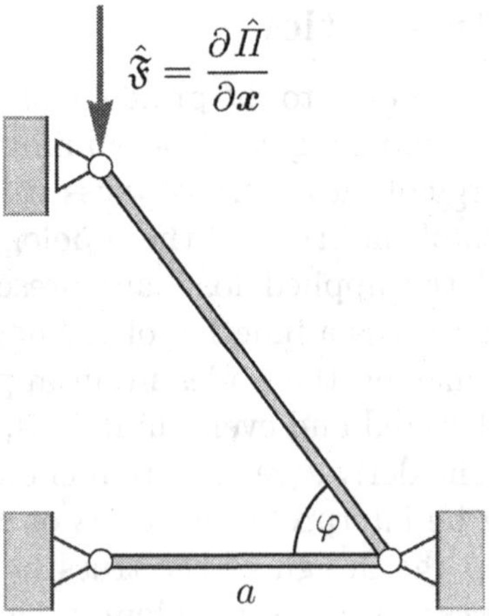

Figure 21.2. Truss in design space

The strain energy stored in the loaded truss,

$$\hat{\Pi} = \frac{aP^2}{2EA} \cdot \frac{1 + \cos^3 \varphi}{\sin^2 \varphi \cos \varphi}, \tag{3}$$

depends on the load P and on the data of the truss, i. e., on the elastic properties of the members and on the geometry of the truss expressed by the angle φ. The dependence of strain energy on the angle φ is displayed in Figure 21.3. The strain energy attains its minimum at an angle of 60°. This configuration can be regarded as the optimum design, in the sense that the deformation of the truss, measured by strain energy, is lowest.

The optimization can be interpreted in terms of *material* or *configurational forces.* These have to be distinguished from *physical forces,* which act on the real truss in physical space. Strain energy is primarily regarded as a function of the nodal displacements $\boldsymbol{u}_k$ of the truss, but it depends also on the positions $\boldsymbol{x}_i$ of the nodes in the undeformed configuration. It may be represented as a function $\Pi(\boldsymbol{x}_i, \boldsymbol{u}_k)$ and gives rise to two different kinds of partial derivatives,

$$\boldsymbol{P}_k = \frac{\partial \Pi}{\partial \boldsymbol{u}_k} \quad \text{and} \quad \mathfrak{F}_i = \frac{\partial \Pi}{\partial \boldsymbol{x}_i}, \tag{4}$$

describing physical and configurational forces, respectively. While the physical force $\boldsymbol{P}_k$ is associated with the virtual displacement $\delta \boldsymbol{u}_k$ of the

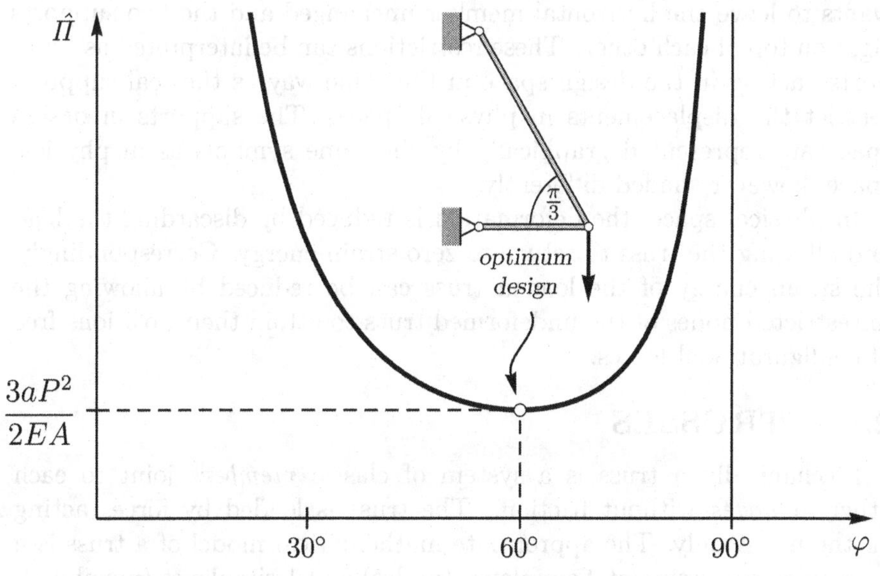

Figure 21.3. Dependence of strain energy on design angle

node in physical space, the configurational or material force $\mathfrak{F}_i$ lives in the "design space" and is associated with a virtual variation $\delta \boldsymbol{x}_i$ of the nodal position in the undeformed configuration of the truss.

Moreover, if the equilibrium conditions are solved for the nodal displacements $\boldsymbol{u}_k$, these depend on the applied forces $\boldsymbol{P}_j$ and again on the nodal coordinates $\boldsymbol{x}_i$. The strain energy of the loaded structure is, therefore, a function

$$\hat{\Pi} = \hat{\Pi}(\boldsymbol{x}_i, \boldsymbol{P}_j) = \Pi(\boldsymbol{x}_i, \boldsymbol{u}_k(\boldsymbol{x}_i, \boldsymbol{P}_j)). \tag{5}$$

The partial derivatives of this function with respect to the nodal coordinates provide a different type of configurational force

$$\hat{\mathfrak{F}}_i = \frac{\partial \hat{\Pi}}{\partial \boldsymbol{x}_i}, \tag{6}$$

which, in the special case of linear elasticity, is related to the configurational force (4) by

$$\hat{\mathfrak{F}}_i = -\mathfrak{F}_i. \tag{7}$$

This simple relation is a consequence of Clapeyron's theorem.

Figure 21.2 shows the configurational force $\hat{\mathfrak{F}}$ acting on the upper node in vertical direction. There are, of course, similar forces on all three nodes in any direction. However, in developing the design, some of the nodes are kept fixed or restricted in their motion. For instance, one wants to leave the horizontal member unchanged and the two supports right on top of each other. These restrictions can be interpreted as "supports" acting in the design space in the same way as the real supports restrict the displacements in physical space. The supports in design space are represented graphically by the same symbols as in physical space, however shaded differently.

In physical space, the deformation is reduced by discarding the load and allowing the truss to return to zero strain energy. Correspondingly, the strain energy of the loaded truss can be reduced by allowing the unrestricted nodes of the undeformed truss to attain their positions free of configurational forces.

2. TRUSSES

Mechanically a truss is a system of elastic *members* joint to each other in *nodes* without friction. The truss is loaded by forces acting on the nodes only. The appropriate mathematical model of a truss is a 1-complex consisting of 0-simplexes (nodes) and 1-simplexes (members), which are properly joined (Braun, 2000). Subsequently, nodes will be designated by Latin letters $i, j, \ldots$ while Greek letters $\alpha, \beta, \ldots$ denote the members. The connectivity of a truss is described by *incidence numbers* $[\alpha, k]$, which are defined as

$$[\alpha, k] = \begin{cases} -1, & \text{if member } \alpha \quad \text{starts} \quad \text{at node } k, \\ +1, & \text{if member } \alpha \text{ terminates at node } k, \\ 0 & \text{otherwise.} \end{cases} \tag{8}$$

The distinction between starting and terminating points of a member introduces an orientation. The matrix of all incidence numbers describes the topological structure of the truss.

The geometry may be specified by prescribing the position vectors $\boldsymbol{x}_k$ of all nodes in the unloaded state of the truss. The edge vector of a member α can then be represented by

$$\boldsymbol{a}_\alpha = \sum_k [\alpha, k] \boldsymbol{x}_k \,, \tag{9}$$

where the summation index may run over all nodes. The incidence numbers single out the proper starting and terminating points, thus reducing the sum to a simple difference. The decomposition

$$\boldsymbol{a}_\alpha = \ell_\alpha \boldsymbol{e}_\alpha \tag{10}$$

yields the length ℓ_α and the direction vector $\boldsymbol{e}_\alpha$ of the member.

It has been tacitly assumed that the truss in its unloaded state is free of stress. In a more general setting, one has to start from the lengths ℓ_α of the undeformed members rather than from a given initial placement of the nodes. This approach within a nonlinear theory is used in a previous paper (Braun, 1999).

When loads are applied to the truss, each node k is displaced by a certain vector $\boldsymbol{u}_k$ from its original position. The linear strain ε_α of a member can be expressed by

$$\varepsilon_\alpha = \frac{1}{\ell_\alpha} \boldsymbol{e}_\alpha \cdot \sum_k [\alpha, k] \boldsymbol{u}_k. \tag{11}$$

As in (9), the incidence numbers single out the end nodes of the member, such that the sum is reduced to a simple difference.

The strain energy of the truss is obtained by summing up the strain energies stored in its members, i. e.,

$$\Pi = \frac{1}{2} \sum_\alpha E A \ell_\alpha \varepsilon_\alpha^2. \tag{12}$$

To simplify the formulation the tensile stiffness EA is assumed to be the same for all members. Inserting (11) and changing the order of summation yields an expression of the form

$$\Pi = \frac{1}{2}\sum_i \sum_k \boldsymbol{u}_i \cdot \boldsymbol{K}_{ik}\boldsymbol{u}_k \tag{13}$$

where the sub-matrices

$$\boldsymbol{K}_{ik} = \sum_\alpha [\alpha, i][\alpha, k]\frac{EA}{\ell_\alpha}\boldsymbol{e}_\alpha \otimes \boldsymbol{e}_\alpha \tag{14}$$

constitute the global stiffness matrix $\boldsymbol{K}$ of the truss. In the linear theory, the strain energy is a quadratic form in the nodal displacements $\boldsymbol{u}_k$, as given by (13). Since the coefficient matrices $\boldsymbol{K}_{ik}$ of this quadratic form depend on the data of the truss, especially on the nodal positions $\boldsymbol{x}_i$, the strain energy is a function $\Pi(\boldsymbol{x}_i, \boldsymbol{u}_k)$ of both nodal positions and nodal displacements.

The dependence on $\boldsymbol{x}_i$ is put forth by the lengths ℓ_α and the direction vectors $\boldsymbol{e}_\alpha$ of the members. The corresponding derivatives are

$$\frac{\partial \ell_\alpha}{\partial \boldsymbol{x}_i} = [\alpha, i]\boldsymbol{e}_\alpha\,, \qquad \frac{\partial \boldsymbol{e}_\alpha}{\partial \boldsymbol{x}_i} = \frac{[\alpha, i]}{\ell_\alpha}(\boldsymbol{I} - \boldsymbol{e}_\alpha \otimes \boldsymbol{e}_\alpha)\,. \tag{15}$$

Using these expression the partial derivative of the strain energy (13) with respect to the nodal position $\boldsymbol{x}_i$ can be calculated. After some straightforward conversions the configurational force

$$\mathfrak{F}_i = \frac{\partial \Pi}{\partial \boldsymbol{x}_i} = \sum_\alpha [\alpha, i]EA\varepsilon_\alpha \left(\frac{1}{\ell_\alpha}\sum_j [\alpha, j]\boldsymbol{u}_j - \frac{3}{2}\varepsilon_\alpha\right) \tag{16}$$

is obtained.

The expression (16), as it stands, does not reveal the connection to the continuum form of the Eshelby stress, as presented by Epstein and Maugin, 1990. This is mainly due to the fact, that the *linear* theory of elasticity is used to describe the deformation of the truss. A nonlinear approach, using the stretch λ instead of the strain ε, provides a more symmetric description and sheds some light on the analogy between continuous and discrete configurational forces (Braun, 1999).

3. APPLICATION

The optimization of truss design using the concept of configurational forces is demonstrated by means of the following example. The truss girder shown in Figure 21.4 is loaded by forces acting on the lower nodes.

The girder shall be optimized in the sense that the strain energy of the loaded structure becomes a minimum. In the design space the undeformed truss is displayed (Figure 21.5). The lower nodes, which are supposed to bear the loads or are supported, should not be moved. These nodes are endowed with supports in the design space. All other nodes can be moved freely. Figure 21.5 shows the configurational forces $\hat{\mathfrak{F}}_k$ acting on the movable nodes. These forces restrain the design in its initial form. If they are released, the design of the undeformed girder will change and eventually assume a shape in which the movable nodes are free of configurational forces.

This optimized truss is displayed in Figure 21.6, again in the deformed state under the same physical loads as in the initial design. The deflection of the girder is much smaller now. Actually the strain energy stored in the deformed truss is reduced to 18% of its initial value.

On the other hand, the slim girder has developed into a bigger truss, which might be unfavorable for other reasons. Also the dead weight of the truss itself has not been taken into account. A more thorough analysis should allow for it as an extra *configuration-dependent* load. This would counteract the expansion of the girder.

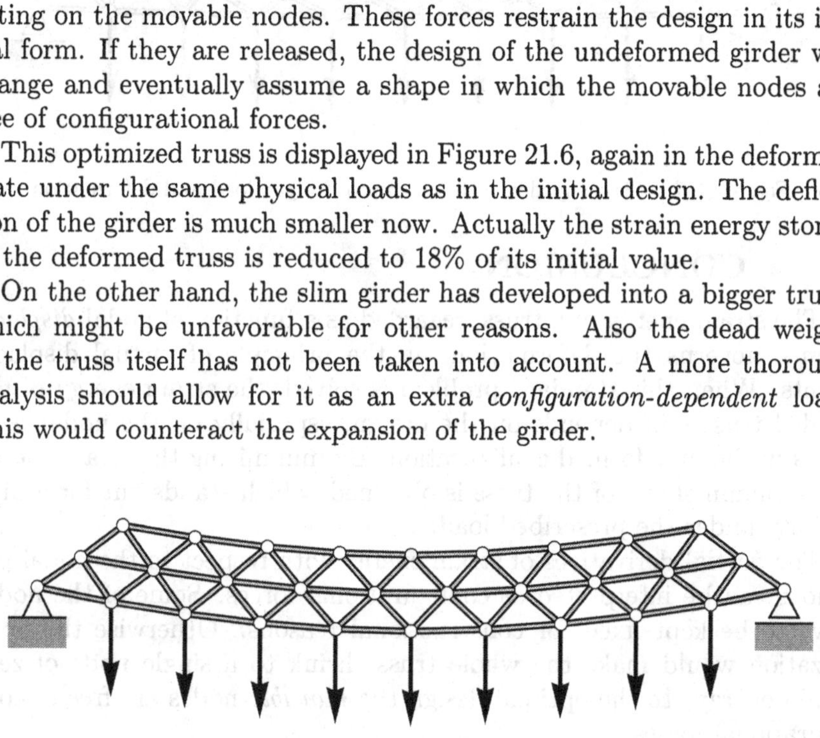

Figure 21.4. Initial design of truss girder: loads and deformation

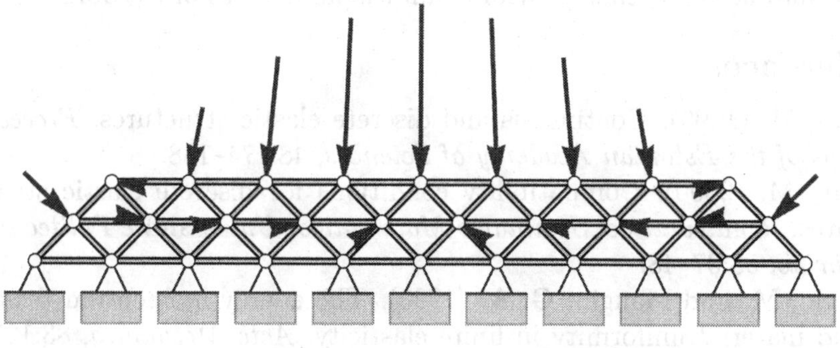

Figure 21.5. Truss girder in design space: configurational forces

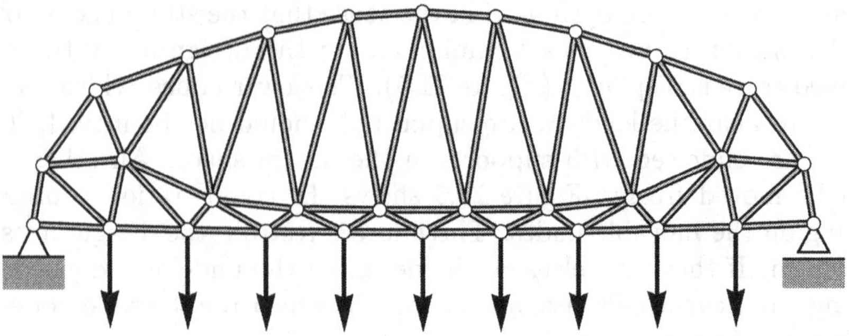

Figure 21.6. Truss girder with optimized design: load and deformation

4. CONCLUSION

The strain energy of a truss, regarded as a function of nodal *displacements*, governs the deformation via the principle of virtual displacements. When this standard problem is solved, the strain energy of the loaded truss still depends on the design, especially on the nodal *positions* in the undeformed configuration. By minimizing the strain energy an optimum shape of the truss is obtained, which stands out for a high rigidity under the prescribed load.

The partial derivatives of strain energy with respect to the nodal positions can be interpreted as configurational forces. Some of the nodes have to be kept fixed for constructional reasons. Otherwise the optimization would make the whole truss shrink to a single node of zero strain energy. In the optimal design the *movable* nodes are free of configurational forces.

The analysis should be extended to include configuration-dependent loads, such as the self weight of the truss. Also initial stresses could be taken into account, thus providing additional degrees of freedom.

References

Braun, M. (1999). Continuous and discrete elastic structures. *Proceedings of the Estonian Academy of Sciences*, 48:174–188.

Braun, M. (2000). Compatibility conditions for discrete elastic structures. *Rendiconti del Seminario Matematico, Università e Politecnico Torino*, 58:37–48.

Epstein, M. and Maugin, G. A. (1990). The energy-momentum tensor and material uniformity in finite elasticity. *Acta Mechanica*, 83:127–133.

Chapter 22

ON STRUCTURAL OPTIMISATION AND CONFIGURATIONAL MECHANICS

Franz-Joseph Barthold

Numerical Methods and Information Processing, University of Dortmund

August-Schmidt-Strasse 6, D-44227 Dortmund, Germany

franz-joseph.barthold@uni-dortmund.de

Abstract Kinematics in structural optimisation and configurational mechanics coincide as long as sufficiently smooth design variations of the material bodies are considered. Thus, variational techniques from design sensitivity analysis can be used to derive the well-known Eshelby tensor. The impact on numerical techniques including computer aided design (cad) and the finite element method (fem) is outlined.

Keywords: Structural optimisation, configurational mechanics, Eshelby tensor

1. INTRODUCTION

Continuum mechanics has been developed to describe the deformation of a single body with fixed geometrical properties under external loads. Applications such as structural design optimisation, see e.g. Kamat, 1993, as well as physical problems such as advancing cracks and surface growth modelled in the framework of configurational mechanics, see e.g. Gurtin, 2000; Kienzler and Hermann, 2000; Kienzler and Maugin, 2001, enforced the consideration of bodies with varying geometry.

This paper proposes an enhancement of the classical kinematical framework of continuum mechanics designed to outline systematically the influence of geometry on continuum mechanical functions and to efficiently perform the variations. This new representation is mainly influenced by ideas from structural design optimisation, i.e. the material derivative approach, see e.g. Arora, 1993, the domain parametrisation approach, see e.g. Tortorelli and Wang, 1993, and subsequent interpretations advocated by the author, see Barthold and Stein, 1996; Barthold, 2002.

2. MATERIAL BODIES AND COORDINATE SYSTEMS

The concept of a material body $\mathcal{B}$ consisting of material points X which continuously fill parts of space with matter is central in continuum mechanics. In general, a material body $\mathcal{B}$ is defined to be a differential manifold. A formulation which highlights their intrinsic properties was published by Noll, 1972, see also Krawietz, 1986 and Bertram, 1989 for additional information.

An intrinsic coordinate system is canonically given by the manifold property of $\mathcal{B}$. The corresponding atlas reads $\mathcal{A} := \{(\mathcal{U}_1, \Phi_1), \ldots, (\mathcal{U}_n, \Phi_n),$ i.e. a finite number of charts $(\mathcal{U}_i, \Phi_i)$ are sufficient to cover $\mathcal{B}$. For each material point $\mathrm{X} \in \mathcal{B}$ there exists an open ball $\mathcal{U}_i \equiv \mathcal{U}_\mathrm{X} \subset \mathcal{B}$ and a one-to-one mapping $\phi_i : \mathcal{U}_i \to D_i \subset \mathrm{R}^3$ into an open ball $D_i \equiv D_\Theta \subset \mathrm{R}^3$, i.e. $\underline{\Theta} = \phi_i(\mathrm{X})$ are the intrinsic coordinates.

The placement of the material body $\mathcal{B}$ into the Euclidean space E^3 lead to the submanifold $\Omega = \chi(\mathcal{B}) \subset \mathrm{E}^3$. The embedding map χ is canonically given by the structure of E^3. The related extrinsic coordinates are $\underline{\mathrm{x}} = \chi(\mathrm{X})$.

The introduced intrinsic and extrinsic parameterisations shall be linked in the sequel. To achieve this, a single chart $(\phi_i, \mathcal{U}_i)$ will be considered. The points of the Euclidean point space E^3 are described by the vector $\mathbf{x}$ of the Euclidean vector space V^3 with Cartesian base system $\{\mathbf{E}_i\}$. The local coordinates as elements of the parameter space R^3 can be described by vectors $\mathbf{\Theta}$ of the vector space Z^3 with Cartesian base system $\{\mathbf{Z}_i\}$. Overall, $\mathbf{\Theta} = \Theta^i \mathbf{Z}_i = \phi_i(\mathrm{X})$ and $\mathrm{X} = \phi_i^{-1}(\mathbf{\Theta})$ as well as $\mathbf{x} = x^i \mathbf{E}_i = \chi(\mathrm{X})$ and $\mathrm{X} = \chi^{-1}(\mathbf{x})$. A composition of these mappings is possible to eliminate the material point X

$$\mathbf{x} = \chi(\mathrm{X}) = \chi(\phi_i^{-1}(\mathbf{\Theta})) = (\chi \circ \phi_i^{-1})(\mathbf{\Theta}) =: \boldsymbol{\kappa}_i(\mathbf{\Theta}). \tag{1}$$

The sufficiently smooth one-to-one mappings $\boldsymbol{\kappa}_i := \chi \circ \phi_i^{-1}$ denote placements $\boldsymbol{\kappa}_i : D_i \to \mathrm{V}$ from the intrinsic coordinate domain D_i into the domain U_i.

The above mappings are valid for a single chart. In order to formulate global mappings, the following assumptions are made without loss of generality, i.e.

$$\mathcal{B} = \bigcup_{i=1}^{m} \mathcal{U}_i \quad \text{and} \quad \Omega = \bigcup_{i=1}^{m} U_i \quad \text{as well as} \quad \mathrm{P}_\Theta = \bigcup_{i=1}^{m} D_i . \tag{2}$$

The atlas $\mathcal{A}_{CM}$ should be chosen such that the necessary balls do not overlap

$$\mathcal{U}_i \cap \mathcal{U}_j = \emptyset \quad \text{and} \quad U_i \cap U_j = \emptyset \quad \text{for } i \neq j. \tag{3}$$

The global mappings $\boldsymbol{\phi} : \mathcal{B} \to \mathrm{P}_\Theta = \boldsymbol{\phi}(\mathcal{B})$ and $\boldsymbol{\kappa} : \mathrm{P}_\Theta \to \Omega = \boldsymbol{\kappa}(\mathrm{P}_\Theta) \subset \mathrm{V}$ are defined by $\boldsymbol{\Theta} = \boldsymbol{\phi}(\mathrm{X})$ and $\mathbf{x} = \boldsymbol{\kappa}(\boldsymbol{\Theta}) := (\chi \circ \boldsymbol{\phi}^{-1})(\boldsymbol{\Theta})$, respectively.

The family of different material bodies $\mathcal{B}$ is parameterised by a time-like design variable s and the deformation is parameterised by time t as usual.

The definitions made above are partly summarised in the subsequent figure.

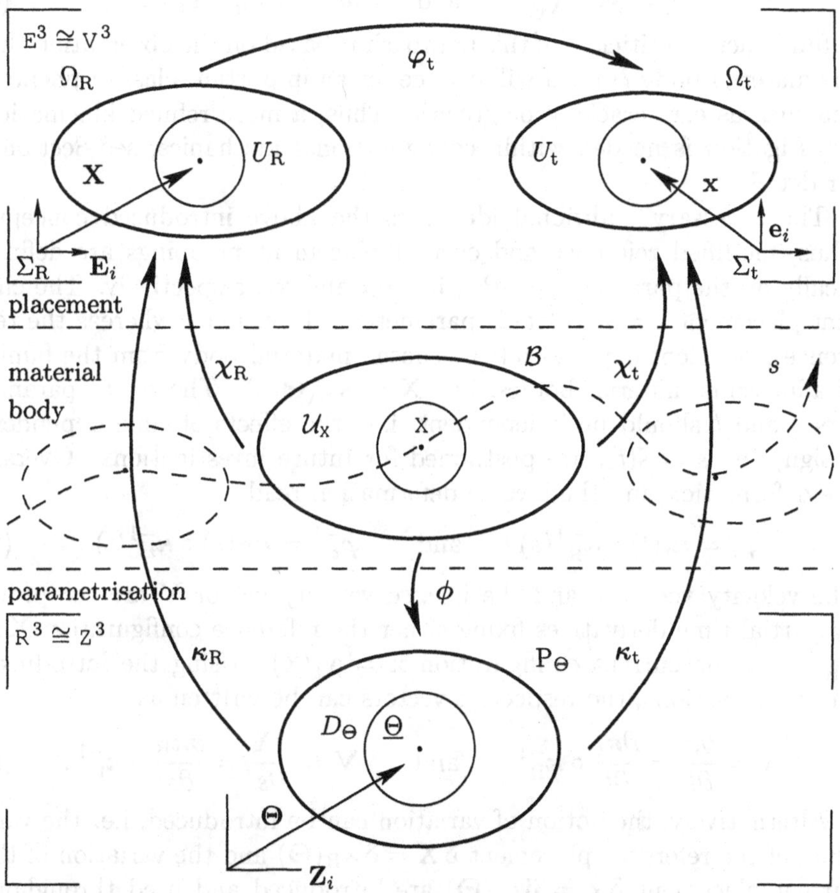

Figure 22.1. **Domains and mappings of a material body**

The structure of parametric mappings as outlined above can be observed in the finite element method and in computer aided geometric design. Both numerical techniques approximate the atlas $\mathcal{A}$ by some finite dimensional model.

3. COMPARISON OF KINEMATICAL MODELS

The inverse deformation mapping φ_t^{-1} for some fixed time t is the starting point of (almost) all kinematical representations in configurational mechanics.

Introducing the reference and current placements of the material body $\mathcal{B}$, i.e. χ_R and χ_t, respectively, the deformation and inverse deformation read

$$\varphi_t = \chi_t \circ \chi_R^{-1} \quad \text{and} \quad \varphi_t^{-1} = \chi_R \circ \chi_t^{-1} . \tag{4}$$

A fundamental criticism of this approach is based on the observation that the material body $\mathcal{B}$ itself will change for an important class of phenomena such as e.g. cracking or growth. Thus, a more refined kinematical investigation is needed within configurational mechanics, see Section 6 for details.

The necessary additional idea uses the above introduced concepts. Thus, modified reference and current placement mappings are defined locally on the parameter set P_Θ, i.e. κ_R and κ_t, respectively. The current placement $\mathbf{x} = \kappa_t(\boldsymbol{\Theta}, t)$ is parameterised by time t whereas the reference placement depends on the chosen material body from the family of admissible material bodies, i.e. $\mathbf{X} = \kappa_R(\boldsymbol{\Theta}, s)$. The scalar parameters s and t should be independent, i.e. the effects of time dependent design, i.e. $s = \hat{s}(t)$, are postponed for future investigations. Overall, the deformation and the inverse deformation read

$$\varphi_t = \kappa_t(t) \circ \kappa_R^{-1}(s) \quad \text{and} \quad \varphi_t^{-1} = \kappa_R(s) \circ \kappa_t^{-1}(t) . \tag{5}$$

The velocity vector $\mathbf{v}$ and the inverse velocity vector $\mathbf{V}$ are computed by partial time derivatives fixing either the reference configuration $\mathbf{X} = \chi_R(\mathsf{X})$ or the current configuration $\mathbf{x} = \chi_t(\mathsf{X})$. Using the introduced parameterisations, the respective vectors can be written as

$$\mathbf{v} = \frac{\partial x}{\partial t} = \frac{\partial \kappa_t}{\partial t} \circ \kappa_R^{-1} \quad \text{and} \quad \mathbf{V} = \frac{\partial X}{\partial s} = \frac{\partial \kappa_R}{\partial s} \circ \kappa_t^{-1} . \tag{6}$$

Alternatively, the notion of variation can be introduced, i.e. the variation of the reference placement $\delta\mathbf{X} = \delta\kappa_R(\boldsymbol{\Theta})$ and the variation of the current placement $\delta\mathbf{x} = \delta\kappa_t(\boldsymbol{\Theta})$ are introduced and used throughout the paper.

4. GRADIENTS AND STRAINS

The material deformation gradient $\mathsf{F} = \mathrm{Grad}_\mathsf{X}\,\varphi_t = \mathbf{g}_i \otimes \mathbf{G}^i$ can be decomposed in form of $\mathsf{F} = \mathsf{M}\,\mathsf{K}^{-1}$ using the local gradients

$$\mathsf{K} = \mathrm{GRAD}_\Theta\,\kappa_R = \mathbf{G}_i \otimes \mathbf{Z}^i \quad \text{and} \quad \mathsf{M} = \mathrm{GRAD}_\Theta\,\kappa_t = \mathbf{g}_i \otimes \mathbf{Z}^i . \tag{7}$$

The (iso-)parametric mappings in fem are the discrete versions of these relationships. The tangent mappings are used to transform all continuum mechanical quantities onto any of the domains $\mathrm{P}_\Theta, \Omega_\mathrm{R}, \Omega_\mathrm{t}$ or the corresponding tangent spaces $\mathtt{T}_\Theta, \mathtt{T}_\mathrm{X}, \mathtt{T}_\mathrm{x}$. All representations are physically equivalent but the parameter set representation using the local coordinates $\boldsymbol{\Theta}$ is advocated to perform all necessary variations. This approach is used throughout the paper.

The fundamental Green strain tensor takes the local form

$$\mathsf{E}_\Theta = \frac{1}{2}\,(\mathsf{g}_\Theta - \mathsf{G}_\Theta) = \frac{1}{2}\,(g_{ij} - G_{ij})\,\mathbf{Z}^i \otimes \mathbf{Z}^j\,, \tag{8}$$

where $\mathsf{G}_\Theta = \mathsf{K}^\mathsf{T}\,\mathsf{K}$ and $\mathsf{g}_\Theta = \mathsf{M}^\mathsf{T}\,\mathsf{M}$ denote the metric tensors and G_{ij}, g_{ij} are the covariant metric coefficients of the reference and current configuration, respectively. All other metric and strain tensors can be obtained by suitable push forward transformations onto the reference and current configurations, see Barthold, 2002 for these and all other details.

5. ENERGY RELEASE RATE AND ESHELBY TENSOR

The Eshelby tensor will be derived by performing the total variation of the potential energy Π. The results will be transformed onto the reference configuration to obtain the known expressions from literature.

5.1. VARIATION OF POTENTIAL ENERGY

The specific free energy Ψ represents the specific strain energy for isothermal processes, i.e. the overall strain energy is given by

$$\Pi := \int_{\mathcal{B}} \Psi\,\mathrm{d}m = \int_{\Omega_\mathrm{t}} \Psi\,\varrho_\mathrm{t}\,\mathrm{d}V_\mathrm{t} = \int_{\Omega_\mathrm{R}} \Psi\,\varrho_\mathrm{R}\,\mathrm{d}V_\mathrm{R} = \int_{\mathrm{P}_\Theta} \Psi\,\varrho_\Theta\,\mathrm{d}V_\Theta\,. \tag{9}$$

The material density is denoted by $\varrho_\mathrm{t}, \varrho_\mathrm{R}$ and ϱ_Θ, respectively. The variation $\delta\Pi$ can be computed using the variation of the mass density

$$\delta\,\varrho_\Theta = \delta(\varrho_\mathrm{R}\,J_\mathrm{K}) = \delta\,\varrho_\mathrm{R}\,J_\mathrm{K} + \varrho_\mathrm{R}\,\delta\,J_\mathrm{K} = \delta\,\varrho_\mathrm{R}\,J_\mathrm{K} + \varrho_\mathrm{R}\,J_\mathrm{K}\,\mathrm{Div}_\mathrm{X}\,\delta\,\mathbf{X} \tag{10}$$

where $J_\mathrm{K} = \det \mathsf{K}$, $\varrho_\Theta = J_\mathrm{K}\,\varrho_\mathrm{R}$ and $\delta\,J_\mathrm{K} = J_\mathrm{K}\,\mathrm{Div}_\mathrm{X}\,\delta\,\mathbf{X}$ has been used. The mass density ϱ_R does not change for closed systems, i.e. $\delta\,\varrho_\mathrm{R}$ vanishes e.g. in case of cracking or for dislocations. Thus, up to now

$$\delta\Pi = \int_{\Omega_\mathrm{R}} [\delta\,\Psi\,\varrho_\mathrm{R} + \Psi\,\varrho_\mathrm{R}\,\mathrm{Div}_\mathrm{X}\,\delta\,\mathbf{X}]\,\mathrm{d}V_\mathrm{R}\,. \tag{11}$$

5.2. VARIATION OF FREE ENERGY

Fundamental considerations from material theory conclude that the free energy function depends on the material deformation gradient $\mathsf{F} = \mathsf{M}\mathsf{K}^{-1}$. Furthermore, material objectivity leads to so-called reduced forms such as

$$\Psi = \Psi(\mathsf{F}) = \Psi(\mathsf{F}^\mathsf{T}\mathsf{F}) = \Psi(\mathsf{K}^{-\mathsf{T}}\mathsf{M}^\mathsf{T}\mathsf{M}\mathsf{K}^{-1}) = \Psi(\mathsf{K}^{-\mathsf{T}}\mathsf{g}_\Theta\mathsf{K}^{-1}) . \tag{12}$$

Finally, limiting the considerations to isotropic materials, we conclude that Ψ depends only on the invariants of $\mathsf{C} = \mathsf{F}^\mathsf{T}\mathsf{F}$. Furthermore, the invariants of $\mathsf{C} = \mathsf{K}^{-\mathsf{T}}\mathsf{g}_\Theta\mathsf{K}^{-1}$ and $\mathsf{B}_\Theta = \mathsf{G}_\Theta^{-1}\,\mathsf{g}_\Theta$ are equal, i.e. $\operatorname{Inv}\mathsf{C} = \operatorname{Inv}(\mathsf{K}^{-1}\mathsf{C}\mathsf{K}) = \operatorname{Inv}(\mathsf{G}_\Theta^{-1}\mathsf{g}_\Theta)$. Overall, we continue with the functional dependency

$$\Psi = \Psi(\mathsf{K},\mathsf{M}) = \Psi(\mathsf{G}_\Theta,\mathsf{g}_\Theta) = \Psi(\mathsf{B}_\Theta) = \Psi(\operatorname{Inv}(\mathsf{B}_\Theta)) \tag{13}$$

The total variation of the free energy split up into two partial variations with respect to time t and design s, i.e. $\delta\,\Psi = \delta_t\,\Psi + \delta_s\,\Psi$. Using the above introduced functional dependencies the variation yield

$$\delta\,\Psi = \frac{\partial\Psi}{\partial\mathsf{M}} : \delta\,\mathsf{M} + \frac{\partial\Psi}{\partial\mathsf{K}} : \delta\,\mathsf{K} = \frac{\partial\Psi}{\partial\mathsf{g}_\Theta} : \delta\,\mathsf{g}_\Theta + \frac{\partial\Psi}{\partial\mathsf{G}_\Theta} : \delta\,\mathsf{G}_\Theta . \tag{14}$$

The partial derivatives lead to second order tensors

$$\tilde{\mathsf{S}}_\Theta := 2\,\frac{\partial\Psi}{\partial\mathsf{g}_\Theta} = 2\,\frac{\partial\Psi}{\partial g_{ij}}\,\mathbf{Z}^i\otimes\mathbf{Z}^j , \tag{15}$$

i.e. the local version of the 2nd Piola-Kirchhoff stress tensor and

$$\tilde{\Sigma}_\Theta := 2\,\frac{\partial\Psi}{\partial\mathsf{G}_\Theta} = 2\,\frac{\partial\Psi}{\partial G_{ij}}\,\mathbf{Z}^i\otimes\mathbf{Z}^j , \tag{16}$$

which is the local version of its 'configurational' counterpart. The physical stress tensor $\tilde{\mathsf{S}}_\Theta$ describes the change of the free energy Ψ for varying metric coefficients g_{ij} of the current configuration in case of physical deformations. Similarly, the configurational stress tensor $\tilde{\Sigma}_\Theta$ describes the change of free energy Ψ for varying metric coefficients G_{ij} of the reference configuration in case of configurational modifications between different material bodies.

Thus, the total variation of the free energy function leads to

$$\delta\,\Psi = \frac{1}{2}\,\tilde{\mathsf{S}}_\Theta : \delta\,\mathsf{g}_\Theta + \frac{1}{2}\,\tilde{\Sigma}_\Theta : \delta\,\mathsf{G}_\Theta = \tilde{\mathsf{S}}_\Theta : \mathsf{M}^\mathsf{T}\,\delta\,\mathsf{M} + \tilde{\Sigma}_\Theta : \mathsf{K}^\mathsf{T}\,\delta\,\mathsf{K}, \tag{17}$$

where the last form is valid due to the symmetry of $\tilde{\mathsf{S}}_\Theta$ and $\tilde{\Sigma}_\Theta$. The configurational tensor $\tilde{\Sigma}_\Theta$ can be eliminated since the partial derivatives of B_Θ w.r.t. G_Θ and g_Θ are related by

$$\frac{\partial \mathsf{B}_\Theta}{\partial \mathsf{G}_\Theta} = -\frac{\partial \mathsf{B}_\Theta}{\partial \mathsf{g}_\Theta} \mathsf{B}_\Theta^T . \tag{18}$$

Therefore, $\tilde{\Sigma}_\Theta = -\mathsf{B}_\Theta \tilde{\mathsf{S}}_\Theta = -\tilde{\mathsf{S}}_\Theta \mathsf{B}_\Theta^T$ and we finally obtain

$$\delta_t \Psi = \tilde{\mathsf{S}}_\Theta : \mathsf{M}^\mathsf{T} \delta \mathsf{M} \quad \text{and} \quad \delta_s \Psi = -\mathsf{B}_\Theta \tilde{\mathsf{S}}_\Theta : \mathsf{K}^\mathsf{T} \delta \mathsf{K} . \tag{19}$$

5.3. THE ESHELBY TENSOR

The partial variation $\delta_t \Pi$ contributes to the weak form which vanishes in case of equilibrium. The remaining partial variation $\delta_s \Pi$, i.e. those parts corresponding to $\delta \mathsf{G}_\Theta$ or $\delta \mathsf{K}$, describe the energy release rate ($\delta_s \Pi_{ext}$ neglected)

$$\mathcal{G} = \delta_s \Pi = \int_{\mathrm{P}_\Theta} \left[\Psi \, \mathrm{Div}_\mathrm{X} \, \delta \mathbf{X} - \mathsf{B}_\Theta \tilde{\mathsf{S}}_\Theta : \mathsf{K} \, \delta \mathsf{K} \right] \varrho_\Theta \, \mathrm{d}V_\Theta . \tag{20}$$

The Eshelby tensor can be obtained from the expression of the energy release rate by some tensor algebra manipulations. Using $\mathrm{Div}_\mathrm{X} \, \delta \mathbf{X} = 1_\mathrm{X} : \mathrm{Grad}_\mathrm{X} \, \delta \mathbf{X}$, $\mathsf{S}_\Theta = \varrho_\Theta \tilde{\mathsf{S}}_\Theta$, $\delta \mathsf{K} = \mathrm{GRAD}_\Theta \, \delta \mathbf{X}$ and $\mathrm{Grad}_\mathrm{X} \, \delta \mathbf{X} = \delta \mathsf{K} \, \mathsf{K}^{-1}$ we obtain

$$\mathcal{G} = \int_{\mathrm{P}_\Theta} \mathsf{b}_\Theta : \mathrm{GRAD}_\Theta \, \delta \mathbf{X} \, \mathrm{d}V_\Theta , \tag{21}$$

where the local Eshelby tensor is defined by

$$\mathsf{b}_\Theta := \varrho_\Theta \Psi \, \mathsf{K}^{-\mathsf{T}} + \mathsf{K} \Sigma_\Theta = \varrho_\Theta \Psi \, \mathsf{K}^{-\mathsf{T}} - \mathsf{K} \, \mathsf{B}_\Theta \mathsf{S}_\Theta .$$

Using the push forward transformations between the tangent spaces we obtain material and spatial versions of the energy release rate

$$\mathcal{G} = \int_{\Omega_\mathrm{R}} \mathsf{b}_\mathrm{X} : \mathrm{Grad}_\mathrm{X} \, \delta \mathbf{X} \, \mathrm{d}V_\mathrm{R} = \int_{\Omega_\mathrm{t}} \mathsf{b}_\mathrm{x} : \mathrm{grad}_\mathrm{x} \, \delta \mathbf{X} \, \mathrm{d}V_\mathrm{t} , \tag{22}$$

where $\mathsf{b}_\mathrm{X} := \mathsf{b}_\Theta \mathsf{K}^\mathsf{T}$ and $\mathsf{b}_\mathrm{x} := \mathsf{b}_\Theta \mathsf{M}^\mathsf{T}$ denote the material and spatial Eshelby tensors, respectively. The well-known expressions can be directly obtained using $W := \varrho_\mathrm{R} \Psi$ as specific potential energy per unit volume

$$\mathsf{b}_\mathrm{X} := W \, \Psi \, 1_\mathrm{X} - \mathsf{F}^\mathsf{T} \mathsf{P}_\mathrm{X} = W \, 1_\mathrm{X} - \mathsf{C} \, \mathsf{S}_\mathrm{X}$$

where $\mathsf{P_X}, \mathsf{S_X}$ denote the 1st and 2nd Piola-Kirchhoff stress tensor, respectively, and $\mathsf{C\,S_X}$ is known as Mandel stress tensor.

Discrete versions of the Eshelby tensor in form of nodal point fictitious forces as well as a discrete design velocity field can be easily obtained by a straightforward analysis in line with standard finite element technology.

6. A CLASSIFICATION SCHEME BASED ON KINEMATICS

An important result of configurational mechanics states that physical and configurational stresses are obtained by variation of the potential energy w.r.t. kinematical quantities defined in terms of the physical or the configurational domain, respectively. Thus, a more refined classification could be derived by analysing the underlying kinematical models on the configurational domain. This idea is highlighted by comparing this paper with investigations by Epstein and Maugin, 2000.

Therein, the existence of a 'linear "transplant" $\mathsf{K}(\mathbf{X})$ from the reference crystal to the tangent neighbourhood of $\mathbf{X}$' for each point $\mathbf{X}$ is assumed. The integrability of the transplant, i.e. the existence of some vector $\mathbf{u}(\mathbf{X})$, is linked with the notation of inhomogeneity of the material. The Eshelby tensor is expressed by the derivative of the potential energy w.r.t. the transplant K.

Both approaches are formally equivalent for results which are solely based on the linear mapping $\mathbf{K}$. This observation is not astonishing because both kinematical models are linked as outlined in Section 3.

Nevertheless, the mapping K models independent phenomena. The smooth design variation of the material body discussed in this paper describes macroscopic modifications of a material body by e.g. engineers in the design process or e.g. advancing cracks. Here, a smooth design is naturally available and the mapping K is the gradient of the geometry mapping as outlined above. On the other hand, non-integrable inhomogeneities of the material body are modelled.

Further investigations may clarify whether the linear mapping $\mathbf{K}$ could be the common basis in configurational mechanics. Consequently, the Eshelby tensor would be a derived quantity and a refined classification of phenomena could be obtained by considering the underlying kinematics in configurational mechanics.

7. DESIGN-SPACE-TIME FRAMEWORK

The continuum mechanical space-time framework should be enlarged in order to describe modification of the material body $\mathcal{B}$. The appropriate sufficiently smooth parametrisation of the family of admissible material bodies by some scalar-valued design variable s has been assumed to exist throughout this paper. This assumption is valid for a brought class of problems in structural optimisation as well as in configurational mechanics.

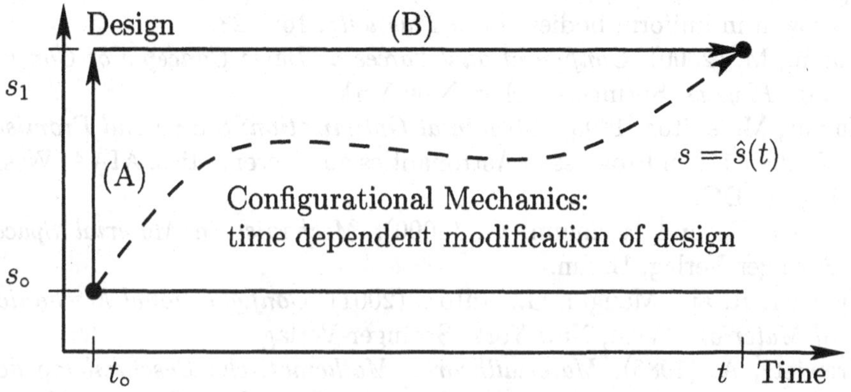

Figure 22.2. **Design-Space-Time Framework**

Future research should focus on time dependent design variations, i.e. a linkage $s = \hat{s}(t)$ between design and time has to be formulated. The figure above highlights the operator split in configurational mechanics, i.e. physical phenomena split up into a pure design part without interaction with the deformation (A) and a pure analysis part for fixed design (B).

8. CONCLUSION

The usefulness and applicability of the local formulation using convected coordinates has been proved in the framework of structural design optimisation, see Barthold, 2002. The outlined concepts can be applied to configurational mechanics as shown above for the derivation of the Eshelby tensor.

References

Arora, J. (1993). An exposition of the material derivative approach for structural shape sensitivity analysis. *Computer Methods in Applied Mechanics and Engineering*, 105:41–62.

Barthold, F.-J. (2002). Zur Kontinuumsmechanik inverser Geometrieprobleme. Habilitationsschrift, Braunschweiger Schriften zur Mechanik 44-2002, TU Braunschweig.

Barthold, F.-J. and Stein, E. (1996). A continuum mechanical based formulation of the variational sensitivity analysis in structural optimization. Part I: analysis. *Structural Optimization*, 11(1/2):29–42.

Bertram, A. (1989). *Axiomatische Einführung in die Kontinuumsmechanik.* BI-Wissenschaftsverlag, Mannheim.

Epstein, M. and Maugin, G. (2000). Thermomechanics of volumetric growth in uniform bodies. *Int.J.Plasticity*, 16:1–28.

Gurtin, M. (2000). *Configurational Forces as Basic Concepts of Continuum Physics.* Springer-Verlag, New York.

Kamat, M., editor (1993). *Structural Optimization: Status and Promise.* Number 150 in Progress in Astronautics and Aeronautics. AIAA, Washington, DC.

Kienzler, R. and Hermann, G. (2000). *Mechanics in Material Space.* Springer-Verlag, Berlin.

Kienzler, R. and Maugin, G., editors (2001). *Configurational Mechanics of Materials*, Wien, New York. Springer-Verlag.

Krawietz, A. (1986). *Materialtheorie. Mathematische Beschreibung des phänomenologischen thermomechanischen Verhaltens.* Springer-Verlag, Berlin.

Noll, W. (1972). A new mathematical theory of simple materials. *Archives of Rational Mechanics*, 48(1):1–50.

Tortorelli, D. and Wang, Z. (1993). A systematic approach to shape sensitivity analysis. *International Journal of Solids & Structures*, 30(9):1181–1212.

VIII

PATH INTEGRALS

Chapter 23

CONFIGURATIONAL FORCES AND THE PROPAGATION OF A CIRCULAR CRACK IN AN ELASTIC BODY

Vassilios K. Kalpakides
University of Ioannina, Department of Mathematics, Ioannina, GR-45110, Greece

Eleni K. Agiasofitou
University of Ioannina, Department of Mathematics, Ioannina, GR-45110, Greece

Abstract The concept of a balance law for an elastic fractured body, in Euclidean and material space, is used to investigate the propagation of a circular crack in an elastic body. In the spirit of modern continuum mechanics, a rigorous localization process results in the local equations in the smooth parts of the body and, in addition, in the relations holding at the crack tip. The latter are used in establishing relations concerning the energy release rates.

1. INTRODUCTION

This paper aims at the investigation of the propagation of a circular crack in an elastic body in the framework of configurational mechanics. To develop the configurational setting, we start with the balance laws for the physical and configurational fields (Maugin, 1993; Dascalu and Maugin, 1995; Steinmann, 2000) and we continue with the derivation of the relevant local equations in a rigorous manner.

Among others, the localization process results in what we call configurational force and moment at the crack tip. These concepts are closely connected with the J and L–integrals of dynamic fracture mechanics.

Finally, our attempt is directed at the energy release rates and their connections with the configurational force and moment. Particularly, introducing the geometry of a circular crack, an elegant relationship between the configurational moment at the crack tip and the rotational energy release rate arises.

The lack of space does not allow us to give the complete proofs of our analysis. One can find all the missing technical details and proofs in a future paper of the same authors.

2. KINEMATICS

We designate with $\mathcal{B}_R$ the reference configuration containing an initial crack which is described by a smooth, non-intersecting curve C_R with the one of its end points to lie on the boundary of the body and the other one to be the crack tip, $\mathbf{Z}_0$. We interpret the evolving crack at the time t with a smooth curve $C(t)$ belonging to a material configuration $\mathcal{B}_t$ thus, we consider infinitely many material configurations $\mathcal{B}_t$, each one assigning to a time t, $t \in I \subset I\!R$. It is required that for $t_2 > t_1 > t_0$ to imply $C_R \subset C(t_1) \subset C(t_2)$. The position of the crack tip at time t is given by the smooth function $\mathbf{Z}(t)$, the derivative of which provides the crack propagation velocity, $\mathbf{V}(t) = d\mathbf{Z}/dt$.

A tip disc $D_\epsilon(t)$ at time t is defined to be a disc of radius ϵ centered at the crack tip $\mathbf{Z}(t)$, that is

$$D_\epsilon(t) = \{\mathbf{X} \in \mathcal{B}_t : |\mathbf{X} - \mathbf{Z}(t)| \leq \epsilon\}. \tag{1}$$

At the time t_0, the tip disc is given by:

$$D_{\epsilon 0} = \{\mathbf{Y} \in \mathcal{B}_R : |\mathbf{Y} - \mathbf{Z}_0| \leq \epsilon\}. \tag{2}$$

Noting that $D_\epsilon(t)$ evolves following the crack tip, we can consider that it arises from a translation of the initial tip disc $D_{\epsilon 0}$, that is, for every $\mathbf{X} \in D_\epsilon(t)$, we can write

$$\mathbf{X} = X(\mathbf{Y}, t) = \mathbf{Y} + \mathbf{Z}(t) - \mathbf{Z}_0, \quad \mathbf{Y} \in D_{\epsilon 0}. \tag{3}$$

Obviously, every point of $D_{\epsilon 0}$ evolves with the velocity of the crack tip

$$\mathbf{V}(\mathbf{Y}, t) = \frac{\partial X}{\partial t}(\mathbf{Y}, t) = \frac{d\mathbf{Z}}{dt} = \mathbf{V}(t), \quad \mathbf{Y} \in D_{\epsilon 0}. \tag{4}$$

Denoting with B_t the current configuration, we introduce the deformational motion, $\chi : \ \mathcal{B}_t \to B_t$

$$\mathbf{x} = \chi(\mathbf{X}, t), \quad \mathbf{x} \in B_t, \quad \mathbf{X} \in \mathcal{B}_t, \ t \in I \subset I\!R, \tag{5}$$

which is a C^2 function for all $(\mathbf{X}, t) \in (\mathcal{B}_t \setminus C(t)) \times I$. Also, it is continuous across the crack curve $C(t)$, as we assume that the crack faces are in perfect contact.

Notice that the material points $\mathbf{X} \in D_\epsilon(t)$ depend on t via the mapping X consequently, we can compose the mappings X and χ for all $\mathbf{X} \in D_\epsilon(t)$ to interpret both the disc motion following the crack evolution and the deformational motion of the body

$$\widetilde{\chi} = \chi \circ X, \quad \mathbf{x} = \widetilde{\chi}(\mathbf{Y}, t) = \chi(X(\mathbf{Y}, t), t), \quad \mathbf{Y} \in D_{\epsilon_0}. \tag{6}$$

Denoting with $\overset{\circ}{\mathbf{x}}$ the partial derivative $\partial\widetilde{\chi}/\partial t$, we can write

$$\begin{aligned} \overset{\circ}{\mathbf{x}} &= \frac{\partial \chi}{\partial \mathbf{X}}(\mathbf{X}, t)\, \frac{\partial X}{\partial t}(\mathbf{Y}, t) + \frac{\partial \chi}{\partial t}(\mathbf{X}, t), \\ &= \mathbf{F}(\mathbf{X}, t)\mathbf{V}(t) + \dot{\mathbf{x}}(\mathbf{X}, t) =: \widetilde{\mathbf{V}}(\mathbf{X}, t), \quad \mathbf{X} \in D_\epsilon(t) \setminus \gamma_D, \end{aligned} \tag{7}$$

where $\mathbf{F} = \partial\chi/\partial\mathbf{X}$, $\dot{\mathbf{x}} = \partial\chi/\partial t$ and $\gamma_D = D_\epsilon(t) \cap C(t)$.

From the assumptions about the smoothness of x, it is implied that $\mathbf{F}$ and $\dot{\mathbf{x}}$ are continuous for $\mathbf{X} \in D_\epsilon(t) \setminus \gamma_D$. However, both $\mathbf{F}$ and $\dot{\mathbf{x}}$ are generally singular at the crack tip $\mathbf{Z}(t)$. The quantity $\widetilde{\mathbf{V}}$ represents the velocity of the deformed disc accounting for the crack evolution velocity as well. Though $\widetilde{\mathbf{V}}$ is defined in terms of $\mathbf{F}$ and $\dot{\mathbf{x}}$ which are singular at the crack tip, we would like $\widetilde{\mathbf{V}}$ to be smooth at the crack tip. Thus, taking the view of (Gurtin, 2000; Gurtin, Podio-Guidugli, 1996), we assume the existence of a bounded, time–dependent function $\widetilde{\mathbf{U}}(t)$ such that

$$\lim_{\mathbf{X} \to \mathbf{Z}(t)} \widetilde{\mathbf{V}}(\mathbf{X}, t) = \widetilde{\mathbf{U}}(t), \quad \text{uniformly in } I. \tag{8}$$

Obviously, the quantity $\widetilde{\mathbf{U}}(t)$ represents the velocity of the deformed crack tip.

3. A BALANCE LAW FOR A FRACTURED BODY

Let Ω be any part of the body in the material configuration $\mathcal{B}_t$. If the crack tip $\mathbf{Z}(t)$ is an interior point of Ω, then there exists a radius ϵ such that $D_\epsilon(t) \subset \Omega$. We denote with Ω_ϵ the subset of Ω which is defined as $\Omega_\epsilon(t) = \Omega \setminus D_\epsilon(t)$. The parts of the crack $C(t)$ contained in Ω_ϵ and Ω will be denoted by γ_ϵ and γ_Ω, respectively , that is, $\gamma_\epsilon = C(t) \cap \Omega_\epsilon(t)$, $\quad \gamma_\Omega = C(t) \cap \Omega$.

Let $\phi(\mathbf{X}, t)$ be a scalar valued function defined in $\mathcal{B}_t$, representing some physical quantity, which may have a singularity at the crack tip and be discontinuous with finite jump along $C(t) \setminus \{\mathbf{Z}(t)\}$. Taking the view of (Gurtin, 2000), we will assume *the integrability of ϕ in the sense* *of Cauchy principal value*, i.e., for all $\Omega \in \mathcal{B}_t$

$$\int_\Omega \phi(\mathbf{X},t)\, dV = \lim_{\epsilon\to 0} \int_{\Omega_\epsilon} \phi(\mathbf{X},t)\, dV. \tag{9}$$

Also, *the line integral of a vector valued function* $\mathbf{g}(\mathbf{X},t)$ *along the curve* γ_Ω *will be meant in the sense*

$$\int_{\gamma_\Omega} \mathbf{g}(\mathbf{X},t)\cdot\mathbf{n}\, dl = \lim_{\epsilon\to 0} \int_{\gamma_\epsilon} \mathbf{g}(\mathbf{X},t)\cdot\mathbf{n}\, dl, \tag{10}$$

where $\mathbf{n}$ is the unit normal to $C(t)$.

The basic axiom connecting the rate of ϕ with its sources, will be a *global balance law* asserting that for every part $\Omega \subset \mathcal{B}_t$ and $t \in I$, it holds

$$\frac{d}{dt}\int_\Omega \phi(\mathbf{X},t)\, dV = \int_{\partial\Omega} \mathbf{f}(\mathbf{X},t)\cdot\mathbf{N}\, dS + \int_\Omega h(\mathbf{X},t)\, dV + g(t), \tag{11}$$

where $\mathbf{N}$ is the outward unit normal to the boundary $\partial\Omega$ and $\mathbf{f}$, h are the flux and the source of ϕ, respectively. *The function g represents the source of ϕ due to the crack evolution.*

Obviously, the integrability of ϕ is not enough to make eq. (11) meaningful so, we must pose extra smoothness on the integrands. We generally assume:

C1: h, ϕ are integrable over Ω.

C2: $\lim_{\epsilon\to 0}\int_{\Omega_\epsilon} \frac{\partial\phi}{\partial t}(\mathbf{X},t)\, dV = \int_\Omega \frac{\partial\phi}{\partial t}(\mathbf{X},t)\, dV$, uniformly in I.

C3: $\int_{\partial D_\epsilon} \phi(\mathbf{X},t)(\mathbf{V}\cdot\mathbf{N})\, dS$ converges uniformly in I, as $\epsilon \to 0$.

C4: Div $\mathbf{f}$, $[\mathbf{f}]\cdot\mathbf{n}$ are integrable over Ω and γ_Ω, respectively.

C5: $\int_{\partial D_\epsilon} \mathbf{f}(\mathbf{X},t)\cdot\mathbf{N}\, dS$ converges, as $\epsilon \to 0$.

One can prove the following

PROPOSITION 1 : *Assume that the Conditions* C1, C2 *and* C3 *hold. Then,* $\int_\Omega \phi(\mathbf{X},t)\, dV$ *is a differentiable function of* t. *In addition, its derivative will be given by the relation*

$$\frac{d}{dt}\int_\Omega \phi(\mathbf{X},t)\, dV = \int_\Omega \frac{\partial\phi}{\partial t}(\mathbf{X},t)\, dV - \lim_{\epsilon\to 0}\int_{\partial D_\epsilon} \phi(\mathbf{X},t)(\mathbf{V}\cdot\mathbf{N})\, dS. \tag{12}$$

Notice that eq. (12) is a version of the *Transport Theorem* appropriate for the needs of the problem under study. Using the Conditions C4–C5, one can prove also the following version for the *Divergence Theorem*

$$\int_{\partial\Omega} \mathbf{f}\cdot\mathbf{N}\, dS = \int_{\Omega} \text{Div}\,\mathbf{f}\, dV + \lim_{\epsilon\to 0}\int_{\partial D_\epsilon} \mathbf{f}(\mathbf{X},t)\cdot\mathbf{N}\, dS + \int_{\gamma\Omega} [\mathbf{f}(\mathbf{X},t)]\cdot\mathbf{n}\, dl. \tag{13}$$

Using all the previous results one can obtain the main result of this section.

PROPOSITION 2 : *The postulation that the balance law* (11) *holds for all* $\Omega \in \mathcal{B}_t$ *and all* $t \in I$ *has as a consequence the following local equations*

$$\begin{aligned} &\frac{\partial\phi}{\partial t} - \text{Div}\,\mathbf{f} - h = 0, \qquad \text{for all } t\in I,\ \mathbf{X}\in\mathcal{B}_t\setminus C(t),\\ &[\mathbf{f}]\cdot\mathbf{n} = 0, \qquad \text{for all } t\in I,\ \mathbf{X}\in C(t)\setminus\{\mathbf{Z}(t)\},\\ &g(t) = -\lim_{\epsilon\to 0}\int_{\partial D_\epsilon} (\phi(\mathbf{V}\cdot\mathbf{N}) + \mathbf{f}\cdot\mathbf{N})\, dS, \qquad \text{for all } t\in I. \end{aligned} \tag{14}$$

Notice that the formulas (14) hold without any constitutive assumption. However, the constitutive assumptions enter into the balance laws indirectly through the definitions of the Eshelby stress tensor and the pseudomomentum (see section 4.2). For this reason, in what follows we confine ourselves in the framework of elasticity.

4. THE BALANCE LAWS FOR THE PHYSICAL AND CONFIGURATIONAL FIELDS

Throughout this section, we assume that each field inserted in a balance law at the position of the abstract functions Φ, $\mathbf{f}$ and h enjoy the corresponding smoothness specified in the previous section.

4.1. THE BALANCES FOR THE PHYSICAL FIELDS

It is reasonable to assume that there are no sources of momentum and angular momentum, due to the crack evolution. On the contrary, the energy balance is directly influenced by the crack growth. Thus, we postulate that the following balances hold for every part $\Omega \in \mathcal{B}_t$ and every time $t \in I$

$$\frac{d}{dt}\int_{\Omega} \rho\dot{\mathbf{x}}\, dV = \int_{\partial\Omega} \mathbf{TN}\, dS, \tag{15}$$

$$\frac{d}{dt}\int_{\Omega} \mathbf{r}\times\rho\dot{\mathbf{x}}\, dV = \int_{\partial\Omega} \mathbf{r}\times\mathbf{TN}\, dS, \tag{16}$$

$$\frac{d}{dt}\int_{\Omega} (W+K)\, dV = \int_{\partial\Omega} \mathbf{TN}\cdot\dot{\mathbf{x}}\, dS - \Phi(t), \tag{17}$$

where ρ, $\mathbf{T}$, W, K and Φ are the mass density, the Piola-Kirchhoff stress tensor, the elastic energy density, the kinetic energy density and the rate of energy dissipation, respectively. All the densities are defined per unit volume in the material configuration. Also, the position vector of $\mathbf{x}$ is denoted with $\mathbf{r} = \mathbf{x} - \mathbf{0}$.

Recalling the Proposition 2, we obtain the local equations

$$\frac{\partial}{\partial t}(\rho\dot{\mathbf{x}}) - \mathrm{Div}\mathbf{T} = 0, \tag{18}$$

$$\frac{\partial}{\partial t}(\mathbf{r} \times \rho\dot{\mathbf{x}}) - \mathrm{Div}(\mathbf{r} \times \mathbf{T}) = 0, \tag{19}$$

$$\frac{\partial}{\partial t}(W + K) - \mathrm{Div}(\mathbf{T}^T\dot{\mathbf{x}}) = 0, \tag{20}$$

for all $t \in I$, $\mathbf{X} \in \mathcal{B}_t \setminus C(t)$ and

$$\lim_{\epsilon\to 0}\int_{\partial D_\epsilon} (\rho\dot{\mathbf{x}}(\mathbf{V}\cdot\mathbf{N}) + \mathbf{TN})\, dS = 0, \tag{21}$$

$$\lim_{\epsilon\to 0}\int_{\partial D_\epsilon} \mathbf{r} \times (\rho\dot{\mathbf{x}}(\mathbf{V}\cdot\mathbf{N}) + \mathbf{TN})\, dS = 0, \tag{22}$$

$$\lim_{\epsilon\to 0}\int_{\partial D_\epsilon} \big((W + K)(\mathbf{V}\cdot\mathbf{N}) + \mathbf{T}^T\dot{\mathbf{x}}\cdot\mathbf{N}\big) dS = \Phi(t), \tag{23}$$

for all $t \in I$.

4.2. THE BALANCES FOR THE CONFIGURATIONAL FIELDS

The balances, which we are dealt with, in the last section do not exhaust all the relevant quantities involved in our problem. We must further consider balances for the configurational fields, that is, the pseudomomentum and the material angular momentum. In particular, we postulate the following balances holding for every part $\Omega \in \mathcal{B}_t$ and every time $t \in I$

$$\frac{d}{dt}\int_\Omega \boldsymbol{\mathcal{P}}\, dV = \int_{\partial\Omega} \mathbf{b}^T\mathbf{N}\, dS + \int_\Omega \tilde{\mathbf{f}}\, dV + \boldsymbol{\mathcal{F}}(t), \tag{24}$$

$$\frac{d}{dt}\int_\Omega \mathbf{R} \times \boldsymbol{\mathcal{P}}\, dV = \int_{\partial\Omega} \mathbf{R} \times \mathbf{b}^T\mathbf{N}\, dS + \int_\Omega \mathbf{R} \times \tilde{\mathbf{f}}\, dV + \boldsymbol{\mathcal{M}}(t), \tag{25}$$

where $\boldsymbol{\mathcal{P}}(\mathbf{X}, t) = -\rho\mathbf{F}^T\dot{\mathbf{x}}$, $\mathbf{b}^T = L\mathbf{I} - \mathbf{F}^T\mathbf{T}$, $\tilde{\mathbf{f}} = \partial L/\partial\mathbf{X}$, (Maugin, 1993). With L is denoted the Langrangian and $\mathbf{R}$ is the position vector of $\mathbf{X}$.

According to Proposition 2, the localization of eqs. (24–25) provide

$$\frac{\partial \mathcal{P}}{\partial t} - \mathrm{Div}\mathbf{b}^T - \tilde{\mathbf{f}} = 0, \tag{26}$$

$$\frac{\partial\left(\mathbf{R}\times\mathcal{P}\right)}{\partial t} - \mathrm{Div}\left(\mathbf{R}\times\mathbf{b}^T\right) - \mathbf{R}\times\tilde{\mathbf{f}} = 0, \tag{27}$$

for all $t \in I$, $\mathbf{X} \in \mathcal{B}_t \setminus C(t)$ and

$$\mathcal{F}(t) = -\lim_{\epsilon\to 0}\int_{\partial D_\epsilon}\left(\mathcal{P}(\mathbf{V}\cdot\mathbf{N}) + \mathbf{b}^T\mathbf{N}\right)dS, \tag{28}$$

$$\mathcal{M}(t) = -\lim_{\epsilon\to 0}\int_{\partial D_\epsilon}\left(\mathbf{R}\times\left(\mathcal{P}(\mathbf{V}\cdot\mathbf{N}) + \mathbf{b}^T\mathbf{N}\right)\right)dS, \tag{29}$$

for all $t \in I$. We will call the quantities $\mathcal{F}(t)$ and $\mathcal{M}(t)$ *configurational force and moment at the crack tip*, respectively. From these quantities, one can extract the integrals

$$\mathbf{F}_\epsilon(t) = \int_{\partial D_\epsilon}\left(\mathcal{P}(\mathbf{V}\cdot\mathbf{N}) + \mathbf{b}^T\mathbf{N}\right)dS,$$

$$\mathbf{M}_\epsilon(t) = \int_{\partial D_\epsilon}\mathbf{R}\times\left(\mathcal{P}(\mathbf{V}\cdot\mathbf{N}) + \mathbf{b}^T\mathbf{N}\right)dS,$$

which can be connected with the well–known J and L–integrals of dynamic fracture, respectively (Golebiewska Herrmann and Herrmann, 1981).

5. THE ENERGY RELEASE RATES

The *dynamic energy release rate* is defined (Freund, 1989) as $G = \Phi/V$, where V is the magnitude of the crack velocity vector ($\mathbf{V} = V\mathbf{t}$). Using the energy and the pseudomomentum equations as well as eq. (8), one can prove that

$$\Phi = -\mathbf{V}\cdot\mathcal{F} \quad \text{or} \quad G = -\mathcal{F}\cdot\mathbf{t}. \tag{30}$$

We examine now the case where the crack evolves along a circular curve of radius R_0. Putting the origin of the coordinates' system at the center of the circle, the position vector of the crack tip $\mathbf{Z}$ can be written as $\mathbf{R}(\mathbf{Z}) = \mathbf{R}_Z = R_0\mathbf{n}$ (see Figure 1.1). Also, we introduce the *angular velocity of the crack tip* as

$$\boldsymbol{\omega} = \omega\mathbf{m} = \frac{V}{R_0}\mathbf{m}, \tag{31}$$

where $\mathbf{m} = \mathbf{t}\times\mathbf{n}$ is the unit normal vector to the plane of the crack.

For each $\mathbf{X} \in \partial D_\epsilon$ it holds that

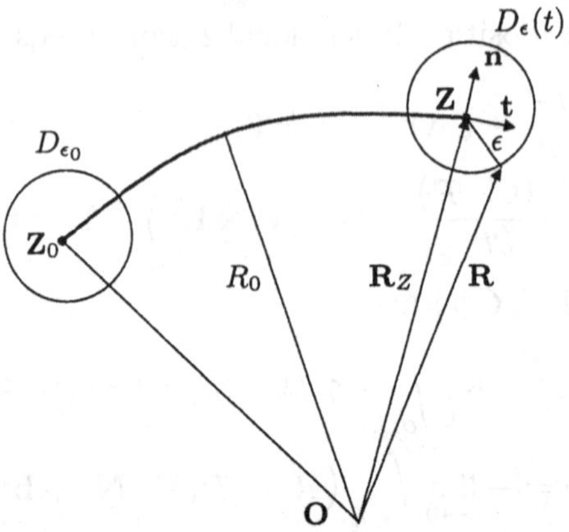

Figure 23.1. A circular crack

$$\mathbf{R}_Z = \mathbf{R} + \boldsymbol{\epsilon}, \tag{32}$$

where $\boldsymbol{\epsilon} = -\epsilon \mathbf{N}$. Then, denoting with $\mathcal{C} = \boldsymbol{\mathcal{P}} \otimes \mathbf{V} + \mathbf{b}^T$, the configurational moment $\boldsymbol{\mathcal{M}}$ (eq. (29)) can be written

$$\begin{aligned} \boldsymbol{\mathcal{M}} &= -\lim_{\epsilon\to 0}\int_{\partial D_\epsilon} \mathbf{R} \times \mathcal{C}\mathbf{N}\, dS \\ &= -\lim_{\epsilon\to 0}\int_{\partial D_\epsilon} \mathbf{R}_Z \times \mathcal{C}\mathbf{N}\, dS + \lim_{\epsilon\to 0}\int_{\partial D_\epsilon} \boldsymbol{\epsilon} \times \mathcal{C}\mathbf{N}\, dS. \end{aligned} \tag{33}$$

Recalling that $-\int_{\partial D_\epsilon} \mathcal{C}\mathbf{N}\, dS$ converges to $\boldsymbol{\mathcal{F}}(t)$, as $\epsilon \to 0$, one can prove that the last term of relation (33) vanishes as $\epsilon \to 0$, therefore

$$\boldsymbol{\mathcal{M}} = -\lim_{\epsilon\to 0}\int_{\partial D_\epsilon} \mathbf{R}_Z \times \mathcal{C}\mathbf{N}\, dS. \tag{34}$$

Next, we calculate the quantity $\boldsymbol{\omega} \cdot \boldsymbol{\mathcal{M}}$

$$\begin{aligned} \boldsymbol{\omega} \cdot \boldsymbol{\mathcal{M}} &= -\frac{V}{R_0}\mathbf{m} \cdot \lim_{\epsilon\to 0}\int_{\partial D_\epsilon} \mathbf{R}_Z \times \mathcal{C}\mathbf{N}\, dS \\ &= -V \lim_{\epsilon\to 0}\int_{\partial D_\epsilon} \mathbf{m}\cdot\mathbf{n} \times \mathcal{C}\mathbf{N}\, dS = -V \lim_{\epsilon\to 0}\int_{\partial D_\epsilon} \mathbf{m}\times\mathbf{n} \cdot \mathcal{C}\mathbf{N}\, dS \\ &= V \lim_{\epsilon\to 0}\int_{\partial D_\epsilon} \mathbf{t}\cdot \mathcal{C}\mathbf{N}\, dS = \mathbf{V}\cdot \lim_{\epsilon\to 0}\int_{\partial D_\epsilon} \mathcal{C}\mathbf{N}\, dS = -\mathbf{V}\cdot\boldsymbol{\mathcal{F}}. \end{aligned} \tag{35}$$

Thus, we have obtained that $\Phi = \boldsymbol{\omega} \cdot \boldsymbol{\mathcal{M}}$. Furthermore, denoting with $G_r = \Phi/\omega$ the *rotational energy release rate*, we can write $G_r = \mathbf{m} \cdot \boldsymbol{\mathcal{M}}$. Please, notice that $G = G_r/R_0$.

Results, similar to $\Phi = \boldsymbol{\omega} \cdot \boldsymbol{\mathcal{M}}$, have been provided by Budiansky and Rice, 1973 and Maugin and Trimarco, 1995 for cavities and disclinations, respectively.

6. CONCLUSIONS

In this work, we have proposed a generalization of the integral balance laws for the case of an elastic body containing a propagating crack. According to the presented analysis, standard concepts of fracture mechanics, like J and L–integrals, have been generalized and their connection with the configurational force and moment at the crack tip has been established. The case of a circular crack has been particularly studied and the notion of the configurational moment at the crack tip has been introduced. Moreover, a new relation between the configurational moment and the energy release rate has been derived.

References

Budiansky, B. and Rice, J. R. (1973). "Conservation Laws and Energy-Release Rates," J. Appl. Mech. 40, 201-203.

Dascalu, C. and Maugin, G. A. (1995). "The Thermoelastic Material-momentum Equation," J. Elast. 39, 201–212.

Freund, L. B. (1989). *Dynamic Fracture Mechanics.* Cambridge: Cambridge University Press.

Golebiewska Herrmann, A. and Herrmann, G. (1981). "On Energy-Release Rates of a plane Crack," J. Appl. Mech. 48, 525–528.

Gurtin, M. E. and Podio-Guidugli, P. (1996). "Configurational Forces and the Basic Laws for Crack Propagation," J. Mech. Phys. Solids 44, 905–927.

Gurtin, M. E. (2000). *Configurational Forces as Basic Concepts of Continuum Physics.* New York: Springer.

Maugin, G. A. (1993) *Material Inhomogeneities in Elasticity.* London: Chapman and Hall.

Maugin, G. A. and Trimarco, C. (1995). "Dissipation of configurational forces indefective elastic solids," Arch. Mech. 47, 81–99.

Steinmann, P. (2000). "Application of material forces to hyperelastostatic fracture mechanics. I. Continuum mechanical setting," Int. J. Solids Struct. 37, 7371–7391.

Chapter 24

THERMOPLASTIC M INTEGRAL AND PATH DOMAIN DEPENDENCE

Pascal Sansen

ESIEE Ecole Supérieure d'Ingénieurs en Electrotechnique et Electronique, Amiens, France

sansen@esiee-amiens.fr

Philippe Dufrénoy

Laboratoire de Mécanique de Lille CNRS UMR 8107, Polytech' Lille, Villeneuve dAscq, France

Philippe.Dufrénoy@polytech-lille.fr

Dieter Weichert

RWTH, Institut fur Allgemeine Mechanik, Aachen, Germany and INSA Rouen, France

Dieter.Weichert@insa-rouen.fr

Abstract The aim of this study is to establish a criteria based on an energy threshold corresponding to crack initiation in the case of dilatational symmetry of a structure under thermomechanical loading. On the base of the J, L, M integrals presented by Knowles and Sternberg, respectively in the case of translational, rotational and scaling symmetry in the light of Noether's theory, many path independent integrals have been defined with applications in fracture mechanics. By the formulation of an adequate Lagrangian formulation, the thermoplastic behaviour is introduced in the M integral through the null Lagrangian theorem. Such an integral finds a physical interpretation as an energy release rate in analogy to the J integral. It equals the change in total available energy due to expansion of an existing cavity. A modification of the M* integral is proposed, in order to have the path-domain independence. A new path-domain independent is described in order to obtain the path domain independence for any transformation of the dilatation group.

Keywords: Crack initiation, dilatational symmetry, M integral, thermoplasticity.

Context and available results

Many industrial applications have for subject the thermomechanical loading. For this kind of application the initiation and growth of cracks must be account for. One example is cracks appearing on brake discs if permanent large circular "hot spots" occur on the friction surface. As shown by Dufrénoy (2003), six hot spots generally appears on the surface of the brake disc of the TGV, with a temperature near one thousand degrees. The thermomechanical solicitations due to these "fixed hot spots" induce cyclic plastic strains leading to thermal fatigue. In this case, as loading is characterized by fields of circular thermal gradients due to hot spots, dilatation symmetry is observed. The consequence of the fatigue cycling is the appearance of a macroscopic crack on the disc. The aim of this paper is to obtain an energetic integral in the case of the scaling symmetry with thermoplasticity. Such an integral finds a physical interpretation as an energy release rate in analogy to the J integral. It equals the change in total available energy due to expansion of an existing cavity, leading to energetic failure criterion.

1. THE M INTEGRAL IN THE CASE OF ELASTICITY

Knowles and Sternberg (1972) wrote different models of crack propagation, in translational symmetry the J integral, a less known integral in rotational symmetry and the M integral (1) in scaling symmetry :

$$M = \int_{\Gamma} x^j (W n_j - \sigma_j^i n_i \frac{\partial u^j}{\partial x_i}) d\Gamma \tag{1}$$

With

x : different points on the contour
W : strain energy
n : normal vector to the contour
σ : stress tensor
u : displacement

According to Olver (1993) the following terms (2) are called P.

$$P = x^j (W n_j - \sigma_j^i n_i \frac{\partial u^j}{\partial x_i}) \tag{2}$$

By Noether's theorem

$$Div(P) = 0 \tag{3}$$

is a conservation law only if we have a variational symmetry. A group of transformation is a variational symmetry group if

$$pr^{(n)}v(L) + Ldiv(\xi) = 0 \tag{4}$$

with v defined in the scaling symmetry by :

$$v = x^i \frac{\partial}{\partial x^i} + \frac{\mathcal{P} - \mathcal{N}}{\mathcal{P}} + u^j \frac{\partial}{\partial u^j} \tag{5}$$

and

$$x \rightarrow \lambda x \qquad u \rightarrow \lambda^{\frac{\mathcal{P}-\mathcal{N}}{\mathcal{P}}} u \tag{6}$$

The conservation law P is given by

$$Div(\vec{P}) = 0 \tag{7}$$

and

$$P_i = \sum_{\alpha=1}^{q} \phi_\alpha \frac{\partial L}{\partial u^\alpha_{,i}} + \xi^i L - \sum_{\alpha=1}^{q} \sum_{j=1}^{p} \xi_j u^\alpha_{,j} \frac{\partial L}{\partial u^\alpha_{,i}} \tag{8}$$

then P, called here M^i, has the form

$$P = M^i = x^j T^i_j + \frac{\mathcal{P} - \mathcal{N}}{\mathcal{P}} u^j \frac{\partial \hat{L}}{\partial u^j_{,i}} \tag{9}$$

with

$$\vec{T} = \hat{L}_n(\chi_{n+1})I - \nabla(\vec{u}_{n+1})^T \sigma_{n+1} \tag{10}$$

$\mathcal{P} = 2$ is the homogeneous strain degree, $\mathcal{N}$ is the space dimension, in the case of a surface integral $\mathcal{N} = 2$. By condition to have a variational symmetry one can write the M integral as the sum of a contour integral and a surface integral. After calculation, the general form of the M integral is

$$M = \int_\Gamma M^i n d\Gamma - \int_\Omega M^i_{,i} d\Omega \tag{11}$$

If we replace M by the previous formulation then

$$M = \int_\Gamma (x^j T^i_j + \frac{\mathcal{P} - \mathcal{N}}{\mathcal{P}} u^j \frac{\partial \hat{L}}{\partial u^j_{,i}}) n_i d\Gamma \tag{12}$$

$$- \int_\Omega (x^j T^i_{j,i} + \mathcal{N}(\hat{L} - W) - \frac{\mathcal{P} - \mathcal{N}}{\mathcal{P}} (\vec{u}_{n+1} \cdot \rho \vec{b}_{n+1})) d\Omega$$

Without singularity the M integral is equal to zero. In the elastic case the lagrangian formulation is equal to the strain energy.

$$\hat{L} = W \tag{13}$$

and the surface integral is null.

2. THE M INTEGRAL IN THE THERMOPLASTIC CASE

In thermoplasticity Wagner and Simo (1992) developped the following Lagrangian formulation

$$\hat{L}_n(\chi_{n+1}) = V(\epsilon_{n+1} - \epsilon^p_{n+1}) + \frac{1}{2}(q_{n+1} : D^{-1} : q_{n+1}) - \rho\vec{b}_{n+1} \cdot \vec{u}_{n+1}$$
$$+ \frac{1}{2T_r C_\epsilon}[T_r\beta : (\epsilon_{n+1} - \epsilon^p_{n+1}) + div(\vec{H}_{n+1}) - \sigma_{n+1} : \epsilon^p_{n+1}$$
$$+ q_{n+1} : D^{-1} : q_{n+1} + m : q_{n+1}]^2 - \lambda_{n+1} f_{n+1} + (\epsilon^p_{n+1} - \epsilon^p_n) : \sigma_{n+1}$$
$$-(q_{n+1} - q_n) : D^{-1} : q_{n+1} + \frac{1}{2T_r\Delta t}(\vec{H}_{n+1} - \vec{H}_n)^T k^{-1}(\vec{H}_{n+1} - \vec{H}_n) \quad (14)$$

With

V	:	elastic strain energy at constant temperature
ϵ	:	strain tensor
ϵ^p	:	plastic strain tensor
q	:	hardening parameter
D	:	hardening coefficient tensor
ρ	:	volume coefficient
b	:	body forces
u	:	displacement
T_r	:	reference temperature
C_ϵ	:	specific heat at constant strain
β	:	tensor coupling strain and temperature
H	:	entropy flux
σ	:	stress tensor
m	:	material property tensor relating temperature and hardening variables
λ	:	inelastic consistency parameter
f	:	yield function
k	:	thermal conductivity tensor
$\bar{t}$	:	boundary traction on the surface.

and

$$\hat{L}^c_n(\chi_{n+1}) = -\vec{\bar{t}}_{n+1} \cdot \vec{u}_{n+1} + \frac{1}{T_r}\bar{\theta}\vec{H}_{n+1} \cdot \vec{n} \quad (15)$$

The last equation (11) corresponds to the calculation of the Lagrangian densities on the contour integral. To use this lagrangian densities it is necessary to verify :

$$\frac{\partial \hat{L}}{\partial \chi} = 0 \qquad \chi = (u_{n+1}, H_{n+1}, \epsilon^p_{n+1}, q_{n+1}\lambda_{n+1}) \tag{16}$$

In the elastic case the variational symmetry is known. The aim is to introduce terms into the lagrangian formulation of the elastic case. We have to check that the added terms are null lagrangian terms. They have to be identically null, following the equation

$$0 \equiv \frac{\partial \hat{L}}{\partial \chi} - D_x \frac{\partial \hat{L}}{\partial \chi_{,x}} = \frac{\partial \hat{L}}{\partial \chi} - \frac{\partial^2 \hat{L}}{\partial x \partial \chi_{,x}} - \chi_{,x} \frac{\partial^2 \hat{L}}{\partial \chi \partial \chi_{,x}} - \chi_{,xx} \frac{\partial^2 \hat{L}}{\partial \chi_{,x}^2} \tag{17}$$

In the elasticic case χ is equal to the displacement and the Lagrangian formulation is equal to the strain energy. Then

$$\chi_{n+1} = \vec{u}_{n+1} \tag{18}$$

and

$$\frac{\partial \hat{L}(\chi_{n+1})}{\partial(\frac{\partial \chi_{n+1}}{\partial x})} = \frac{\partial w(\chi_{n+1})}{\partial(\frac{\partial u_{n+1}}{\partial x})} = \sigma_{n+1} \tag{19}$$

Now by application of the same method in thermoplasticity it is necessary to check the following equation

$$\frac{\partial \hat{L}}{\partial(\frac{\partial \chi_{n+1}}{\partial x})} = (\sigma_{n+1}) \tag{20}$$

After calculation :

$$\frac{\partial \hat{L}}{\partial(\frac{\partial \chi_{n+1}}{\partial x})} = (\sigma_{n+1}, -\frac{\theta_{n+1}}{T_r}, 0, 0, 0)^T \tag{21}$$

The terms added in the lagrangian formulation of the elastic case to obtain the lagrangian formulation in the thermoplastic case are not null lagrangian terms. In this case the following hypothesis has to be done :

$$Div(\vec{H}) = 0 \tag{22}$$

with the following definition for $\vec{H}$ from the temperature Fourier's propagation law :

$$\dot{H} = h = -k \bigtriangledown \theta \tag{23}$$

In the thermoplastic case the result of the M* integral calculation is then

$$M^* = \int_\Gamma (M^i) n d\Gamma - \int_\Omega (M^i_{,i}) d\Omega \tag{24}$$

By introducing equations (6), (7), (8), and if the body forces are neglected

$$\rho \vec{b}_{n+1} \cdot \vec{u}_{n+1} = 0 \tag{25}$$

The result of the M* integral calculation is :

$$\begin{aligned}
M^* = & \int_\Gamma (x((V(\epsilon_{n+1} - \epsilon^p_{n+1}) + \frac{1}{2}(q_{n+1} : D^{-1} : q_{n+1}) - \lambda_{n+1} f_{n+1} \\
& + \frac{1}{2T_r C_\epsilon}(T_r \beta : (\epsilon_{n+1} - \epsilon^p_{n+1}) - \sigma_{n+1} : \epsilon^p_{n+1} + q_{n+1} : D^{-1} : q_{n+1} \\
& + m : q_{n+1})^2 + (\epsilon^p_{n+1} - \epsilon^p_n) : \sigma_{n+1} - (q_{n+1} - q_n) : D^{-1} : q_{n+1} \\
& + \frac{1}{2T_r \Delta t}(\vec{H}_{n+1} - \vec{H}_n)^T k^{-1} (\vec{H}_{n+1} - \vec{H}_n) 1) - (\nabla \vec{u}_{n+1})^T \sigma_{n+1}) \\
& + \frac{\mathcal{P} - \mathcal{N}}{\mathcal{P}} (\vec{u}_{n+1} \sigma_{n+1})) n d\Gamma \\
& - \int_\Omega (x(q_{n+1} : D^{-1} : \nabla q_{n+1} - \sigma_{n+1} : \nabla \epsilon^p_{n+1} + \frac{1}{T_r \Delta t}(\nabla \vec{H}_{n+1} \\
& - \nabla \vec{H}_n)^T k^{-1} (\vec{H}_{n+1} - \vec{H}_n) - \frac{\theta_{n+1}}{T_r}[T_r \beta : (\nabla \epsilon_{n+1} - \nabla \epsilon^p_{n+1}) \\
& - \sigma_{n+1} : \nabla \epsilon^p_{n+1} + q_{n+1} : D^{-1} : \nabla q_{n+1} + m : \nabla q_{n+1}]) \\
& + \mathcal{N}(\frac{1}{2}(q_{n+1} : D^{-1} : q_{n+1}) + (\epsilon^p_{n+1} - \epsilon^p_n) : \sigma_{n+1} - \lambda_{n+1} f_{n+1} \\
& + \frac{1}{2T_r C_\epsilon}(T_r \beta : (\epsilon_{n+1} - \epsilon^p_{n+1}) - \sigma_{n+1} : \epsilon^p_{n+1} + q_{n+1} : D^{-1} : q_{n+1})^2 \\
& - (q_{n+1} - q_n) : D^{-1} : q_{n+1} + \frac{1}{2T_r \Delta t}(\vec{H}_{n+1} - \vec{H}_n)^T k^{-1} \\
& (\vec{H}_{n+1} - \vec{H}_n)) d\Omega
\end{aligned} \tag{26}$$

3. MODIFICATION OF THE M INTEGRAL TO OBTAIN PATH DOMAIN INDEPENDENCE

According to the conditions of variational symmetry (6), the condition to obtain the path domain independence in the elastic case is to have u constant (24), then

$$W(\nabla \vec{u}) \cong W(\frac{1}{r}) \tag{27}$$

Studying the general case of $u \cong r^{\omega}$, $\mathcal{P} = 2$ and $\mathcal{N} = 2$ the displacement, the strain, the stress and the deformation energy are written in the form

$$u \cong r^{-\omega} \qquad \sigma \cong r^{-(\omega+1)} \qquad \epsilon \cong r^{-(\omega+1)} \qquad W \cong r^{-2(\omega+1)} \tag{28}$$

In the case of elasticity if one examines the dimension of the M integral

$$M^{\omega} \cong (r^{-2(\omega+1)}r)r = r^{-2\omega} \tag{29}$$

Then path-domain independence is obtained only if $\omega = 0$, this is the solution of Flamand's problem. As the deformation is a two degree homogeneous function, one can write

$$W(\frac{1}{r}\frac{r^{\omega}}{r^{\omega}}) = r^{2\omega}W(\frac{1}{r^{\omega+1}}) \qquad with \qquad r^{2\omega} = (x_i n_i)^{2\omega} \tag{30}$$

Then :

$$M^{\omega} = \int_{\Gamma}(x_i n_i)^{2\omega}(Wx^i n_i - \sigma^i_j n_i \frac{\partial u^j}{\partial x_k}x^k)d\Gamma \tag{31}$$

The dimension of this integral is then :

$$M^{\omega} \cong r^{2\omega}(r^{-2(\omega+1)}r)r = r^0 \tag{32}$$

thus one has path domain independence. In the thermoplastic case one proceeds in the same way as previously to have a new M^{ω^*} integral.

$$M^{\omega^*} = \int_{\Gamma} M^i n d\Gamma - \int_{\Omega} M^i_{,i} d\Omega \tag{33}$$

$$M^{\omega^*} = \int_{\Gamma}((x_i n_i)^{2\omega}(x^j T^i_j)n_i)d\Gamma$$
$$- \int_{\Omega}((x_i n_i)^{2\omega}(x^j T^i_{j,i}) + 2((x_i n_i)^{2\omega} + 2\omega(x_i n_i)^{2\omega-1}x^j)(\hat{L} - W))d\Omega \tag{34}$$

About the dimension of $M^{\omega *}$ integral one has path domain independence.

$$M^{\omega^*} \cong r^{2\omega}(r^{-2(\omega+1)}r)r = r^0 \tag{35}$$

4. PHYSICAL INTERPRETATION

$$\Pi_n(\chi_n) = \Pi_{n+1}(\chi_{n+1}) + D(\chi_n, \chi_{n+1}) + F(\chi_{n+1}) \tag{36}$$

With

Π_n	:	total energy at time n
Π_{n+1}	:	total energy at time n+1
D	:	the dissipated energy between time n to time n+1
F	:	the dissipated energy in the crack propagation.

Then after calculation

$$\frac{\partial F(\chi_{n+1})}{\partial t} = -M^{\omega *} \tag{37}$$

The $M^{\omega *}$ integral is an energy release rates for the scaling symmetry.

5. CONCLUSION

The study of the brake disc allows us to have an interesting case of scaling symmetry with thermoplasticity. The M integral has been extended to this case with discussion of path domain independence. Experimental investigation would had to confirm the existence of an energy threshold and cavity expansion.

References

Dufrénoy, P. and Weichert, D. (2003). *A thermomechanical model for the analysis of disc brakes fracture mechanisms.* J. of Thermal Stresses 26(8), 815-828.

Dufrénoy, P., Bodovillé, G. and Degallaix, G. (2002). *Damage mechanisms and thermomechanical loadings of brake discs.* Temperature-Fatigue Interaction, ESIS Publication 29, Elsevier Science, 167-176.

Knowles, J.K. and Sternberg, E. (1972). *On a class of conservation laws in linearized and finite elastostatics.* Archive of Rational Mechanics and Analysis 44, 187-211.

Noether, E. (1918). *Göttinger Nachrichten,* Mathematisch-Physikalische Klasse, 2, 235, English translation by Tavel, M.A. (1971). Transport theory and statistical physics 1, 183-207.

Olver, P.J. (1993). *Applications of Lie groups to differential equations,* Springer Verlag, New York.

Sansen, P. (1999a). *Formulation energetique d'un critère de rupture locale d'un solide en thermoplasticité.* Ph.D. thesis, Université des Sciences et Technologies de Lille - Polytech Lille.

Sansen, P, Dufrénoy, P., Weichert, D. (1999b). *Indépendance de l'intégrale de contour pour une symétrie de dilatation.* C. R. Acad. Sci Paris 327, série II b, 1351-1354.

Sansen, P, Dufrénoy, P., Weichert, D. (2000). *Fracture parameter for thermoplasticity relative to material scaling.* Zamm, Berlin 80 Suppl.2, 495-496.

Sansen, P, Dufrénoy, P., Weichert, D. (2001). *Fracture parameter for thermoplasticity in the case of dilatation symmetry.* Int. J. of fracture vol. 111, 61-66.

Sansen, P, Dufrénoy, P., Weichert, D. (2002). *Path-independent integral for the dilatation symmetry group in thermoplasticity.* Int. J. of fracture vol. 117, 337-346.

Simo, J.C. and Honein, T. (1990). *Variational formulation, discrete conservation laws, and path-domain independent integrals for elasto-viscoplasticity.* Journal of applied mechanics 57, 488-497.

Wagner, D.A. and Simo, J.C. (1992). *Fracture parameter for thermoinelasticity.* International Journal of Fracture 56, 159-187.

IX

DELAMINATION & DISCONTINUITIES

Chapter 25

PEELING TAPES

Paolo Podio-Guidugli
Dipartimento di Ingegneria Civile, Università di Roma "Tor Vergata"
Via del Politecnico, 1 - 00133 Roma, Italy
ppg@uniroma2.it

Abstract Two basic peeling problems formulated by Ericksen (1991) are studied, both in statics and in dynamics. While equilibrium is treated variationally, the evolution of the tape tip is modeled as the result of a configurational force balance.

Keywords: Adhesion, static and dynamic peeling models, peel tests.

1. INTRODUCTION

Peeling a scotch tape off your desk, opening up the velcron flaps of your wind jacket or, if you happen to be a gecko, to lift one of your paw up to move on when hanging from a ceiling, all this requires expenditure of mechanical power. That power may be thought of as needed to overcome the peculiar resistance commonly referred to as the "adhesion force", or to compensate for the time rate of change of some "adhesion energy". Intuitively, a force should act in response to the breaking of microscopical bonds occurring when the tape is peeled or the flaps are pulled apart; and an energy would be required to form new free surface, since a peculiar system of forces and stresses conceivably maintains the structural integrity of the contact bond. Be it in terms of "force" or "energy" or both, a concept of *adhesion* replaces, at the macroscopic length scale typical of continuum mechanics, for a detailed description of the bonding mechanism, such description being ascribed to a multiple-scale analysis.

Phenomenological theories of adhesion have been proposed, intended for a wide range of applications: see, e.g., Frémond (1988) and the literature quoted therein. As to peeling in particular, a simple variational theory is found in Sections 7.3 and 7.4 of Ericksen (1991); Ericksen's theory deals with two prototype problems, and captures the essentials of the quasi-static phenomenology of peeling. Burridge and Keller (1978) have attempted to a dynamical theory; their model is under many respects questionable, but they deserve credit for singling out peeling as an interesting example of a one-dimensional *moving free-boundary*, whose associated evolution problem has a mixed "parabolic-hyperbolic" nature.

I here claim that recent work on two-dimensional dynamical fracture of Gurtin and Podio-Guidugli (1996a and, especially, 1998) may serve as a basis for a sound and general treatment of dynamical peeling. Indeed, there is a striking kinematic similarity between the motion of a crack tip and the motion of a *tape tip*, that is, a point separating the peeled part of a tape from the part adhering to the substrate. But peeling problems are inherently easier, since the trajectory of a tape tip is a prescribed curve, whereas one of the major difficulties of dynamical fracture is to predict the shape of an evolving crack (and cracks may kink, bifurcate, self-intersect, *etc.*).[1]

In this paper, which updates the contents of a talk with the same title given a few years ago (Podio-Guidugli, 1996), I discuss the prototype problems of Ericksen from the point of view of gaining information on modeling adhesion. First, in Section 2, I confine myself to statics, as Ericksen did. Then, in Section 3, I briefly indicate how the ideas of Gurtin and Podio-Guidugli, 1998 may be adapted to model dynamical peeling. In the last section I touch very briefly on some modeling issues that seem to deserve further attention.

2. THE STATICS OF SOFT AND HARD PEELING

Peeling experiments are intended to test and measure adhesion: neither the tape should break nor pieces should come off the substrate (this is why depilation would require a more complicated model than peeling). One expects the mechanical responses of the tape and the substrate to have an influence on the outcome of a peeling experiment: to focus on adhesion, it would seem best to begin with the simplest assumptions possible. In my presentation, as is done in the quoted sections of Ericksen's book (1991), I assume the tape to be infinitely strong and the substrate to be flat and rigid.

2.1. SOFT PEELING

Let the tape to be peeled occupy the interval $[0, L]$ in its reference configuration, and let a given dead load $\mathbf{f} = f\mathbf{e}$ be applied at the tape's right end, with $f > 0$, $\mathbf{e} = \cos\varphi\, \mathbf{c}_1 + \sin\varphi\, \mathbf{c}_2$, $\{\mathbf{c}_1, \mathbf{c}_2\}$ an orthonormal basis, and $\varphi \in [0, \pi]$. The goal is to find whether there are peeled configurations of equilibrium, with the tape tip at $Z \in [0, L]$.

Since the load is dead, the load potential W_l has the expression

$$W_l = -\mathbf{f} \cdot \mathbf{u}, \tag{1}$$

where $\mathbf{u}$ is the equilibrium displacement of the right end. For δ the current length of the detached portion of the tape, one has that

$$\mathbf{u} = -(L - Z)\mathbf{c}_1 + \delta \mathbf{e}, \tag{2}$$

and that

$$\delta = \lambda(L - Z), \quad \lambda - 1 = \frac{f}{\beta}, \quad \beta > 0, \tag{3}$$

where λ is stretch and β the elastic stiffness of the tape, assumed for semplicity to be linearly elastic. Then, the load potential is

$$W_l = -(L - Z) f (\lambda - \cos\varphi), \tag{4}$$

and the elastic energy stored in the tape is

$$W_e = (L - Z) \frac{1}{2} \beta (\lambda - 1)^2. \tag{5}$$

As to the adhesion potential W_a, I take it with Ericksen proportional to the length of the detached portion through an *adhesion modulus* α:

$$W_a = (L - Z)\alpha, \quad \alpha > 0. \tag{6}$$

In conclusion, the total potential is

$$W_t^{(s)} = W_a + W_e + W_l = (L - Z)\Big(\alpha + \frac{1}{2}\beta(\lambda - 1)^2 - f(\lambda - \cos\varphi)\Big). \tag{7}$$

2.1.1 Inextensible tape. For $\lambda = 1$, the equilibrium analysis is easily carried out: there is a critical value

$$f_c^{(i)}(\alpha, \varphi) = \alpha \frac{1}{1 - \cos\varphi} \tag{8}$$

of the applied load at which $W_t^{(s)}(Z) \equiv 0$, and hence equilibrium is possible at any $Z \in [0, L]$ (*arbitrary peeling*); for $f > f_c^{(i)}$, $W_t^{(s)}(Z) \leq 0$ attains its minimum value at $Z = 0$ (*complete peeling*); for $f < f_c^{(i)}$, $W_t^{(s)}(Z) \geq 0$, and hence minimum for $Z = L$ (*no peeling*). These conclusions seem to agree fairly well with experience (Federico *et al.*, 1996). In fact, in principle at least, (8) suggests a simple procedure to validate the model (and then measure the adhesion modulus): one should verify that $f_c^{(i)}(\alpha, \frac{\pi}{2}) = 2 f_c^{(i)}(\alpha, \pi)$.

2.1.2 Extensible tape. In this case, relation (8) is replaced by

$$f_c^{(e)}(\alpha,\beta,\varphi)=\beta\Big(\sqrt{(1-\cos\,\varphi)^2+2\,\frac{\alpha}{\beta}}-1+\cos\,\varphi\Big)\,. \qquad (9)$$

It is easy to see that

$$f_c^{(e)}<f_c^{(i)} \quad \text{and} \quad \lim_{\beta\to+\infty} f_c^{(e)}=f_c^{(i)}\,. \qquad (10)$$

Moreover, in a 90°-degree peeling test,

$$f_c^{(e)}\simeq\alpha\Big(1-\frac{1}{2}\frac{\alpha}{\beta}\Big)\,. \qquad (11)$$

2.2. HARD PEELING

Let now the vertical displacement of right end of the tape be assigned:

$$\mathbf{u}\cdot\mathbf{c}_2=D>0\,. \qquad (12)$$

Then, the tape tip position Z, the stretch λ in the peeled part of the tape and its angle φ to the substrate must satisfy the constraint

$$\lambda(L-Z)\sin\,\varphi=D. \qquad (13)$$

The total potential is

$$W_t^{(h)}=(L-Z)\Big(\alpha+\frac{1}{2}\,\beta(\lambda-1)^2\Big). \qquad (14)$$

There is one triplet (Z,λ,φ) satisfying (13) which minimizes $W_t^{(h)}$,

$$Z^{(e)}=L-\frac{D}{\lambda^{(e)}},\quad \lambda^{(e)}=\Big(1+2\,\frac{\alpha}{\beta}\Big)^{\frac{1}{2}},\quad \varphi=\frac{\pi}{2}\,. \qquad (15)$$

Hence, at equilibrium, the peeled portion of the tape must be perpendicular to the substrate and have length equal to the imposed vertical displacement, a finding that should not be difficult to confirm or contradict.[2] For an inextensible tape,

$$L-D=:Z^{(i)}<Z^{(e)},\quad \lim_{\beta\to+\infty} Z^{(e)}=Z^{(i)}\,. \qquad (16)$$

3. TAPE-TIP DYNAMICS

I begin by a condensed summary of the contents of Gurtin and Podio-Guidugli (1998), with a view toward suggesting how the format there proposed to characterize the evolution of crack tips might be applied, with adaptations, to study the evolution of tape tips.

Gurtin and Podio-Guidugli postulate the balance equations basic to a theory of dynamical fracture; these equations involve, in addition to the standard force balances of classical continuum mechanics, the balances of the nonstandard *configurational forces* (Gurtin (1995, 2000) performing work in the evolution of macroscopic structural defects such as cracks. The evolution equations obtain from the postulated balance equations when explicit choices are made for the inertial forces acting at the crack tip, both standard and configurational; these constitutive choices are motivated along lines put forward in Podio-Guidugli (1997).[3] Ground for further constitutive characterization of crack kinetics is provided by a purely mechanical dissipation inequality holding for an evolving referential control volume of arbitrary shape.[4] Roughly, this inequality requires that the time rate of the sum of bulk and surface free energies be not greater than the working performed inside the control volume and at its boundary. By localization at the crack tip, this inequality establishes the basic dissipative nature of the crack's kinetics.

For simplicity, I here restrict attention to the case of soft peeling of an inextensible tape and, due to lack of space, proceed by formal analogy with the case of a propagating crack treated in Gurtin and Podio-Guidugli (1996a, 1998).

Let then $\mathbf{v} = V\mathbf{c}_1$ be the tip velocity, with $V = \dot{Z}(t)$ the tip speed, and let the evolving referential control volume be a *tip disk*, that is to say, a disk $\mathcal{D}_\varepsilon$ of radius ε centered at the tip and moving with the tip velocity.

For a straight crack, the *normal configurational force balance at the tip* has the form:

$$\Big(\lim_{\varepsilon\to 0}\int_{\partial\mathcal{D}_\varepsilon} \mathbf{C}^d\mathbf{n} + \mathbf{g}\Big)\cdot\mathbf{c}_1 = 0\,, \tag{17}$$

where

$$\mathbf{C}^d = (\psi + \kappa_{rel})\mathbf{1} - \mathbf{F}^T\mathbf{S} \tag{18}$$

is the *dynamic Eshelby tensor*, with ψ the free energy density per unit referential volume, κ_{rel} the density of kinetic energy relative to the tip, $\mathbf{F}$ the deformation gradient, and $\mathbf{S}$ the Piola stress; and where $\mathbf{n}$ is the outer unit normal to the contour of the tip disk.

For an inextensible tape of mass density ρ per unit length, I take

$$\psi \sim \alpha, \quad 2\,\kappa_{rel} \sim \rho\,|\dot{\mathbf{u}} - \mathbf{v}|^2 = \rho\,\dot{Z}^2, \quad \mathbf{F} \sim \mathbf{1} \tag{19}$$

(note that, for $\lambda = 1$, (2) and $(3)_1$ imply that $\mathbf{u} = (L - Z)(\mathbf{e} - \mathbf{c}_1)$, whence $\dot{\mathbf{u}} = -\dot{Z}\mathbf{e} + \mathbf{v}$), and I assume that

$$\lim_{\varepsilon \to 0} \int_{\partial R_\varepsilon} (\Psi + \kappa_{rel})\mathbf{n} \sim \Big(\alpha + \frac{1}{2}\rho \dot{Z}^2\Big)\mathbf{e}, \quad \lim_{\varepsilon \to 0} \int_{\partial R_\varepsilon} \mathbf{F}^T \mathbf{S}\mathbf{n} \sim f\mathbf{e}. \qquad (20)$$

Furthermore, I stipulate that the following dissipation inequality restricts the constitutive choice of the *internal configurational force* $\mathbf{g}$ at the tip:

$$\mathbf{g}(V) \cdot \mathbf{v} = V\,\mathbf{g}(V) \cdot \mathbf{c}_1 \leq 0 \quad \text{for all } V, \quad \mathbf{g}(0) \cdot \mathbf{c}_1 = 0, \qquad (21)$$

(so that $\mathbf{g}$ does oppose the motion of the tip; note that the component $\mathbf{g} \cdot \mathbf{c}_2$ of $\mathbf{g}$ has reactive nature: it does not enter the dissipation inequality (21) and hence needs not a constitutive specification). All in all, I lay down the following evolution equation for the tape tip:

$$\dot{Z}^2 + \hat{g}(Z, \dot{Z}) = \hat{f}(t), \qquad (22)$$

where the function $\hat{g}$ must be chosen consistent with (21), and where the forcing term $\hat{f}(t)$ is proportional to $(f(t) - \alpha)\cos\varphi$. This equation can read as an expression of balance involving: a driving force, measured by f; an adhesion force, measured by α in statics and $\hat{g}$ in dynamics; and an external distance force, assumed to have solely inertial nature.

4. HINTS FOR FURTHER DEVELOPMENTS

Certain classes of smooth solutions of equation (22) have been considered by Ansini (2000). In the experiments of peeling under constant applied load described by Barquins and Ciccotti (1997), a jerky peeling mode is observed for sufficiently large loads, accompanied by emission of both sound and light waves; the jerky motion of the tape tip offers an interesting example of deterministic chaotic dynamics. It would be interesting to see whether *slip-stick peeling regimes* would be induced solely by a suitable choice of the constitutive function $\hat{g}$. As a matter of fact, the works by Tsai and Kim (1993), Hong and Yue (1995), Barquins and Ciccotti (1997), Ciccotti, Giorgini and Barquins (1998) call attention to the role of the tape's elastic energy, which is jerkily stored/released as the tape tip slips/sticks. While in these papers the elastic energy is taken quadratic in the tape's stretch, I surmise that a nonconvex elastic energy would introduce another potential source of chaotic behavior.

But the simple peeling model here presented could be also implemented in other ways. For example, it seems sensible to imagine that the tape is made to adhere to the substrate after some stretching. Or, the tape might be assigned some bending stiffness, in which case one could dream of something like a "plastic hinge" forming at the tape tip. Finally and, in my mind, more importantly, it would be interesting to look for a *cohesive zone model* of tape-tip motion. Models based on standard wisdom should in principle require modest adaptations of what one learns, e.g., in the papers by Chowdury and Narasimhan (2000) and Chandra *et al.* (2002); to adapt the ideas in Wu (1992) would be a different, perhaps worth-telling story.

Acknowledgments

This work has been supported by Progetto Cofinanziato 2002 "Modelli Matematici per la Scienza dei Materiali" and by TMR Contract FMRX-CT98-0229 "Phase Transitions in Crystalline Solids".

Notes

1. The idea that peeling can be regarded as a particular type of fracture is not new; see Hong and Yue, 1995 and the literature they quote.

2. See Federico *et al.* (1996) where peeling problems are examined with a view toward applying the results to junctions in "geosynthetic" coatings for earth or concrete dams. What brought me back to peeling problems, many years after I read with much interest Burridge and Keller's 1978 paper, was precisely the temporal coincidence between my theoretical work with Gurtin on crack propagation in a configurational framework (Gurtin and Podio-Guidugli, 1996a, 1996b, 1998) and the practical interest for geosynthetics of Federico and other colleagues in my Department.

3. See also Gurtin and Podio-Guidugli (1996b).

4. See Gurtin and Struthers (1990); Gurtin (1995); Gurtin and Podio-Guidugli (1996a).

References

Ansini, L. (2000). "Modellazione e analisi del peeling," Atti Sem. Mat. Fis. Univ. Modena **48**, 217-223.

Barquins, M., and M. Ciccotti. (1997). "On the kinetics of peeling of an adhesive tape under a constant imposed load," Int. J. Adhesion and Adhesives **17**, 66-68.

Burridge, R. and J.B. Keller. (1978). "Peeling, slipping and cracking - Some one-dimensional free-boundary problems in mechanics," SIAM Rev. **20**, 31-61.

Chandra, N., H. Li, C. Shet, and H. Ghonem. (2002). "Some issues in the application of cohesive zone models for metallic-ceramic interfaces," Int. J. Solids Structures **39**, 2827-2855.

Chowdhury, S.R. and R. Narasimhan. (2000). "A cohesive finite element formulation for modelling fracture and delamination of solids," Sadhana **25**, 561-587.

Ciccotti, M., Giorgini, B., and M. Barquins. (1998). "Stick-slip in the peeling of an adhesive tape: evolution of theoretical model," Int. J. Adhesion and Adhesives **18**, 35-40.

Ericksen, J.L. (1991). *Introduction to the Thermodynamics of Solids.* Chapman & Hall: London·New York·Tokyo·Melbourne·Madras.

Federico, F., Lupoi, A., and P. Podio-Guidugli. (1996). "Il peeling dei geosintetici. Aspetti teorici, sperimentali e normativi," Res. Rep. 23.6.96, Department of Civil Engineering, University of Rome "Tor Vergata".

Frémond, M. (1988). "Contact with adhesion," Chapter IV of *Topics in Nonsmooth Mechanics*, Birkhäuser Verlag: Basel·Boston·Berlin.

Gurtin, M.E. (1995). "The nature of configurational forces," Arch. Rational Mech. Anal. **131**, 67-100.

Gurtin, M.E. (2000). *Configurational Forces as Basic Concepts of Continuum Physics.* Springer-Verlag: Berlin.

Gurtin, M.E., and Podio-Guidugli, P. (1996a). "Configurational forces and the basic laws for crack propagation," J. Mech. Phys. Solids **44**, 905-927.

Gurtin, M.E., and Podio-Guidugli, P. (1996b). "On configurational inertial forces at a phase interface," J. Elasticity **44**, 255-269.

Gurtin, M.E., and Podio-Guidugli, P. (1998). "Configurational forces and a constitutive theory for crack propagation that allows for kinking and curving," J. Mech. Phys. Solids **46**, 1-36.

Gurtin, M.E., and A. Struthers. (1990). "Multiphase thermomechanics with interfacial structure. 3. Evolving phase boundaries in the presence of bulk deformation," Arch. Rational Mech. Anal. **112**, 97-160.

Hong, D.C., and S. Yue. (1995). "Deterministic chaos in failure dynamics: dynamics of peeling of adhesive tapes," Phys. Rev. Letters **74**, 254-257.

Podio-Guidugli, P. (1996). "Peeling tapes," invited talk at the conference on "Recent Developments in Solid Mechanics" in honor of Wolf Altman, Rio de Janeiro, July 31st, 1996.

Podio-Guidugli, P. (1997). "Inertia and invariance," Ann. Mat. Pura. Appl. (IV) **172**, 103-124.

Tsai, K.-H., and K.-S. Kim. (1993). "Stick-slip in the thin film peel test - I. The 90° peel test," Int. J. Solids Structures **30**, 1789-1806.

Wu, C.H. (1992). "Cohesive elasticity and surface phenomena," Q. Appl. Math. **50**, 73-103.

Chapter 26

STABILITY AND BIFURCATION WITH MOVING DISCONTINUITIES

Claude Stolz

CNRS UMR7649, Laboratoire de Mécanique des Solides

Ecole Polytechnique, 91128 Palaiseau cedex

stolz@lms.polytechnique.fr

Rachel-Marie Pradeilles-Duval

CNRS UMR7649, Laboratoire de Mécanique des Solides

Ecole Polytechnique, 91128 Palaiseau cedex

rachel@lms.polytechnique.fr

Abstract The propagation of moving surface inside a body is analysed in the framework of thermodynamics, when the moving surface is associated with an irreversible change of mechanical properties. The thermodynamical force associated to the propagation has the form of an energy release rate. Quasistatic rate boundary value problem is given when the propagation of the interface is governed by a normality rule. Extension to generalised media to study delamination is also investigated.

1. INTRODUCTION

This paper is concerned mostly with the description of damage involved on the evolution of a moving interface along which mechanical transformation occurs, (Pradeilles-Duval and Stolz, 1995). Some connection can be made with the notion of configurational forces, (Gurtin, 1995 ; Maugin, 1995; Truskinovsky, 1987).

A domain Ω is composed of two distinct volumes Ω_1, Ω_2 of two linear elastic materials with different characteristics. The bounding between the two phases is perfect and the interface is denoted by Γ, ($\Gamma = \partial\Omega_1 \cap \partial\Omega_2$). The external surface $\partial\Omega$ is decomposed in two parts $\partial\Omega_{\rm u}$ and $\partial\Omega_{\rm T}$ on which the displacement $\underline{u}^d$ and the loading $\underline{\rm T}^d$ are prescribed respectively. The material 1 changes into material 2 along the interface Γ by an irreversible process. Hence Γ moves with the normal velocity $\underline{c} = \phi \underline{\nu}$ in the reference state, where $\underline{\nu}$ is the outward Ω_2 normal vector and ϕ is positive. When the surface Γ is moving volume average of any mechanical quantity f has a rate defined by:

$$\frac{\mathrm{d}}{\mathrm{d}t}\int_{\Omega(\Gamma)} f \,\mathrm{d}\Omega = \int_{\Omega(\Gamma)} \dot{f} \,\mathrm{d}\Omega - \int_{\Gamma} [\, f \,]_{\Gamma}\, \underline{c}.\underline{\nu}\,\mathrm{d}S, \tag{1}$$

where $[\, f\,]_{\Gamma} = f_1 - f_2$ represents the jump of f accross Γ. The state of the system is characterized by the displacement field $\underline{u}$, from which the strain field $\boldsymbol{\varepsilon}$ is derived, and by the spatial distribution of the two phases defined by the position of Γ. The two phases have the same mass density ρ. The free energy density w_i is a quadratic function of the strain $\boldsymbol{\varepsilon}$, $\rho w_i = \frac{1}{2}\boldsymbol{\varepsilon} : \boldsymbol{C}^i : \boldsymbol{\varepsilon}$ in Ω_i. The potential energy $\mathcal{E}$ of the structure Ω $(\Omega_1 \cup \Omega_2)$ has the following form

$$\mathcal{E}(\underline{\tilde{u}}, \Gamma, \underline{\tilde{\mathrm{T}}}^d) = \sum_{i=1,2}\int_{\Omega_i(\Gamma)} \rho\, w_i(\boldsymbol{\varepsilon}(u))\,\mathrm{d}\Omega - \int_{\partial\Omega_{\mathrm{T}}} \underline{\mathrm{T}}^d.\underline{u}\,\mathrm{d}S.$$

When Γ is known, the body is a heterogeneous medium with two elastic phases, and under prescribed loading an equilibrium state $\underline{u}^{sol}$ is given by the stationnarity of the potential energy written as

$$\frac{\partial \mathcal{E}}{\partial \underline{u}}\cdot\delta\underline{u} = \sum_{i=1,2}\int_{\Omega_i} \rho\frac{\partial w_i}{\partial \boldsymbol{\varepsilon}} : \boldsymbol{\varepsilon}(\delta\underline{u})\,\mathrm{d}\Omega - \int_{\partial\Omega_{\mathrm{T}}} \underline{\mathrm{T}}^d.\delta\underline{u}\,\mathrm{d}S = 0, \tag{2}$$

for all $\delta\underline{u}$ kinematically admissible field satisfying $\delta\underline{u} = 0$ over $\partial\Omega_{\mathrm{u}}$. Then the solution $\underline{u}^{sol}$ satisfies (PB1):

- the local constitutive relations: $\boldsymbol{\sigma} = \rho\frac{\partial w_i}{\partial \boldsymbol{\varepsilon}}$, on Ω_i,
- the momentum equations: $\mathrm{div}\,\boldsymbol{\sigma} = 0$, on Ω , $\boldsymbol{\sigma}.\underline{n} = \underline{\mathrm{T}}^d$ over $\partial\Omega_{\mathrm{T}}$,
- the compatibility relations: $2\boldsymbol{\varepsilon} = \nabla\underline{u} + \nabla^t\underline{u}$, in Ω, $\underline{u} = \underline{u}^d$ over $\partial\Omega_{\mathrm{u}}$,
- the perfect bounding over Γ: $[\,\boldsymbol{\sigma}\,]_{\Gamma}.\underline{\nu} = 0$ and $[\,\underline{u}\,]_{\Gamma} = 0$.

For a prescribed history of the loading, we must determine the rate of all mechanical fields and the normal propagation ϕ to characterize the position of the interface Γ at each time. Let us introduce the convected derivative $\mathcal{D}_\phi$ of any function $f(\underline{X}_\Gamma, t)$ defined by

$$\mathcal{D}_\phi f = \lim_{\tau\to 0}\frac{f(\underline{X}_\Gamma + \phi\underline{\nu}\tau, t+\tau) - f(\underline{X}_\Gamma, t)}{\tau}. \tag{3}$$

For example, the equation of a moving surface is $S(\underline{X}_{\Gamma}), t) = 0$, then $\mathcal{D}_\phi S = 0$. This defines the normal velocity ϕ as $\phi = -\frac{\partial S}{\partial t} / \parallel \frac{\partial S}{\partial \underline{X}} \parallel$ and $\underline{\nu} = \frac{\partial S}{\partial \underline{X}} / \parallel \frac{\partial S}{\partial \underline{X}} \parallel$.

Hadamard's relations. The bounding being perfect between the phases, the displacement and the stress vector are continuous along Γ. Their rates $(\underline{v}, \dot{\boldsymbol{\sigma}})$ have discontinuities according to the compatibility conditions of Hadamard, rewritten with the convected derivative: the continuity of displacement induces

$$[\underline{u}]_\Gamma = 0 \Rightarrow \mathcal{D}_\phi([\underline{u}]_\Gamma) = [\underline{v}]_\Gamma + \phi[\nabla \underline{u}]_\Gamma . \underline{\nu} = 0, \tag{4}$$

and continuity of stress vector implies:

$$[\boldsymbol{\sigma}]_\Gamma . \underline{\nu} = 0 \Rightarrow \mathcal{D}_\phi([\boldsymbol{\sigma}]_\Gamma . \underline{\nu}) = [\dot{\boldsymbol{\sigma}}]_\Gamma . \underline{\nu} - \mathrm{div}_\Gamma([\boldsymbol{\sigma}]_\Gamma \, \phi) = 0. \tag{5}$$

The last equation is obtained taking the equilibrium equation into account and the surface divergence defined by $\mathrm{div}_\Gamma F = \mathrm{div} F - \underline{\nu}.\nabla F.\underline{\nu}$.

Orthogonality property for discontinuities. Since the displacement and the stress vector are continuous along the interface,

$$[\underline{u}]_\Gamma = 0, \Rightarrow [\nabla \underline{u}]_\Gamma . \underline{e}_\alpha = 0, \quad [\boldsymbol{\sigma}]_\Gamma . \underline{\nu} = 0, \tag{6}$$

the discontinuities of the stress $\boldsymbol{\sigma}$ and of $\nabla \underline{u}$ have the property of orthogonality:

$$[\boldsymbol{\sigma}]_\Gamma : [\nabla \underline{u}]_\Gamma = 0. \tag{7}$$

Dissipation analysis. The total dissipation of the system is

$$\mathrm{D} = \int_{\Omega(\Gamma)} \boldsymbol{\sigma} : \dot{\boldsymbol{\varepsilon}} \, \mathrm{d}\Omega - \frac{\mathrm{d}}{\mathrm{d}t} \int_{\Omega(\Gamma)} \rho w \, \mathrm{d}\Omega = -\frac{\partial \mathcal{E}}{\partial \Gamma} . \dot{\Gamma} = \int_\Gamma \mathcal{G} \phi \, \mathrm{d}S \geq 0, \tag{8}$$

taking the momentum conservation and the boundary conditions into account. The quantity $\mathcal{G}(\underline{X}_\Gamma, t) = [w]_\Gamma - \boldsymbol{\sigma} : [\boldsymbol{\varepsilon}]_\Gamma$ has an analogous form to the driving traction force acting on a surface of strain discontinuity proposed by Abeyratne and Knowles, 1990.

2. EVOLUTION OF THE INTERFACE

Let denotes $\Gamma^+ = \{x \in \Gamma / \mathcal{G}(x) = G_c\}$ and let considers the rule defined as a generalization of Griffith's law: $\phi \geq 0$, on $\Gamma^+, \phi = 0$, otherwise. At the point $x_\Gamma(t) \in \Gamma^+$, we have $\mathcal{G}(x_\Gamma(t), t) = G_c$, this is an implicit equation for the position of the interface, and the derivative of $\mathcal{G}$ following the moving surface vanishes: $\mathcal{D}_\phi \mathcal{G} = 0$. This leads to the consistency condition written for all point on Γ^+

$$(\phi - \phi^*)\mathcal{D}_\phi \mathcal{G} \geq 0, \forall \phi^* \geq 0, \text{ over } \Gamma^+, \tag{9}$$

and ϕ and ϕ^* are in $\mathcal{K} = \{\beta/\beta \geq 0 \text{ on } \Gamma^+, \beta = 0 \text{ otherwise}\}$. The evaluation of $\mathcal{D}_\phi \mathcal{G}$ is obtained easily

$$\mathcal{D}_\phi \mathcal{G} = \mathcal{D}_\phi [w]_\Gamma - \mathcal{D}_\phi \boldsymbol{\sigma}_2 : [\nabla \underline{u}]_\Gamma - \boldsymbol{\sigma}_2 : [\mathcal{D}_\phi \nabla \underline{u}]_\Gamma \tag{10}$$

$$= [\boldsymbol{\sigma}]_\Gamma : \nabla \underline{v}_1 - \dot{\boldsymbol{\sigma}}_2 : [\nabla \underline{u}]_\Gamma - \phi G_n, \tag{11}$$

where $G_n = -[\boldsymbol{\sigma}]_\Gamma : (\nabla\nabla \underline{u}_1 . \underline{\nu}) + \nabla \boldsymbol{\sigma}_2 . \underline{\nu} : [\nabla \underline{u}]_\Gamma$.

The rate boundary value problem. The solution is governed by the derivation of (PB1) relatively to time, taking account of Hadamard's relations (4,5) along Γ and of the propagation law (9). Defining the functional $\mathcal{F}$

$$\mathcal{F}(\tilde{\underline{v}}, \tilde{\phi}, \dot{\underline{\mathrm{T}}}^d) = \int_\Omega \frac{1}{2} \boldsymbol{\varepsilon}(\underline{v}) : \boldsymbol{C} : \boldsymbol{\varepsilon}(\underline{v}) \, \mathrm{d}\Omega - \int_{\partial\Omega_\mathrm{T}} \dot{\underline{\mathrm{T}}}^d . \underline{v} \, \mathrm{d}S \tag{12}$$

$$- \int_\Gamma \phi [\boldsymbol{\sigma}]_\Gamma : \nabla \underline{v}_1 \, \mathrm{d}S + \int_\Gamma \frac{1}{2} \phi^2 G_n \, \mathrm{d}S, \tag{13}$$

the solution satisfies the inequality

$$0 \leq \frac{\partial \mathcal{F}}{\partial \underline{v}} (\underline{v} - \underline{v}^*) + \frac{\partial \mathcal{F}}{\partial \phi} (\beta - \phi), \tag{14}$$

amoung the set $\mathcal{K}.\mathcal{A}$ of admissible fields $(\underline{v}^*, \phi^*)$:

$$\mathcal{K}.\mathcal{A} = \Big\{ (\underline{v}, \phi) / \, \underline{v} = \underline{v}^d \text{ over } \partial\Omega_\mathrm{u}, \mathcal{D}_\phi \underline{u} = 0, \text{ on } \Gamma, \; \phi \in \mathcal{K} \Big\}. \tag{15}$$

Some typical examples are presented in Pradeilles-Duval and Stolz, 1995.

Stability and bifurcation. Consider the velocity $\underline{v}$ solution of the rate boundary value problem for any given velocity ϕ. This field $\underline{v}$ is solution of a classical problem of heterogeneous elasticity with non classical boundary conditions on Γ : $\mathcal{D}_\phi([\boldsymbol{\sigma}]_\Gamma . \underline{\nu}) = 0, \;\; \mathcal{D}_\phi [\underline{u}]_\Gamma = 0$. Consider the value W of $\mathcal{F}$ for this solution $\underline{v}(\phi, \, \underline{v}^d, \dot{\underline{\mathrm{T}}}^d)$

$$W(\phi, \, \underline{v}^d, \dot{\underline{\mathrm{T}}}^d) = \mathcal{F}(\underline{v}(\phi, \, \underline{v}^d, \dot{\underline{\mathrm{T}}}^d), \phi, \dot{\underline{\mathrm{T}}}^d). \tag{16}$$

The stability of the actual state is determined by the condition of the existence of a solution

$$\delta\phi \frac{\partial^2 W}{\partial\phi\partial\phi} \delta\phi \geq 0, \; \delta\phi \in \mathcal{K} - \{0\}, \tag{17}$$

and the uniqueness and non bifurcation is characterized by

$$\delta\phi \frac{\partial^2 W}{\partial\phi\partial\phi}\delta\phi \geq 0,\ \delta\phi \in \{\beta/\beta = 0 \text{ over } \Gamma - \Gamma^+\} - \{0\}. \tag{18}$$

3. DELAMINATION OF LAMINATES

The change of mechanical characteristics along moving front is used for the study of the degradation of laminates. We consider that a sound laminate (domain 0) with known characteristics is transformed into two laminates (1 and 2), separated by the crack of delamination as shown is the figure 26.1. Each laminate is described by an homogeneous plate or by an homogeneous beam, whose middle surface is denoted by S. The sound part has suffix 0, the two others parts are designed by suffices 1, 2.

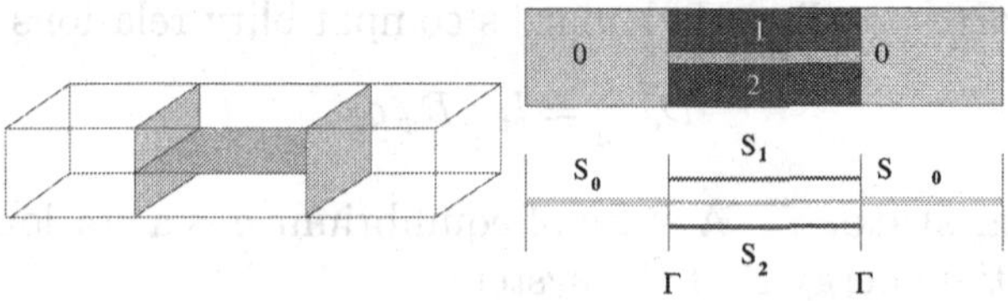

Figure 26.1. Modelization with beams.

The kinematic modeling of the beams or of the plates has a great influence on the behaviour of the delamination as well as the modeling of the continuity relations on the displacement along the delamination front which induces specific value for the energy release rate.

Kinematics of plates. A point of the middle surface S has curvilinear coordinates (x_1, x_2). The normal to this surface is denoted by $\underline{e}_3$. The normal coordinate is x_3. The displacement of the point (x_1, x_2, x_3) is defined by :

$$\underline{\xi} = \underline{u}(x_1, x_2) + w(x_1, x_2)\underline{e}_3 + \underline{\theta}(x_1, x_2)x_3, \quad (x_1, x_2) \in S. \tag{19}$$

where $\underline{u}$ is the plane displacement, w is the normal displacement and θ is the local rotation of the normal vector to the middle surface. With these fields, the strain inside the plate has the following form :

$$\boldsymbol{\varepsilon}(\underline{\xi}) = \boldsymbol{\varepsilon}(\underline{u}) + \frac{1}{2}(\underline{e}_3 \otimes \underline{\gamma} + \underline{\gamma} \otimes \underline{e}_3) - x_3\boldsymbol{\kappa}. \tag{20}$$

The distortion γ is defined by $\underline{\gamma} = \nabla w - \underline{\theta}$; the local rotation $\underline{\theta}$ gives $\boldsymbol{\kappa} = \frac{1}{2}(\nabla\underline{\theta} + \nabla^T\underline{\theta})$, and the membrane strain $\boldsymbol{\varepsilon}(\underline{u})$ satisfies $2\boldsymbol{\varepsilon}(\underline{u}) = (\nabla\underline{u} + \nabla^T\underline{u})$. The free energy of the plate is choosen naturally as a function of the generalized strains : $W = W(\boldsymbol{\varepsilon}(\underline{u}), \boldsymbol{\kappa}, \underline{\gamma})$.

The assemblage of plates. Consider the generalized parameters $\underline{q}_i = (\underline{u}_i, w_i, \underline{\theta}_i)$ and the generalized strains $\nabla \underline{q}_i = (\nabla \underline{u}_i, \nabla w_i, \nabla \underline{\theta}_i)$. We obtain the generalized stresses derived from the free energy W_i:

$$W_i(\varepsilon(\underline{u}_i), \boldsymbol{\kappa}_i, \gamma_i) = F_i(\underline{q}_i, \nabla \underline{q}_i), \quad \boldsymbol{\sigma}_i = \frac{\partial F_i}{\partial \nabla \underline{q}_i}, \quad \underline{T}_i = \frac{\partial F_i}{\partial \underline{q}_i}. \tag{21}$$

Along the front, the section of the sound plate ($i = 0$) imposes its motion to the two others plates ($i = 1, 2$) :

$$\underline{u}_i = \underline{u}_o + h_i \underline{\theta}_o, \quad w_i = w_o, \quad \underline{\theta}_i = \underline{\theta}_o. \tag{22}$$

These continuity conditions are easily rewritten with the set of generalized parameters :

$$\underline{q}_i = \boldsymbol{l}_i . \underline{q}_o, \quad \underline{x} \in \Gamma. \tag{23}$$

Hence the corresponding Hadamard's compatiblity relations are

$$\mathcal{D}_\phi \, \underline{q}_i = \boldsymbol{l}_i : \mathcal{D}_\phi \, \underline{q}_o. \tag{24}$$

Equilibrium state. A state of equilibrium $\underline{q}$ is a stationnary value of the potential energy $\mathcal{E}$ of the system:

$$\mathcal{E}(\tilde{\underline{q}}, \underline{\mathrm{T}}^d) = \sum_{i=0}^{2} \int_{S_i} F_i(\underline{q}, \nabla \underline{q}) \, \mathrm{d}S - \int_{\partial S_T} \underline{\mathrm{T}}^d . \underline{q} \, \mathrm{d}s. \tag{25}$$

The variation of $\mathcal{E}$ amoung the set of kinematically admissible fields

$$\mathcal{K}.\mathcal{A}. = \{ \underline{q} / \, \underline{q} = \underline{q}^d \text{ over } \partial S_q, \, \underline{q}_i = \boldsymbol{l}_i . \underline{q}_o . \text{ over } \Gamma \},$$

gives rize to the equilibrium equations :

$$0 = \frac{\partial F_i}{\partial \underline{q}_i} - \mathrm{div} \boldsymbol{\sigma}_i = \underline{T}_i - \mathrm{div} \boldsymbol{\sigma}_i, \text{ on } S_i, \tag{26}$$

$$0 = \underline{\nu} . (-\sum_{i=1}^{2} \boldsymbol{\sigma}_i . \boldsymbol{l}_i + \boldsymbol{\sigma}_o) = \underline{\nu} . [|\boldsymbol{\sigma}|]_\Gamma, \text{along } \Gamma, \tag{27}$$

$$\underline{\mathrm{T}}^d = \boldsymbol{\sigma}_o . \underline{n}, \text{ along } \partial S_T. \tag{28}$$

where the tractions $\underline{\mathrm{T}}^d$ are imposed on ∂S_T complementary part of ∂S_q. We have introduced the usefull notation $[|f|]_\Gamma = f_o - \sum_{i=1}^{2} f_i l_i$.

Analysis of the dissipation. The dissipation is given by the balance of the power of external loading and the reversible stored energy :

$$\mathrm{D} = \mathcal{P}_e - \frac{\mathrm{d}}{\mathrm{d}t} \sum_{i=0}^{2} \int_{S_i} F_i \, \mathrm{d}S = \int_\Gamma \mathcal{G} \phi \, \mathrm{d}S = -\frac{\partial \mathcal{E}}{\partial \Gamma} . \dot{\Gamma} \geq 0, \tag{29}$$

where the thermodynamical force $\mathcal{G}$ is the field :

$$\mathcal{G}(s) = -\frac{\partial \mathcal{E}}{\partial \Gamma}(s) = F_o - \underline{\nu}.\sigma_o.\nabla \underline{q}_o.\underline{\nu} - \sum_{i=1}^{2}(F_i - \underline{\nu}.\sigma_i.\nabla \underline{q}_i.\underline{\nu}). \quad (30)$$

Propagation law. The propagation law is defined as before :

$$\mathcal{G}(s,t) < G_c, \quad \phi = 0 \text{ and } \quad \mathcal{G}(s,t) = G_c, \quad \phi \geq 0. \quad (31)$$

The propagation is then possible when the critical value is reached. The velocity ϕ is included in the set of admissible propagation :

$$\mathcal{K} = \{\phi^* / \phi^*(s) \geq 0, \mathcal{G}(s) = G_c, \phi^* = 0, \text{otherwise}\}.$$

The rate boundary value problem. The rate boundary value problem is written in terms of rate of displacement. The solution $(\dot{\underline{q}}, \phi)$ of the rate boundary value satisfies the set of local equations obtained by the derivation of the constitutive law (21) and of the conservation of the momentum (26) with respect to time, the continuity relations along the front (24), the continuity relations on the stress vector :

$$\mathcal{D}_\phi(\underline{\nu}.[|\sigma|]_\Gamma) = \underline{\nu}.[|\dot{\sigma}|]_\Gamma - \mathrm{div}_\Gamma(\phi[|\sigma|]_\Gamma) + \phi[|\underline{T}|]_\Gamma = 0, \quad (32)$$

the propagation law:

$$(\phi - \phi^*)\,\mathcal{D}_\phi \mathcal{G} \geq 0, \quad \phi \in \mathcal{K}, \quad \forall \phi^* \in \mathcal{K}, \quad (33)$$

and the boundary conditions: $\dot{\underline{q}} = \dot{\underline{q}}^d$ over ∂S_q and $\dot{\sigma}.\underline{n} = \dot{\underline{\mathrm{T}}}^d$ over ∂S_T. The effective expression of the consistency condition gives a relation between the rate of the displacement and the velocity of propagation. The variation of $\mathcal{G}$ is given by :

$$\mathcal{D}_\phi \mathcal{G} = [|\sigma|]_\Gamma : \nabla \dot{\underline{q}}_o + [|\underline{T}|]_\Gamma.\dot{\underline{q}}_o + \sum_{i=1}^{2} \dot{\sigma}_i : (\nabla \underline{q}_i - l_i.\nabla \underline{q}_o) - \phi G_n,$$

$$G_n = -[|\sigma|]_\Gamma : \nabla\nabla \underline{q}_o.\underline{\nu} - [|\underline{T}|]_\Gamma.\nabla \underline{q}_o.\underline{\nu} + \sum_{i=1}^{2}(\nabla \sigma_i.\underline{\nu}) : (\nabla \underline{q}_i - l_i.\nabla \underline{q}_o).$$

Introducing now the potential $\mathcal{F}(\dot{\underline{q}}, \phi, \dot{\underline{\mathrm{T}}}^d)$:

$$\mathcal{F} = \sum_{i=0}^{2} \int_{S_i} \frac{1}{2}(\dot{\underline{q}}_i.\frac{\partial^2 F_i}{\partial \underline{q}_i \partial \underline{q}_i}.\dot{\underline{q}}_i + \dot{\underline{q}}_i.\frac{\partial^2 F_i}{\partial \underline{q}_i \partial \nabla \underline{q}_i}.\nabla \dot{\underline{q}}_i$$
$$+ \nabla \dot{\underline{q}}_i.\frac{\partial^2 F_i}{\partial \nabla \underline{q}_i \partial \underline{q}_i}.\dot{\underline{q}}_i + \nabla \dot{\underline{q}}_i.\frac{\partial^2 F_i}{\partial \nabla \underline{q}_i \partial \nabla \underline{q}_i}.\nabla \dot{\underline{q}}_i)\,\mathrm{d}S$$
$$- \int_\Gamma \left([|\sigma|]_\Gamma : \nabla \dot{\underline{q}}_o + [|\underline{T}|]_\Gamma.\dot{\underline{q}}_o\right)\phi\,\mathrm{d}S + \int_\Gamma \frac{\phi^2}{2} G_n\,\mathrm{d}S - \int_{\partial S_T} \dot{\underline{\mathrm{T}}}^d.\dot{\underline{q}}\,\mathrm{d}s,$$

the solution of the rate boundary value problem $(\tilde{\dot{\underline{q}}}, \tilde{\phi}) \in \mathcal{K}.\mathcal{A}$ is a solution of the variational inequality:

$$\frac{\partial \mathcal{F}}{\partial \tilde{\dot{\underline{q}}}}.(\tilde{\dot{\underline{q}}} - \tilde{\dot{\underline{q}}}^*) + \frac{\partial \mathcal{F}}{\partial \tilde{\phi}}.(\tilde{\phi} - \tilde{\phi}^*) \geq 0, \tag{34}$$

amongs the set $(\tilde{\dot{q}}^*, \tilde{\phi}^*) \in \mathcal{K}.\mathcal{A}$.

$$\begin{aligned} \mathcal{K}.\mathcal{A}. \;=\; & \Big\{(\tilde{\dot{q}}, \tilde{\phi}) / \; \dot{\underline{q}}_i + \phi \nabla \, \underline{q}_i.\underline{\nu} = l_i.(\dot{\underline{q}}_o + \phi \nabla \, \underline{q}_o.\underline{\nu}), \text{ over } \Gamma, \\ & \underline{q}_o = \underline{q}^d \text{ over } \partial S_q, \quad \phi \in \mathcal{K}\Big\}. \end{aligned}$$

This framework can be extended to different models of beams or plates and to dynamics using of kinetic energy and hamiltonian formalism, (Stolz, 1995, 2000).

References

Abeyratne, R., Knowles, J., On the driving traction acting on a surface of strain discontinuity in a continuum, *J. Mech. Phys. Solids*, **38**, 3, pp. 345-360, 1990.

Gurtin, M. E., The nature of configurational forces, *Arch. Rat. Mech. Anal.*, **131**, pp. 67-100, 1995.

Maugin, G. A., Material forces : concept and applications, *ASME, Appl. Mech. Rev.*, **48**, pp. 213-245, 1995.

Pradeilles-Duval, R.M., Stolz, C., Mechanical transformations and discontinuities along a moving surface, *J. Mech. Phys. Solids*, **43**, 1, pp. 91-121, 1995.

Stolz, C., Functional approach in non linear dynamics, *Arch. Mech*, **47**, 3, 421-435, 1995.

Stolz, C., Thermodynamical description of running discontinuities, in *Continuum Thermodynamics*, pp 410-412, G.A. Maugin et al. (eds), Kluwer Academic Publishers, 2000.

Truskinovsky, L. M., Dynamics of non equilibrium phases boundaries in a heat conducting non linearly elastic medium, *P.M.M*, **51**, pp. 777-784, 1987.

Chapter 27

ON FRACTURE MODELLING BASED ON INVERSE STRONG DISCONTINUITIES

Ragnar Larsson and Martin Fagerström
Department of Applied Mechanics
Chalmers University of Technology, Sweden
ragnar.larsson@me.chalmers.se

1. INTRODUCTION

The concept of material forces is widely used in the fracture mechanics area due to the close connection between the material forces, represented by the Eshelby stress tensor (or energy momentum tensor) [1], and material inhomogeneities and defects. Traditionally, the use of material forces within fracture mechanics has mostly concerned Linear Elastic Fracture Mechanics (LEFM). For a survey of the continuum mechanical framework, see [2] and [3] considering the numerical aspects.

The goal of the present paper is to formulate the momentum balance such that concepts of LEFM and Non-Linear Fracture Mechanics (NLFM) can be unified in a natural way and solved within the X-FEM concept. It turns out that we arrive at a Lagrangian/Eulerian description of the fracture mechanical problem, where the inverse deformation involving the discontinuity is related to the Eulerian frame. As to the separation of the deformation continuous and discontinuous portions, it is noted that X-FEM, originally introduced by Belytschko and Black [4] and Moës et.al. [5], has advantages in the possibility to introduce strong discontinuities by enrichment near the crack tip. In practice, this is done by considering the continuous displacement and the discontinuous counterpart (with additional enrichment in the vicinity of the crack tip) as two independent fields, superimposed to give the total displacements.

2. STRONG DISCONTINUITY CONCEPT

The objective of the present section is to define the kinematics of the strong discontinuity in relation to a consequent inverse discontinuity.

Let us first consider the formulation in terms of the direct deformation map defined as

$$\varphi(\boldsymbol{X},t) = \varphi_c(\boldsymbol{X},t) + H_S(S(\boldsymbol{X}))\boldsymbol{d}(\boldsymbol{X},t) \text{ with } \boldsymbol{d} = \boldsymbol{x} - \boldsymbol{x}_c \tag{1}$$

where the total deformation map involves a continuous portion φ_c and a discontinuous portion $\boldsymbol{d}$ as shown in Figure 27.1. The Heaviside function, H_S is considered centered at the internal discontinuity boundary Γ_S, subdividing the solid into a minus side B_0^- and a plus side B_0^+.

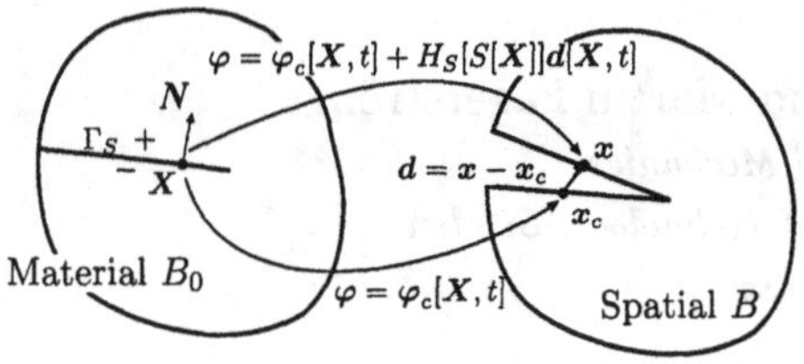

Figure 27.1. Total direct deformation map consisting of one continuous and one discontinuous portion

The pertinent discontinuous deformation gradient becomes

$$\boldsymbol{F} = \varphi \otimes \boldsymbol{\nabla}_X = \boldsymbol{F}_c + H_S \boldsymbol{F}_d + \delta_S\, \boldsymbol{d} \otimes \boldsymbol{N} \tag{2}$$

where $\boldsymbol{F}_c = \varphi_c \otimes \boldsymbol{\nabla}_X$ and $\boldsymbol{F}_d = \boldsymbol{d} \otimes \boldsymbol{\nabla}_X$ and $\boldsymbol{N}$ is the outward normal vector of Γ_S (pointing from minus to plus side) and δ_S is the Dirac delta function.

Let us next consider the strong discontinuity in terms of the inverse deformation maps. We then consider a material point $\boldsymbol{X}$, which due to separation of the material is specified in two separate deformation maps representing total and continuous placement of the component defined as

$$\boldsymbol{X} = \boldsymbol{\phi}_c(\boldsymbol{x}_c, t) = \boldsymbol{\phi}(\boldsymbol{x}, t) \text{ with } \boldsymbol{\phi}_c = \varphi_c^{-1} \text{ and } \boldsymbol{\phi} = \varphi^{-1}. \tag{3}$$

It is also of significant interest to consider the relation between material and spatial velocities due to the kinematic consideration. To this end, we introduce the direct velocities $\boldsymbol{v} = \dot{\varphi} = \boldsymbol{v}_c + \dot{\boldsymbol{d}}$ and $\boldsymbol{v}_c = \dot{\varphi}_c$, and use (3) to establish the relations

$$\dot{\boldsymbol{X}} = \boldsymbol{f}_c \cdot \boldsymbol{v}_c + \boldsymbol{V}_c = \boldsymbol{0};\ \dot{\boldsymbol{X}} = \boldsymbol{f} \cdot (\dot{\varphi}_c + H_S \dot{\boldsymbol{d}}) + \boldsymbol{V} = \boldsymbol{0} \tag{4}$$

with

$$\boldsymbol{f}_c = \boldsymbol{F}_c^{-1} \circ \boldsymbol{\phi}_c(\boldsymbol{x}_c);\ \boldsymbol{f} = \boldsymbol{F}^{-1} \circ \boldsymbol{\phi}(\boldsymbol{x});\ \boldsymbol{V}_c = \frac{\partial \boldsymbol{\phi}_c}{\partial t};\ \boldsymbol{V} = \frac{\partial \boldsymbol{\phi}}{\partial t} \tag{5}$$

where $\circ$ denotes composition. It may be remarked that the inverse deformation gradients, i.e. $\boldsymbol{f}$ and $\boldsymbol{f}_c$, are the deformation gradients considering the current (fixed) spatial configuration B as reference for the material configuration B_0. Likewise, the material velocities $\boldsymbol{V}$ and $\boldsymbol{V}_c$ are measured relative to the current (fixed) spatial configurations B and B_0, respectively. Note that $\dot{\bullet}$ denotes the time derivative with respect to fixed material configuration whereas $\bullet_v$ corresponds to the counterpart with fixed spatial configuration.

Upon combining the equations in (4) we arrive at

$$H_S \boldsymbol{f} \cdot \dot{\boldsymbol{d}} = - [\![\boldsymbol{f}]\!] \cdot \boldsymbol{v}_c - \boldsymbol{D}_v \tag{6}$$

where we have introduced the rate of material jump $\boldsymbol{D}_v = \boldsymbol{V} - \boldsymbol{V}_c$ and the jump $[\![\boldsymbol{f}]\!] = \boldsymbol{f} - \boldsymbol{f}_c$ of the inverse deformation gradient. On the basis of $\boldsymbol{D}_v$ relating the material jump with respect to a fixed B, we define the total material discontinuity from the temporal integration

$$\boldsymbol{D}(\boldsymbol{X}, t) = \int_0^t \boldsymbol{D}_v \, dt. \tag{7}$$

In the following we shall associate "variations" with the material rate (relative to the material configuration B_0) denoted by Δ, whereas variation associated with changes relative to the spatial configuration is denoted by δ. Hence, the relation (6) relating the direct and inverse discontinuities may be expressed in the material and spatial variations as

$$H_S \boldsymbol{f} \cdot \Delta \boldsymbol{d} = - [\![\boldsymbol{f}]\!] \cdot \Delta \boldsymbol{\varphi}_c - \delta \boldsymbol{D}. \tag{8}$$

3. WEAK FORM, DIRECT/INVERSE DEFORMATION

For the deformation gradient pertinent to the direct deformation map in (1), it is possible to establish the weak form of the momentum balance in two separate problems, as discussed in e.g. [6], written in the first Piola-Kirchhoff stress $\boldsymbol{\Sigma}_1^t$, and the corresponding nominal traction vector $\boldsymbol{t}_1 = \boldsymbol{\Sigma}_1^t \cdot \boldsymbol{N}$, and the (mechanical) body force $\boldsymbol{b}^{mec}$ as

$$\int_{B_0} \boldsymbol{\Sigma}_1^t : \Delta \boldsymbol{F}_c \, dV = \int_{\Gamma_0} \Delta \boldsymbol{\varphi}_c \cdot \boldsymbol{t}_1 d\Gamma + \int_{B_0} \Delta \boldsymbol{\varphi}_c \cdot \boldsymbol{b}^{mec} dV \tag{9a}$$

$$\int_{B_0} H_S \boldsymbol{\Sigma}_1^t : \Delta \boldsymbol{F}_d \, dV + \int_{\Gamma_S} \Delta \boldsymbol{d} \cdot \boldsymbol{t}_1 \, d\Gamma = \int_{B_0} H_S \, \Delta \boldsymbol{d} \cdot \boldsymbol{b}^{mec} dV \tag{9b}$$

where the direct continuous deformation map $\boldsymbol{\varphi}_c$ and the direct discontinuity $\boldsymbol{d}$ are considered as two independent fields.

Since the object is to rewrite (9ab) in terms of the inverse deformation map, we first introduce the transformation $\boldsymbol{\Sigma}_1^t = \boldsymbol{f}^t \cdot \boldsymbol{T}$ between the first Piola-Kirchhoff stress and the Mandel stress tensor $\boldsymbol{T}$. Moreover, we also introduce the traction vector $\boldsymbol{Q} = -\boldsymbol{T} \cdot \boldsymbol{N}$, kinematically associated with the material discontinuity $\delta\boldsymbol{D}$. This yields

$$\begin{aligned}&\int_{B_0}\boldsymbol{\Sigma}_1^t : \Delta\boldsymbol{F}_c dV - \int_{B_0}(H_S\boldsymbol{T} : (\boldsymbol{f}_c \cdot (\boldsymbol{F}_c \otimes \boldsymbol{\nabla}_X) \cdot \boldsymbol{f}_c \cdot \Delta\boldsymbol{\varphi}_c))\, dV + \\ &\int_{B_0}(H_S\boldsymbol{T} : (\boldsymbol{f} \cdot (\boldsymbol{F} \otimes \boldsymbol{\nabla}_X) \cdot \boldsymbol{f}_c \cdot \Delta\boldsymbol{\varphi}_c))\, dV + \int_{\Gamma_S}\Delta\boldsymbol{\varphi}_c \cdot [\![\boldsymbol{f}^t]\!] \cdot \boldsymbol{Q}\, d\Gamma - \\ &\int_{B_0}H_S([\![\boldsymbol{f}^t]\!] \cdot \boldsymbol{T}) : \Delta\boldsymbol{F}_c dV = \\ &\int_{\Gamma_0}\Delta\boldsymbol{\varphi}_c \cdot \boldsymbol{t}_1 d\Gamma + \int_{B_0}\Delta\boldsymbol{\varphi}_c \cdot \boldsymbol{b}^{mec} dV + \int_{B_0}\Delta\boldsymbol{\varphi}_c \cdot [\![\boldsymbol{f}^t]\!] \cdot \boldsymbol{B}^{mec} dV\end{aligned} \tag{10a}$$

$$\begin{aligned}&-\int_{B_0}H_S\boldsymbol{T} : \delta\boldsymbol{L}_d\, dV + \int_{\Gamma_S}\delta\boldsymbol{D} \cdot \boldsymbol{Q} d\Gamma - \int_{B_0}H_S\boldsymbol{\Sigma}_1^t : (\boldsymbol{F} \otimes \boldsymbol{\nabla}_X) \cdot \delta\boldsymbol{D}\, dV \\ &= \int_{B_0}H_S\, \delta\boldsymbol{D} \cdot \boldsymbol{B}^{mec} dV\end{aligned} \tag{10b}$$

respectively, introducing the material body force $\boldsymbol{B}^{mec} = -\boldsymbol{F}^t \cdot \boldsymbol{b}^{mec}$.

Now, realizing that $[\![\boldsymbol{f}]\!]$ is only defined on a small subregion D_0^+ within B_0 with $[\![\boldsymbol{f}]\!] = \boldsymbol{0}$ on parts of the boundary, shown in Figure 27.2, and using the divergence theorem for D_0^+ results in a simplified continuous equation. We make the following substitution:

$$\int_{\Gamma_S}H_S\Delta\boldsymbol{\varphi}_c \cdot [\![\boldsymbol{f}^t]\!] \cdot \boldsymbol{Q}\, d\Gamma = \int_{D_0^+}H_S(\Delta\boldsymbol{\varphi}_c \cdot [\![\boldsymbol{f}^t]\!] \cdot \boldsymbol{T}) \cdot \boldsymbol{\nabla}_X. \tag{11}$$

The rest of the boundary terms vanish either due to zero jump in inverse deformation gradient or due to zero traction on a free surface. It turns out that the divergence term, i.e the RHS, in (11) cancels all terms in (10a) containing a discontinuity in inverse deformation gradient which finally leads to the continuous formulation

$$\int_{B_0}\boldsymbol{\Sigma}_1^t : \Delta\boldsymbol{F}_c dV = \int_{\Gamma_0}\Delta\boldsymbol{\varphi}_c \cdot \boldsymbol{t}_1 d\Gamma + \int_{B_0}\Delta\boldsymbol{\varphi}_c \cdot \boldsymbol{b}^{mec} dV. \tag{12}$$

4. WEAK FORM EXPRESSED IN ESHELBY STRESS TENSOR

It is of significant interest to advance the discontinuous weak form (10b) to include also the Eshelby stress tensor $\boldsymbol{M}^t = \rho_0\,\psi\boldsymbol{1} - \boldsymbol{T}$ with the corresponding traction reformulated as $\boldsymbol{Q} = \boldsymbol{M}^t \cdot \boldsymbol{N} - \rho_0\psi\boldsymbol{N}$. We remark that the Eshelby stress is the balance stress tensor in the pseudo-momentum relation for the inverse deformation problem, as discussed in e.g. refs. [2] and [3].

Let us consider the variation of the inverse discontinuous deformation $\delta\boldsymbol{D}$ defined on a subdomain D_0 of B_0 with the internal crack boundary Γ_S, as shown in Figure 27.2. Based on the subdivision of the subdomain D_0 in Figure 27.2 we now apply the divergence theorem to D_0^+ as

$$\begin{aligned}&-\int_{\Gamma_S}\rho_0\,\psi\,\delta\boldsymbol{D} \cdot \boldsymbol{N} d\Gamma + \int_{\partial D_0^+}\rho_0\,\psi\,\delta\boldsymbol{D} \cdot \boldsymbol{N}\, d\Gamma = \\ &\int_{D_0}H_S\,\rho_0\,\psi\boldsymbol{1} : \boldsymbol{L}_d\, dV + \int_{D_0}H_S\,\delta\boldsymbol{D} \cdot \boldsymbol{\nabla}_X(\rho_0\,\psi)\, dV.\end{aligned} \tag{13}$$

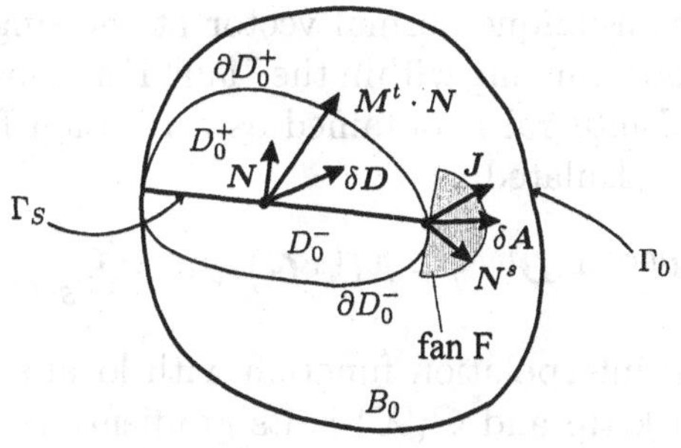

Figure 27.2. The material discontinuity defined only on a subdomain D_0 of B_0. Also note the reaction force $\boldsymbol{J}$ at the crack tip in the discontinuous problem with corresponding virtual crack extension $\delta\boldsymbol{A}$.

Moreover, upon assuming Dirichlet boundary conditions for $\delta\boldsymbol{D}$ along ∂D_0^+ and combining (10b) and (13) we obtain the discontinuous problem in terms of the Eshelby stress (leaving the continuous problem unchanged) as

$$\begin{aligned}&\int_{D_0} H_S \boldsymbol{M}^t : \delta\boldsymbol{L}_d \, dV + \int_{\Gamma_S} \delta\boldsymbol{D} \cdot \boldsymbol{M}^t \cdot \boldsymbol{N} d\Gamma = \\ &\int_{D_0} H_S \, \delta\boldsymbol{D} \cdot (\boldsymbol{B}^{mec} + \boldsymbol{B}^{inh}) \, dV\end{aligned} \tag{14}$$

with $\boldsymbol{B}^{inh} = -\boldsymbol{\nabla}_X(\rho_0\psi) + \boldsymbol{\Sigma}_1^t : (\boldsymbol{F} \otimes \boldsymbol{\nabla}_X) = -\left.\frac{\partial(\rho_o\psi)}{\partial \boldsymbol{X}}\right|_{expl}$.

5. LEFM: J-INTEGRAL APPROACH

The regular part of the boundary is now considered as a free surface with orientation $\boldsymbol{N}$ (rather than a cohesive zone) along which the physical stress is zero ($\boldsymbol{Q} = -\boldsymbol{F}^t \cdot \boldsymbol{t}_1 = \boldsymbol{0}$), whereby the traction vector pertinent to the Eshelby stress along Γ_S becomes $\boldsymbol{M}^t \cdot \boldsymbol{N} = \boldsymbol{Q} + \rho_0 \, \psi \boldsymbol{N} = \rho_0 \, \psi \boldsymbol{N}$.

We are now in position to formulate the J-integral approach to elastic fracture based on the discontinuous problem (14). Upon subdividing Γ_S into a regular part Γ_{S^r} and a singular part Γ_{S^s}, we may rewrite the boundary term as

$$\int_{\Gamma_S} \rho_0 \, \psi \delta\boldsymbol{D} \cdot \boldsymbol{N} d\Gamma = \int_{\Gamma_{S^r}} \rho_0 \, \psi \delta\boldsymbol{D} \cdot \boldsymbol{N} d\Gamma - \delta\boldsymbol{A} \cdot \boldsymbol{J} \tag{15}$$

where we included also the crack tip extension $\delta\boldsymbol{A} = -\delta\boldsymbol{D}(\boldsymbol{X}_A)$, even though we have a Dirichlet boundary condition at the crack tip. It appears that the corresponding force variable $\boldsymbol{J}$ to $\delta\boldsymbol{A}$ is the vectorial form of the J-integral, first introduced by Rice [7], defined as

$$\boldsymbol{J} = \lim_{\Gamma_{S^s} \to 0} \int_{\Gamma_{S^s}} \rho_0 \, \psi \boldsymbol{N}^s \, d\Gamma \tag{16}$$

where N^s is the non-unique normal vector at the singular point of the discontinuity surface, ranging within the "fan" F as shown in Figure 27.2.

Evidently, the J-integral is obtained as a reaction force at the crack tip, which can be calculated as

$$J = \int_{D_0} \left(N(\boldsymbol{B}^{mec} + \boldsymbol{B}^{inh}) - \boldsymbol{M}^t \cdot \boldsymbol{G} \right) dV - \int_{\Gamma_{Sr}} \rho_0 \, \psi N \boldsymbol{N} d\Gamma. \quad (17)$$

where $N(\boldsymbol{X})$ is an interpolation function with local support D_{0l} in the vicinity of the crack tip and $\boldsymbol{G}(\boldsymbol{X})$ is its gradient. Note that using the fact that the normal vector $\boldsymbol{N}^s$ varies within the fan F, the Heaviside function H_S can be removed within the integrands in (17).

6. NLFM: COHESIVE ZONE MODEL

In the case of NLFM we focus on the specification of a constitutive interface model describing the successive degradation of the cohesive stresses along Γ_S as a function of the developed discontinuity. We propose a constitutive relation formulated in the crack closing traction $\boldsymbol{Q} = -\boldsymbol{T} \cdot \boldsymbol{N}$, as a function of the material inverse discontinuity $\boldsymbol{D}$, i.e. $\boldsymbol{Q}(\boldsymbol{D})$. Pertinent to a fracture process the idea is to consider the relaxation of $\boldsymbol{Q}(\boldsymbol{D})$ as a damage-plasticity process within the interface, formulated in the spirit of the developments in [8]. The nominal traction $\boldsymbol{Q}$ is then defined in terms of an effective traction vector

$$\hat{\boldsymbol{Q}} = (1-\alpha)\hat{\boldsymbol{Q}} \text{ with } \hat{\boldsymbol{Q}} = \boldsymbol{K} \cdot \boldsymbol{D}^e = \boldsymbol{K} \cdot (\boldsymbol{D} - \boldsymbol{D}^p) \quad (18)$$

where $0 \leq \alpha \leq 1$ is the damage variable and $\boldsymbol{D}$ is the total (crack closing) material discontinuity. A simple choice is $\boldsymbol{K} = K\boldsymbol{1}$ where K is an artificial stiffness (or penalty) parameter of the interface. Next we assume an evolution equation for the damage development related to a plastic multiplier λ_v defined so that $\alpha_v = B\lambda_v$. Moreover, λ_v is the plastic multiplier that is controlled by the Karush-Kuhn-Tucker conditions $F \leq 0, \lambda_v \geq 0, F\lambda_v = 0$ where $F(\hat{\boldsymbol{Q}})$ is the condition for fracture loading and unloading corresponding to perfect plasticity.

As to the factor B in the evolution law for α, it is assumed that B is a monotonic function in the argument α so that $B(\alpha) = \frac{1}{S(1-\alpha)} > 0$ where S is the damage parameter. It appears that the damage parameter S can be calibrated for mode I fracture with fracture energy G_f^I as

$$G_f^I = \int_0^{-\infty} \boldsymbol{Q} \cdot \boldsymbol{N} dD_n^p = \int_0^1 2S(1-\alpha)\sigma_f d\alpha = S\sigma_f \Rightarrow S = \frac{G_f^I}{\sigma_f} \quad (19)$$

where σ_f is the failure stress in pure tension. Moreover, to arrive at the last equality in (19) the evolution of the material jump is defined in terms of a damage-plasticity potential $G(\hat{Q})$ as

$$D_v^p = \lambda_v \frac{\partial G}{\partial Q} = \frac{\lambda_v}{1-\alpha} \frac{\partial G}{\partial \hat{Q}}. \tag{20}$$

7. NUMERICAL EXAMPLE

In order to illustrate the relation between LEFM and NLFM, we consider a *Double Cantilever Beam* (DCB) test, cf. Figure 27.3, and compare the results from the J-integral approach in Section 5 and the results from the cohesive zone model approach in Section 6. The discretization is made using constant strain triangular elements representing both the direct displacement field and the inverse discontinuous field. As to the LEFM approach, the choice of local interpolation function when calculating the reaction force (vectorial J-integral) is a delicate matter. In this example, we used the ordinary interpolation function with support only in the crack tip node for the representation of $N(\boldsymbol{X})$ (similar to the material force method), which requires a very fine discretization (see Figure) in order to accurately resolve the singularity at the crack tip, cf. [9]. Hence, remeshing has to be performed as the crack propagates through the material. The Griffith's criterion is used to signal fracture at the crack tip.

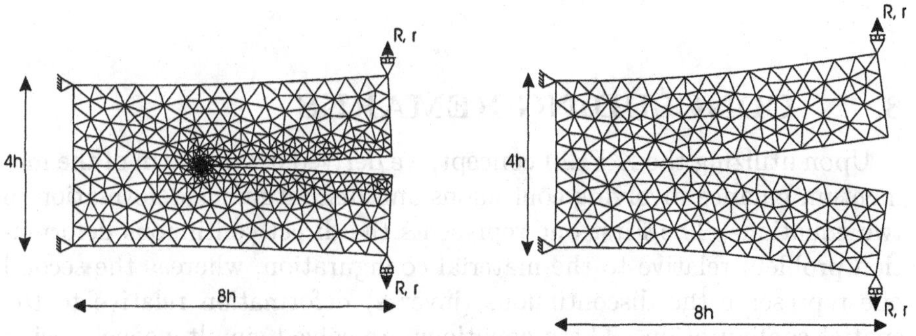

Figure 27.3. Deformed double cantilever beams for J-integral approach (left) and cohesive zone approach (right). The reaction forces (R) and the corresponding node displacements (r) are indicated with arrows.

The analysis considers plane strain conditions and a neo-Hookean hyperelastic model for the continuum response, corresponding to Young's modulus $E = 4$ GPa and Poisson's ratio $\nu = 0.3$. The fracture parameters are taken as $\sigma_f = 40$ MPa, $G_f^I = 1500$ N/m, cohesive zone parameters $\gamma = 3$ and $\mu = 0$ (see [8]); The dimension parameter $h = 0.25$

mm and beam thickness $4h$. The resulting *reaction force - node displacement* curve is shown in Figure 27.4 with the interesting feature of LEFM and NLFM coincide when the crack opening displacement (or the crack length) has reached a certain value.

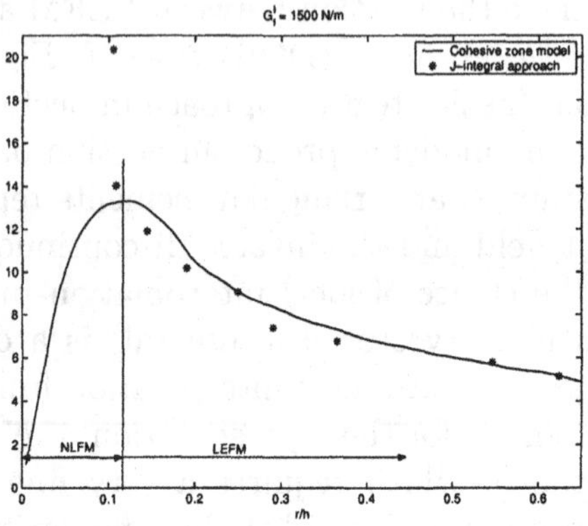

Figure 27.4. Comparison of J-integral approach (stars) and cohesive zone model (solid line) for a DCB-test. Note that LEFM and NLFM (nearly) coincide for $r/h \gtrsim 0.12$.

8. CONCLUDING REMARKS

Upon utilizing the X-FEM concept, we derived a weak form of the momentum balance involving continuous and discontinuous deformation in two equations; The First one represents the direct continuous deformation problem relative to the material configuration, whereas the second one represents the discontinuous (inverse) deformation relative to the spatial configuration. These equations are solved simultaneously using FE interpolation of both fields to obtain the total displacement in a fracture mechanics problem. The key issue in this development is the kinematical consideration of the direct discontinuity in relation to its material counterpart. An interesting feature of the formulation is that a link between LEFM and NLFM was established in terms of the extension of the cohesive zone. It turns out that LEFM is nothing but the special case of NLFM when the cohesive zone is confined entirely to the crack tip. This feature was also verified numerically in a DCB-test, where it appears that LEFM and NLFM coincide for crack lengths larger than a certain value.

References

[1] J.D. Eshelby. The energy-momentum tensor. *J. Elast.*, 5:321–335, 1975.

[2] P. Steinmann. Application of material forces to hyperelastostatic fracture mechanics. I. Continuum mechanical setting. *Int. J. Sol. Struct.*, 37:7371–7391, 2000.

[3] P. Steinmann, D. Ackermann, and F.J. Barth. Application of material forces to hyperelastostatic fracture mechanics. II. Computational setting. *Int. J. Sol. Struct.*, 38:5509–5526, 2001.

[4] T. Belytschko and T. Black. Elastic crack growth in finite elements with minimal remeshing. *Int. J. Num. Meth. Eng.*, 45:601–620, 1999.

[5] N. Moës, J. Dolbow, and T. Belytschko. A finite element method for crack growth without remeshing. *Int. J. Num. Meth. Eng.*, 46:131–150, 1999.

[6] G.N. Wells. *Discontinuous modelling of strain localisation and failure.* PhD thesis, Delft University of Technology, Faculty of Aerospace Engineering, 2001.

[7] J. R. Rice. A path independent integral and the approximate analysis of strain concentrations by notches and cracks. *J. Appl. Mech.*, 35:379–386, 1968.

[8] R. Larsson and N. Jansson. Geometrically non-linear damage interface based on regularized strong discontinuity. *Int. J. Num. Meth. Eng.*, 54:473–497, 2002.

[9] P. Heintz, F. Larsson, P. Hansbo, and K. Runesson. On error control and adaptivity for computing material forces in fracture mechanics. Proceed. WCCM V. Vienna University of Technology, Austria, 2002.

X

INTERFACES & PHASE TRANSITION

Chapter 28

MAXWELL'S RELATION FOR ISOTROPIC BODIES

Miroslav Šilhavý

Mathematical Institute of the AV ČR
Žitná 25
115 67 Prague 1
Czech Republic
Department of Mathematics
University of Pisa
Via F. Buonarroti, 2
56127 Pisa
Italy

Abstract The paper determines the forms of equations of force equilibrium and Maxwell's relation for stable coherent phase interfaces in isotropic two–dimensional solids. If any of the two principal stretches of the first phase differs from the two principal stretches of the second phase, one obtains the equality of two *generalized* scalar forces and of a *generalized* Gibbs function. The forms of these quantities depend on whether the two principal stretches both increase (decrease) when crossing the interface or whether one of the stretches increases and the other decreases. Apart from this nondegenerate case, also the degenerate cases are discussed. The proofs use the rank 1 convexity condition for isotropic materials.

Keywords: Phase transitions, equilibrium, interfaces

Mathematics Subject Classification 2000: 74N10, 49J45

1. INTRODUCTION AND RESULTS

Consider a two-dimensional nonlinear elastic body with a continuously differentiable stored energy $f : \mathbb{M}_+^{2\times 2} \to \mathbb{R}$, defined on the set $\mathbb{M}_+^{2\times 2}$ of 2×2 matrices with positive determinant, and with the Piola–Kirchhoff stress $\mathbf{S}(\mathbf{A}) := \partial f(\mathbf{A})/\partial \mathbf{A}$. If two (homogeneous) phases of deformation

gradients **A**, **B** form a stable state with an interface of referential normal **n** then they satisfy the following conditions (see, e.g., [7]):
(a) the geometrical compatibility (Hadamard's condition)

$$\mathbf{B} - \mathbf{A} = \mathbf{a} \otimes \mathbf{n}, \tag{1}$$

where **a** is some vector;
(b) the balance of forces and Maxwell's relation:

$$\mathbf{S}_A \mathbf{n} = \mathbf{S}_B \mathbf{n}, \quad f_B = f_A + \mathbf{S}_A \cdot (\mathbf{B} - \mathbf{A}); \tag{2}$$

(c) the energy f is rank 1 convex at **A** and **B**.

Here $\mathbf{S}_A, \mathbf{S}_B$ and f_A, f_B are the values of **S** and f at **A**, **B**. Recall that f is said to be rank 1 convex at $\mathbf{A} \in \mathbb{M}_+^{2\times2}$ if ([3], [4])

$$f(\mathbf{B}) \geq f(\mathbf{A}) + \mathbf{S}(\mathbf{A}) \cdot (\mathbf{B} - \mathbf{A}) \tag{3}$$

for every $\mathbf{B} \in \mathbb{M}_+^{2\times2}$ that is rank 1 connected to **A**, i.e., that satisfies $\det(\mathbf{A} - \mathbf{B}) = 0$. Note that the rank 1 convexity follows from stability considerations.

For fluids one can eliminate $\mathbf{A}, \mathbf{B}, \mathbf{S}_A, \mathbf{S}_B, \mathbf{n}$ from (2) to obtain scalar equations: the equality of pressures and Gibbs functions

$$p_\mathbf{A} = p_\mathbf{B}, \quad g_\mathbf{A} = g_\mathbf{B}, \tag{4}$$

where $g = f + pv$ and $v = \det \mathbf{F}$ the specific volume. The goal of this note is to examine the analogs of (4) for two–dimensional isotropic solids. Thus the question is whether one can pass from the tensorial equations (2) to scalar equations involving invariants of deformation.

The ***principal stretches*** (= singular values) $\alpha_1 \geq \alpha_2$ of a deformation gradient **A** are the eigenvalues of $\sqrt{\mathbf{A}\mathbf{A}^T}$; we write $\alpha = (\alpha_1, \alpha_2)$ for the ordered pair of principal stretches. The stored energy of an isotropic body can be expressed also in terms of the singular values: $f(\mathbf{A}) = \tilde{f}(\alpha)$ where $\tilde{f}$ is a continuously differentiable symmetric function defined on the open first quadrant in $\mathbb{R}^2$. We define the ***principal forces*** s_1, s_2 by $s_i(\alpha) = \partial\tilde{f}(\alpha)/\partial\alpha_i$. If **A** is diagonal, $\mathbf{A} = \operatorname{diag}(\alpha_1, \alpha_2)$, then $\mathbf{S} \equiv \mathbf{S}(\mathbf{A})$ is diagonal with $\mathbf{S} = \operatorname{diag}(s_1, s_2)$.

We say that two deformation gradients $\mathbf{A}, \mathbf{B}, \mathbf{A} \neq \mathbf{B}$, form a ***stable interface*** if they satisfy Conditions (a)–(c). If **A**, **B** form a stable interface we call the principal stretches α, β of **A**, **B** the ***data*** of the interface. We say that the data are ***nondegenerate*** if $\alpha_i \neq \beta_j$ for all $i, j \in \{1, 2\}$. Let

$$\epsilon \equiv \epsilon(\alpha, \beta) := \operatorname{sgn}((\beta_1 - \alpha_1)(\beta_2 - \alpha_2)).$$

The following is the analog of (4) for isotropic solids.

Theorem 1. *Let α, β be data of a stable interface. We have the following assertions:*
(i) *if the data are nondegenerate and $\epsilon = 1$ then the quantities*

$$\left.\begin{array}{l} k_+ := \dfrac{\alpha_1 s_1 - \alpha_2 s_2}{\alpha_1 - \alpha_2}, \quad c_+ := -\dfrac{s_1 - s_2}{\alpha_1 - \alpha_2}, \\ g_+ := \tilde{f} - k_+(\alpha_1 + \alpha_2) - c_+\alpha_1\alpha_2 \equiv \tilde{f} - \dfrac{\alpha_1^2 s_1 - \alpha_2^2 s_2}{\alpha_1 - \alpha_2} \end{array}\right\} \tag{5}$$

are equal on the two sides of the interface;
(ii) *if the data are nondegenerate and $\epsilon = -1$ then the quantities*

$$\left.\begin{array}{l} k_- := \dfrac{\alpha_1 s_1 - \alpha_2 s_2}{\alpha_1 + \alpha_2}, \quad c_- := \dfrac{s_1 + s_2}{\alpha_1 + \alpha_2}, \\ g_- := \tilde{f} - k_-(\alpha_1 - \alpha_2) - c_-\alpha_1\alpha_2 \equiv \tilde{f} - \dfrac{\alpha_1^2 s_1 + \alpha_2^2 s_2}{\alpha_1 + \alpha_2} \end{array}\right\} \tag{6}$$

are equal on the two sides of the interface;
(iii) *if $\alpha_i = \beta_j$ for some $i, j \in \{1, 2\}$ then*

$$\left.\begin{array}{c} s_{\bar{i}}(\alpha) = s_{\bar{j}}(\beta), \\ \tilde{f}(\alpha) - s_{\bar{i}}(\alpha)\alpha_{\bar{i}} = \tilde{f}(\beta) - s_{\bar{j}}(\beta)\beta_{\bar{j}}; \end{array}\right\} \tag{7}$$

here $\bar{i}, \bar{j}$ are the complementary indices, i.e., $\bar{i}, \bar{j} \in \{1, 2\}$ and $\bar{i} \neq i, \bar{j} \neq j$.

In the nondegenerate cases (i), (ii) the quantities $k_\pm, c_\pm$ play roles of generalized forces and $g_\pm$ the roles of generalized Gibbs functions. Which of the triples k_+, c_+, g_+ or k_-, c_-, g_- applies is determined by ϵ. Note that $\epsilon = 1$ occurs if the two principal stretches both increase (decrease) when crossing the interface while $\epsilon = -1$ occurs if one of the stretches increases and the other decreases when crossing the interface. In the degenerate case (iii) the conditions of equilibrium take a one-dimensional form (7). Conditions (i)–(iii) collapse to (4) for fluids. Indeed if $\tilde{f}(\alpha) = \varphi(\alpha_1\alpha_2)$ and $s_1 = -p\alpha_2, s_2 = -p\alpha_1$, where $p = -\varphi'$ is the pressure, then

$$k_\pm \equiv 0, \quad c_\pm = -p, \quad g_\pm = f + pv.$$

Next we address ourselves the question of whether the data determine the interface uniquely. We have to exclude arbitrary rotations in the physical space and in the reference configuration. Thus we say that the interfaces $\mathbf{A}, \mathbf{B}$ and $\bar{\mathbf{A}}, \bar{\mathbf{B}}$ are ***equivalent*** if there exist rotations $\mathbf{Q}, \mathbf{R}$ such that $\bar{\mathbf{A}} = \mathbf{QAR}, \bar{\mathbf{B}} = \mathbf{QBR}$. Otherwise we call the interfaces ***distinct.*** The unicity of the interface depends on whether at least one phase is solid or whether both of them are liquid. We say that a pair of principal stretches α is a ***liquid point*** if the principal forces s at α satisfy

$$s_1\alpha_1 = s_2\alpha_2; \tag{8}$$

otherwise we call α a ***solid point.*** Thus α is a liquid point if and only if the stress is hydrostatic. We note that for fluids the equality (8) holds everywhere; for solids (8) clearly holds if $\alpha_1 = \alpha_2$. However, in the absence of ellipticity (8) can hold also if $\alpha_1 \neq \alpha_2$.

Theorem 2. *Let α, β be data of a stable interface.*
(i) *If α, β are liquid points then any two rank 1 connected tensors $\mathbf{A}, \mathbf{B}$ with the principal stretches α, β form a stable interface;*
(ii) *if α is a solid point then there are at most two distinct stable interfaces with data α, β; viz. $\mathbf{A} = \operatorname{diag}(\alpha)$ and $\mathbf{B} = \mathbf{A} + \mathbf{a} \otimes \mathbf{n}$ with*

$$\left.\begin{aligned} n_1 &= \sqrt{\frac{(\alpha_1 - \sigma\beta_2)(\beta_1 - \alpha_1)}{(\alpha_1 - \sigma\alpha_2)(\hat{\beta}_\sigma - \hat{\alpha}_\sigma)}}, & n_2 &= \pm\sqrt{\frac{(\alpha_2 - \sigma\beta_1)(\beta_2 - \alpha_2)}{(\alpha_1 - \sigma\alpha_2)(\hat{\beta}_\sigma - \hat{\alpha}_\sigma)}}, \\ a_1 &= (\hat{\beta}_\sigma - \hat{\alpha}_\sigma)n_1, & a_2 &= \sigma(\hat{\beta}_\sigma - \hat{\alpha}_\sigma)n_2 \end{aligned}\right\} \tag{9}$$

where $\sigma = 1$ if $\epsilon = 0, 1$ and $\sigma = -1$ if $\epsilon = -1$, and $\hat{\alpha}_\sigma = \alpha_1 + \sigma\alpha_2, \hat{\beta}_\sigma = \beta_1 + \sigma\beta_2$.

The set of all matrices $\mathbf{B} = \mathbf{A} + \mathbf{a} \otimes \mathbf{n}$ with singular values β forms two continuous families parametrized by $\mathbf{n}$, [10]. Thus if both phases are liquid, there is a large indeterminacy of the interface; on the contrary, if at least one of the two phases is solid then the normal and amplitude are given by (9). Note that the forms of $\mathbf{n}, \mathbf{a}$ are independent of the material and if $\mathbf{A}$ is chosen to be diagonal then either $\mathbf{B}$ is symmetric (if $\epsilon = 1$) or $\operatorname{diag}(1, -1)\mathbf{B}$ is symmetric (if $\epsilon = -1$).

We require $\mathbf{A} \neq \mathbf{B}$ as a part of the definition of the interface; however, $\alpha = \beta$ is not excluded. Two matrices $\mathbf{A}, \mathbf{B} \in \mathbb{M}_+^{2\times 2}$ are said to be ***twins*** if they are rank 1 connected and have the same singular values. Twins can form a stable interface only if $\mathbf{A}, \mathbf{B}$ are liquid points:

Proposition 3. *If twins $\mathbf{A}, \mathbf{B}$ with the common pair of singular values α form a stable interface then α is a liquid point.*

Maxwell's relation and the force balance hold trivially. The amplitude depends on the normal via Ericksen's twinning formula (19), below.

Of particular interest is a solid/liquid interface:

Proposition 4. *Any stable interface with data α, β, where α is a solid and β a liquid point, is equivalent to $\mathbf{A}, \mathbf{B}$ where $\mathbf{A} = \operatorname{diag}(\alpha), \mathbf{B} =$*

diag(γ) *and* $\alpha_i = \gamma_i$ *for some* $i = 1, 2$; *moreover,*

$$\left.\begin{aligned} s_{\bar{i}}(\alpha) &= s_{\bar{i}}(\gamma), \\ \tilde{f}(\alpha) - s_{\bar{i}}(\alpha)\alpha_{\bar{i}} &= \tilde{f}(\gamma) - s_{\bar{i}}(\gamma)\gamma_{\bar{i}} \end{aligned}\right\} \tag{10}$$

where $\bar{i}$ *is the index complementary to* i.

Thus if the solid phase $\mathbf{A}$ is chosen to be diagonal, the normal and amplitude are proportional and eigenvectors of $\mathbf{A}$. Equation $(10)_2$ can be written as $\tilde{f}(\alpha) - s_{\bar{i}}(\alpha)\alpha_{\bar{i}} = \tilde{f}(\beta) + p_{\mathbf{B}}v_{\mathbf{B}}$ where $p_{\mathbf{B}}$ is the pressure of $\mathbf{B}$ and $v_{\mathbf{B}} = \det \mathbf{B}$ is the specific volume.

Finally note that a liquid/liquid interface can also occur in an isotropic solid; then we obtain the gibbsian thermostatics of fluids (4).

2. PROOFS

Denote by $\mathbb{H}^2$ the set of all pairs $\alpha = (\alpha_1, \alpha_2)$ satisfying $\alpha_1 \geq \alpha_2 > 0$. For any $\alpha \in \mathbb{H}^2$ and any $\epsilon \in \{1, -1\}$ let $\hat{\alpha}_\epsilon = \alpha_1 + \epsilon\alpha_2$ and recall the quantities $c_\pm = c_\pm(\alpha), k_\pm = k_\pm(\alpha)$ given by $(5)_1$, $(6)_1$ for each $\alpha \in \mathbb{H}^2$ for which the corresponding denominators are different from 0.

Theorem 5. ([1], [8]) *The function* f *is rank 1 convex at* diag$(\alpha), \alpha \in \mathbb{H}^2$, *if and only if, with* $s = s(\alpha)$, *we have* $s_1\alpha_1 - s_2\alpha_2 \geq 0$ *and*

$$\tilde{f}(\beta) \geq \tilde{f}(\alpha) + H(\alpha, \beta)$$

for every $\beta \in \mathbb{H}^2$ *that satisfies*

$$(\alpha_1 - \beta_2)(\beta_1 - \alpha_2) \geq 0 \tag{11}$$

where, with the notation $\epsilon := \epsilon(\alpha, \beta)$, *the quantity* H *is defined by*

$$H(\alpha, \beta) = \begin{cases} k_+(\alpha)(\hat{\beta}_+ - \hat{\alpha}_+) + c_+(\alpha)(\beta_1\beta_2 - \alpha_1\alpha_2) & \textit{if } \epsilon = 1, \\ s(\alpha) \cdot (\beta - \alpha) & \textit{if } \epsilon = 0, \\ k_-(\alpha)(\hat{\beta}_- - \hat{\alpha}_-) + c_-(\alpha)(\beta_1\beta_2 - \alpha_1\alpha_2) & \textit{if } \epsilon = -1. \end{cases}$$

The nonlinear function $H(\alpha, \beta)$ plays the role of the Weierstrass excess function. The restriction (11) follows from the fact, [2], that $\beta \in \mathbb{H}^2$ are singular values of some rank 1 perturbation of diag(α) if and only if (11) holds. The following follows from the considerations in [8]:

Proposition 6. *Let* f *be rank 1 convex at* $\mathbf{A} =$ diag$(\alpha), \alpha \in \mathbb{H}^2$ *and let* $\mathbf{B} \in \mathbb{M}_+^{2\times 2}$, *with singular values* β, *be rank 1 connected to* $\mathbf{A}$. *Then:*

(i)

$$H(\alpha, \beta) \geq \mathbf{S}(\mathbf{A}) \cdot (\mathbf{B} - \mathbf{A}); \tag{12}$$

(ii) *if for* $s := s(\alpha)$ *we have* $s_1\alpha_1 - s_2\alpha_2 > 0$ *then the equality in* (12) *holds only if* $\mathbf{B} = \mathbf{A} + \mathbf{a} \otimes \mathbf{n}$ *where* $\mathbf{a}, \mathbf{n}$ *are as in* (9).

Proof of Theorem 1. Write $\epsilon = \epsilon(\alpha, \beta)$. Let $\mathbf{A}, \mathbf{B}$ be a stable interface with data α, β, write $\mathbf{B} = \mathbf{A} + \mathbf{a} \otimes \mathbf{n}$ and assume that $\mathbf{A} = \operatorname{diag}(\alpha)$. Since f is rank 1 convex at $\mathbf{A}$, Theorem 5, Proposition 6(i) and Maxwell's relation imply that

$$\tilde{f}(\beta) \geq \tilde{f}(\alpha) + H(\alpha, \beta) \geq \tilde{f}(\alpha) + \mathbf{S}(\mathbf{A}) \cdot (\mathbf{B} - \mathbf{A}) = \tilde{f}(\beta). \tag{13}$$

Thus we have the equality signs throughout; in particular

$$\tilde{f}(\beta) = \tilde{f}(\alpha) + H(\alpha, \beta). \tag{14}$$

If the data are nondegenerate then (14) takes the form

$$\tilde{f}(\beta) = \tilde{f}(\alpha) + k_\epsilon(\alpha)(\hat{\beta}_\epsilon - \hat{\alpha}_\epsilon) + c_\epsilon(\alpha)(\beta_1\beta_2 - \alpha_1\alpha_2). \tag{15}$$

Proof of (i): If $\epsilon = 1$ then

$$(\gamma_1 - \alpha_1)(\gamma_2 - \alpha_2) > 0, \quad (\alpha_1 - \gamma_2)(\gamma_1 - \alpha_2) > 0, \quad \gamma_1 > \gamma_2 > 0$$

for all $\gamma \in \mathbb{R}^2$ sufficiently close to β. An appeal to Theorem 5 shows that

$$\tilde{f}(\gamma) \geq \tilde{f}(\alpha) + k_+(\alpha)(\hat{\gamma}_+ - \hat{\alpha}_+) + c_+(\alpha)(\gamma_1\gamma_2 - \alpha_1\alpha_2)$$

for all γ sufficiently close to β; moreover, for $\gamma = \beta$ we have the equality by (15). The differentiation with respect to γ at $\gamma = \beta$ provides $c_+(\alpha) = c_+(\beta), k_+(\alpha) = k_+(\beta)$ and (15) gives $g_+(\alpha) = g_+(\beta)$. The Proof of (ii) is similar to (i). (iii): Note first that if $i \in \{1, 2\}$ and if we define $\varphi : (0, \infty) \to \mathbb{R}$ by $\varphi(\tau) = \tilde{f}(\alpha_i, \tau)$, $\tau \in (0, \infty)$, then the rank 1 convexity of f at $\operatorname{diag}(\alpha)$ implies

$$\varphi(\tau) \geq \varphi(\alpha_{\bar{i}}) + s_{\bar{i}}(\alpha)(\tau - \alpha_{\bar{i}}) \tag{16}$$

for all $\tau > 0$. This is the separate convexity of rank 1 convex functions [4] which also follows from Theorem 5. To prove (iii), consider first the case $i = j = 1$ so that $\bar{i} = \bar{j} = 2$ and $\beta_1 = \alpha_1$. Using that either $\alpha_1 > \alpha_2$ or $\beta_1 > \beta_2$ we assume the latter. Since (16) holds for each $\tau > 0$ and for $\tau = \beta_2$ we have the equality by (14), the differentiation provides $\varphi'(\beta_2) = s_2(\alpha)$. The definition of φ and the use of $\beta_2 < \beta_1 = \alpha_1$ gives that $\varphi'(\beta_2) = \tilde{f}_2(\alpha_1, \beta_2) = \tilde{f}_2(\beta) = s_2(\beta)$. Thus we have (7) and with that equation the equality (16) at $\tau = \beta_2$ gives $(7)_2$. The remaining cases are similar. □

Proof of Theorem 2. (i): If α, β are liquid points then by Theorem 1,

$$p_\alpha = p_\beta, \quad g_\alpha = g_\beta \tag{17}$$

where $p_\alpha = -s_1(\alpha)/\alpha_2 = -s_2(\alpha)/\alpha_1, g_\alpha = \tilde{f}(\alpha) + p_\alpha\alpha_1\alpha_2$, are the pressure and the Gibbs function of α and p_β, g_β have a similar meaning. To show that any two rank 1 connected tensors $\mathbf{A}, \mathbf{B}$ with the singular values α, β form a stable interface we have to verify Maxwell's relation. Noting that $\mathbf{S}_A = -p_\alpha \operatorname{cof} \mathbf{A}$ we find $\mathbf{S}_A \cdot (\mathbf{B} - \mathbf{A}) = -p_\alpha(\beta_1\beta_2 - \alpha_1\alpha_2)$ since $\mathbf{A}, \mathbf{B}$ are rank 1 connected. Thus $(17)_2$ gives $(2)_2$. (ii): As shown in the proof of Theorem 1, we have (see (13))

$$H(\alpha, \beta) = \mathbf{S}(\mathbf{A}) \cdot (\mathbf{B} - \mathbf{A}).$$

By Proposition 6(ii), which applies here since $\mathbf{A}$ is a solid phase and $s_1\alpha_1 - s_2\alpha_2 \geq 0$ by Theorem 5, we have $\mathbf{B} = \mathbf{A} + \mathbf{a} \otimes \mathbf{n}$ where $\mathbf{a}, \mathbf{n}$ are as asserted. □

Proof of Proposition 3. Assume that $\mathbf{A} = \operatorname{diag}(\alpha)$. Since $\alpha = \beta$, we have $\tilde{f}(\alpha) = \tilde{f}(\beta)$, and Maxwell's relation implies

$$\mathbf{S}_A\mathbf{n} \cdot \mathbf{a} = 0; \tag{18}$$

moreover, if we normalize to $|\mathbf{n}| = 1$ then $\mathbf{a}$ is given by [5]

$$\mathbf{a} = 2\left(\frac{\mathbf{A}^{-1}\mathbf{n}}{|\mathbf{A}^{-1}\mathbf{n}|^2} - \mathbf{A}\mathbf{n}\right). \tag{19}$$

Assuming that $\mathbf{A}$ is diagonal, we have $\mathbf{S} = \mathbf{S}_A = \operatorname{diag}(s)$ where $s := s(\alpha)$. The combination of (18) with (19) leads to

$$n_1^2 n_2^2(\alpha_1^2 - \alpha_2^2)(s_1\alpha_1 - s_2\alpha_2) = 0 \tag{20}$$

where n_1, n_2 the components of $\mathbf{n}$. The condition $\mathbf{A} \neq \mathbf{B}$ implies $n_1 \neq 0, n_2 \neq 0$ and thus if $\alpha_1^2 = \alpha_2^2$, i.e., $\alpha_1 = \alpha_2$ then $s_1 = s_2$; so $s_1\alpha_1 - s_2\alpha_2 = 0$, i.e., $\mathbf{A}$ is a liquid point. If $\alpha_1^2 \neq \alpha_2^2$, then $s_1\alpha_1 - s_2\alpha_2 = 0$ by (20) and thus $\mathbf{A}$ is a liquid point again. By symmetry also $\mathbf{B}$ is a liquid point. □

Proof of Proposition 4. Write $\epsilon := \epsilon(\alpha, \beta)$. Note first that necessarily $\alpha_i = \beta_j$ for some $i, j \in \{1, 2\}$, i.e., Case (iii) in Theorem 1 occurs. Indeed, assuming that Case (i) or Case (ii) occurs leads to $k_\epsilon(\alpha) = k_\epsilon(\beta)$ which is impossible since $k_\epsilon(\alpha) \neq 0$ (for $\mathbf{A}$ is a solid point) while $k_\epsilon(\beta) = 0$ (for $\mathbf{B}$ is a liquid point). By Theorem 2(ii), $\mathbf{a}, \mathbf{n}$ are given by (9). We combine this with $\alpha_i = \beta_j$ for some $i, j \in \{1, 2\}$ to infer that either $a_1 = n_1 = 0$ or $a_2 = n_2 = 0$. We distinguish the following cases: (a) $\epsilon = 0$ (and hence either $\alpha_1 = \beta_1$ or $\alpha_2 = \beta_2$), (b) $\epsilon \neq 0, \alpha_1 = \beta_2$, (c) $\epsilon \neq 0, \alpha_2 = \beta_1$. If Case (a) occurs then $\sigma = 1$ and (9) imply that $a_1 = n_1 = 0$ if $\alpha_1 = \beta_1$ while $a_2 = n_2 = 0$ if $\alpha_2 = \beta_2$. If case (b) occurs then we have $\alpha_2 < \alpha_1 = \beta_2 \leq \beta_1$ and thus $\beta_1 \geq \alpha_1, \beta_2 \geq \alpha_2$ so that $\epsilon \geq 0$, which in combination with $\epsilon \neq 0$ implies $\epsilon = 1$ and

$\sigma = 1$. Relations (9) then give $a_1 = n_1 = 0$. Finally if (c) occurs then $\alpha_1 > \alpha_2 = \beta_1 \geq \beta_2$; thus $\epsilon = \sigma = 1$ as in the preceding case and (9) give $a_2 = n_2 = 0$. The relations $a_1 = n_1 = 0$ or $a_2 = n_2 = 0$ imply that $\mathbf{B} = \mathbf{A} + \mathbf{a} \otimes \mathbf{n}$ is diagonal $\mathbf{B} = \operatorname{diag}(\gamma)$ with either $\alpha_1 = \gamma_1 > 0$ or $\alpha_2 = \gamma_2 > 0$. From $\det \mathbf{B} = \gamma_1\gamma_2 > 0$ we conclude that both the two components of γ are positive. Thus $\gamma \in (0, \infty)^2$. This completes the proof of (i). Note that in terms of γ, the ordered singular values β of $\mathbf{B} = \operatorname{diag}(\gamma)$ are either $\beta = \gamma$ or $\beta = (\gamma_2, \gamma_1)$. Referring to Item (ii) of Theorem 1 and considering separately the cases $\beta = \gamma$ or $\beta = (\gamma_2, \gamma_1)$, one finds that (7) reduce to (10). □

Acknowledgments

This research was supported by Grant 201/00/1516 of the Grant Agency of the Czech Republic.

References

[1] Aubert, G. Necessary and sufficient conditions for isotropic rank-one convex functions in dimension 2. *J. Elasticity*, 39:31–46, 1995.

[2] Aubert, G. & Tahraoui, R. Sur la faible fermeture de certains ensembles de contrainte en élasticité non-linéaire plane. *C. R. Acad. Sci. Paris*, 290:537–540, 1980

[3] Ball, J. M. Convexity conditions and existence theorems in nonlinear elasticity *Arch. Rational Mech. Anal.*, 63:337–403, 1977.

[4] Dacorogna, B. Direct methods in the calculus of variations. Springer, Berlin, 1989.

[5] Ericksen, J. L. Continuous martensitic transitions in thermoelastic solids. *J. Thermal Stresses*, 4:107-119, 1981.

[6] Rosakis, P. Characterization of convex isotropic functions. *J. Elasticity*, 49:257–267, 1997.

[7] Šilhavý, M. The mechanics and thermodynamics of continuous media. Springer, Berlin, 1997.

[8] Šilhavý, M. Convexity conditions for rotationally invariant functions in two dimensions. In *Applied nonlinear functional analysis*, A. Sequeira, H. Beirao da Veiga & J. Videman, editors, pp. 513–530. Kluwer Press. New York, 1999.

[9] Šilhavý, M. On isotropic rank 1 convex functions. *Proc. Roy. Soc. Edinburgh*, 129A:1081–1105, 1999.

[10] Šilhavý, M. Rank 1 perturbations of deformation gradients. *Int. J. Solids Structures*, 38:943–965, 2001.

Chapter 29

DRIVING FORCE IN SIMULATION OF PHASE TRANSITION FRONT PROPAGATION

Arkadi Berezovski

Institute of Cybernetics at Tallinn Technical University, Centre for Nonlinear Studies, Akadeemia tee 21, 12618 Tallinn, Estonia

Gérard A. Maugin

Université Pierre et Marie Curie, Laboratoire de modélisation en mécanique, UMR 7607, tour 66, 4 place Jussieu, case 162, 75252, Paris cédex 05, France

Abstract Dynamics of martensitic phase transition fronts in solids is determined by the driving force (a material force acting at the phase boundary). Additional constitutive information needed to describe such a dynamics is introduced by means of non-equilibrium jump conditions at the phase boundary. The relation for the driving force is also used for the modeling of the entropy production at the phase boundary.

Keywords: Martensitic phase transformations, moving phase boundary, thermomechanical modeling.

1. INTRODUCTION

The problem of the macroscopic description of a phase transition front propagation in solids suggests the solution of an initial-boundary value problem for a system of equations, which includes not only conservation laws of continuum mechanics and material constitutive equations but also certain conditions at a moving phase boundary. This problem remains nonlinear even in the small-strain approximation.

For the stress-induced martensitic phase transformations, large speeds of phase boundaries are inconsistent with diffusion during the transformation. This leads to the sharp-interface model of the phase boundary, where interfaces are treated as discontinuity surfaces of zero thickness. Jump conditions following from continuum mechanics are fulfilled at interfaces between two phases. However, the jump relations are not sufficient to determine a speed of the phase-transition front in crystalline solids. What the continuum mechanics is able to determine is the so-called driving force acting on a phase boundary [1, 2, 11, 13]. To select the physically relevant solution from the set of available solutions, the notions of a nucleation criterion and a kinetic relation have been imported from materials science [1, 2]. Additionally, the criterion for the nucleation of an austenite-to-martensite phase transformation is assumed to be the attainment of a critical value of driving force at the phase boundary.

Unfortunately, we have no tools to derive the kinetic relation except simple-minded phenomenology. Therefore, we propose another way to introduce the additional constitutive information taking into account the non-equilibrium nature of the phase transformation process.

We propose a thermomechanical description of the phase transition front propagation based on the material formulation of continuum mechanics and the non-equilibrium thermodynamic consistency conditions at the phase boundary. These conditions are connected with contact quantities, which are used for the description of non-equilibrium states of discrete elements representing a continuous body in the framework of the thermodynamics of discrete systems [12]. Thus, we do not use any explicit expression for the kinetic relation at the phase boundary. The additional constitutive information is introduced by the prescription of entropy production at the phase boundary in the form which is influenced by the expression of the driving force.

2. LINEAR THERMOELASTICITY

The phase transformation is viewed as a deformable thermoelastic phase of a material growing at the expense of another deformable thermoelastic phase. For the sake of simplicity we consider both phases of the material as isotropic thermoelastic heat conductors. Neglecting geometrical nonlinearities, the main equations of thermoelasticity at each regular material point in the absence of body force are the following:

$$\rho_0(\mathbf{x})\frac{\partial v_i}{\partial t} - \frac{\partial \sigma_{ij}}{\partial x_j} = 0, \tag{1}$$

$$\frac{\partial \sigma_{ij}}{\partial t} = \lambda(\mathbf{x})\frac{\partial v_k}{\partial x_k}\delta_{ij} + \mu(\mathbf{x})\left(\frac{\partial v_i}{\partial x_j} + \frac{\partial v_j}{\partial x_i}\right) + m(\mathbf{x})\frac{\partial \theta}{\partial t}\delta_{ij}, \tag{2}$$

$$C(\mathbf{x})\frac{\partial\theta}{\partial t} = \frac{\partial}{\partial x_i}\left(k(\mathbf{x})\frac{\partial\theta}{\partial x_i}\right) + m(\mathbf{x})\frac{\partial v_k}{\partial x_k}. \tag{3}$$

of which the second one is none other than the time derivative of the Duhamel-Neumann thermoelastic constitutive equation [3]. Here t is time, x_j are spatial coordinates, v_i are components of the velocity vector, σ_{ij} is the Cauchy stress tensor, ρ_0 is the density, θ is temperature, λ and μ are the Lamé coefficients, $C(\mathbf{x}) = \rho_0 c$, c is the specific heat at constant stress. The *dilatation coefficient* α is related to the thermoelastic coefficient m, and the Lamé coefficients λ and μ by $m = -\alpha(3\lambda + 2\mu)$. The indicated explicit dependence on the point $\mathbf{x}$ means that the body is materially inhomogeneous in general. The obtained system of equations is a system of conservation laws with source terms. There exist well developed numerical methods for the solution of this system of equations including the case of inhomogeneous media, for example, the wave-propagation algorithm [9]. However, in the case of moving phase transition fronts we need some additional considerations.

To consider the possible irreversible transformation of a phase into another one, the separation between the two phases is idealized as a sharp, discontinuity surface $\mathbb{S}$ across which most of the fields suffer finite discontinuity jumps. Let $[A]$ and $< A >$ denote the jump and mean value of a discontinuous field A across $\mathbb{S}$, the unit normal N_i to $\mathbb{S}$ being oriented from the "minus" to the "plus" side:

$$[A] := A^+ - A^-, \qquad < A > := \frac{1}{2}(A^+ + A^-). \tag{4}$$

The phase transition fronts considered are *homothermal* (no jump in temperature; the two phases coexist at the same temperature) and *coherent* (they present no defects such as dislocations). Consequently, we have the following continuity conditions [11]:

$$[V_j] = 0, \qquad [\theta] = 0 \qquad \text{at} \quad \mathbb{S}, \tag{5}$$

where material velocity V_j is connected with the physical velocity v_i (in the sense of Maugin [10])

$$v_i = -(\delta_{ij} + \frac{\partial u_i}{\partial x_j})V_j. \tag{6}$$

Jump relations associated with balance of linear momentum and entropy inequality read [11]:

$$\tilde{V}_N[\rho_0 v_i] + N_j[\sigma_{ij}] = 0, \quad \tilde{V}_N[S] + N_i\left[\frac{k}{\theta}\frac{\partial\theta}{\partial x_i}\right] = \sigma_{\mathbb{S}} \geq 0. \tag{7}$$

where $\tilde{V}_N = \tilde{V}_i N_i$ is the normal speed of the points of $\mathbb{S}$, and $\sigma_{\mathbb{S}}$ is the entropy production at the interface.

As it was shown in [11], the entropy production can be expressed in terms of the so-called "material" driving force $f_{\mathbb{S}}$ such that

$$f_{\mathbb{S}} \tilde{V}_N = \theta_{\mathbb{S}} \sigma_{\mathbb{S}} \geq 0, \tag{8}$$

where $\theta_{\mathbb{S}}$ is the temperature at $\mathbb{S}$.

In addition, the *balance of "material" forces* at the interface between phases [1, 2, 11, 13] can be specified to the form

$$f_{\mathbb{S}} = -[W] + < \sigma_{ij} > [\varepsilon_{ij}]. \tag{9}$$

However, we cannot use the jump relations (7) until we could determine the value of the velocity of the phase boundary.

A possible solution is the introduction of an additional constitutive relation between the material velocity at the interface and the driving force in the form of kinetic relation [1, 2]. Unfortunately, we have no tools to derive the kinetic relation. Therefore, we are forced to look for another way to introduce the additional constitutive information to be able to describe the motion of a phase boundary. At this point we should examine the situation:

- we cannot use jump relations at the phase boundary because the velocity of the phase boundary is undetermined.
- we have used local equilibrium approximation because it is assumed that all the fields are defined correctly.
- we have not any distinction between the presence and the absence of phase transformation.

We propose to improve the situation by the introduction of non-equilibrium thermodynamic jump conditions at the phase boundary [8].

3. NON-EQUILIBRIUM JUMP CONDITIONS AT THE PHASE BOUNDARY

We start with the classical equilibrium conditions at the phase boundary. The classical equilibrium conditions at the phase boundary consist, for the example of single-component fluid-like systems, of the equality of temperatures, pressures and chemical potentials in the two phases.

In the homothermal case, the continuity of temperature at the phase boundary still holds, and the continuity of the chemical potential can be replaced by the relation (9). What we need is to change the equilibrium condition for pressure.

In non-equilibrium, we expect that the value of energy of an element differs from its equilibrium value so that

$$U = U_{eq} + U_{ex}. \tag{10}$$

To be able to make the distinction between the presence and the absence of the phase transformation, we propose to replace the classical jump relation for pressure by two different non-equilibrium jump relations, which include the excess energy and which are distinct for the processes with and without entropy production [8]

$$\left[\left(\frac{\partial(U_{eq}+U_{ex})}{\partial V}\right)_{\theta}\right] = 0, \quad \left[\left(\frac{\partial(U_{eq}+U_{ex})}{\partial V}\right)_{p}\right] = 0. \tag{11}$$

The last two conditions differ from the equilibrium condition for pressure only by fixing different variables in the corresponding thermodynamic derivatives. This choice of the fixed variables is influenced by the stability conditions for single-component fluid-like systems [8]. To be able to exploit the introduced conditions, we need to have a more detailed description of non-equilibrium states than the only introduction of the energy excess. It seems that the most convenient description of the non-equilibrium states can be obtained by means of the thermodynamics of discrete systems [12]. In such thermodynamics, the thermodynamic state space is extended by means of so-called *contact quantities*. Equations (11) must be rewritten in the context of solid mechanics (see below).

4. CONTACT QUANTITIES AND DRIVING FORCE

Following the main ideas of finite volume numerical methods, we divide the body in a *finite number of identical elements*. The states of discrete elements have not to be necessarily in equilibrium, especially, because of the occurrence of phase transformations. To be able to describe non-equilibrium states of discrete elements, we represent the free energy in each element as the sum of two terms

$$W = \bar{W} + W_{ex}, \tag{12}$$

where $\bar{W}$ is the local equilibrium value of free energy and W_{ex} is the excess free energy. Then we introduce a *contact dynamic stress tensor* and an *excess entropy*

$$\Sigma_{ij} = \frac{\partial W_{ex}}{\partial \varepsilon_{ij}}, \qquad S_{ex} = -\frac{\partial W_{ex}}{\partial \theta}, \tag{13}$$

in addition to local equilibrium stress and entropy

$$\bar{\sigma}_{ij} = \frac{\partial \bar{W}}{\partial \varepsilon_{ij}}, \qquad \bar{S} = -\frac{\partial \bar{W}}{\partial \theta}. \tag{14}$$

In linear thermoelasticity, the thermodynamic consistency conditions (11) at the interface between discrete elements take on the following form in the homothermal case [6]

$$\left[-\bar{\theta} \left(\frac{\partial \bar{\sigma}_{ij}}{\partial \theta} \right)_{\varepsilon_{ij}} + \bar{\sigma}_{ij} - \bar{\theta} \left(\frac{\partial \Sigma_{ij}}{\partial \theta} \right)_{\varepsilon_{ij}} + \Sigma_{ij} \right] N_j = 0, \tag{15}$$

$$\left[\bar{\theta} \left(\frac{\partial \bar{S}}{\partial \varepsilon_{ij}} \right)_{\sigma_{ij}} + \bar{\sigma}_{ij} + \bar{\theta} \left(\frac{\partial S_{ex}}{\partial \varepsilon_{ij}} \right)_{\sigma_{ij}} + \Sigma_{ij} \right] N_j = 0. \tag{16}$$

Just these conditions we apply to determine the values of the contact quantities in the bulk and at the phase boundary [6]. However, we need to take into account whether the phase transition takes place or not.

We propose to expect the initiation of the stress-induced phase transition if both consistency conditions (15), (16) are fulfilled at the phase boundary simultaneously [8]. Eliminating the jumps of stresses from the system of equations (15), (16), we obtain then

$$\left[\bar{\theta} \left(\frac{\partial}{\partial \theta} \bar{\theta} \left(\frac{\partial S_{ex}}{\partial \varepsilon_{ij}} \right)_{\sigma} \right)_{\varepsilon} - \bar{\theta} \left(\frac{\partial S_{ex}}{\partial \varepsilon_{ij}} \right)_{\sigma} \right] N_j = 0. \tag{17}$$

To be able to calculate the jumps of the derivatives of the excess entropy, we propose the following procedure. First, we suppose that the non-equilibrium entropy production during phase transformation is described by means of the jump of excess entropy at the phase boundary

$$[S] = [S_{eq}] + [S_{ex}] = [S_{ex}]. \tag{18}$$

Then we exploit the jump relation corresponding to the balance of entropy $(7)_2$ and the expression for the entropy production in terms of the driving force (8):

$$\tilde{V}_N [S] + \left[\frac{k}{\bar{\theta}} \frac{\partial \bar{\theta}}{\partial N} \right] = \frac{f_{\mathbb{S}} \tilde{V}_N}{\theta_{\mathbb{S}}}. \tag{19}$$

Assuming the continuity of heat flux at the phase boundary in the homothermal case, we have

$$[S_{ex}] = \frac{f_{\mathbb{S}}}{\theta_{\mathbb{S}}}. \tag{20}$$

Further we extend the definition of the excess entropy at every point of the body by similarity to (20):

$$S_{ex} = \frac{f}{\theta}, \quad f = -\bar{W} + < \bar{\sigma}_{ij} > \bar{\varepsilon}_{ij} + f_0, \quad [f_0] = 0. \tag{21}$$

This supposes that at the point where (21) is defined, there exists in thought an oriented surface of unit normal **N**. If there is no discontinuity across this surface, then f is a so-called generating function (the complementary energy changed of sign and up to a constant). If there does exist a discontinuity then the expression becomes meaningful only if the operator $[\cdots]$ is applied to it.

Then we can compute the derivatives of the excess entropy with respect to thermodynamic variables ε_{ij} as usual

$$\left(\frac{\partial S_{ex}}{\partial \varepsilon_{ij}}\right)_\sigma = \left(\frac{\partial}{\partial \varepsilon_{ij}}\left(\frac{f}{\bar{\theta}}\right)\right)_\sigma. \tag{22}$$

The combined consistency condition (17) for the normal component of the stress tensor σ_{11} can be specified to the form [7]

$$[f]\left\langle\frac{2(\lambda+\mu)}{\alpha(3\lambda+2\mu)}\right\rangle + < f > \left[\frac{2(\lambda+\mu)}{\alpha(3\lambda+2\mu)}\right] = \theta_0^2[\alpha(3\lambda+2\mu)]. \tag{23}$$

Remember that the value of f_0 is undetermined yet. One of the simplest possibilities is to choose this value in such a way that

$$< f >= 0. \tag{24}$$

Therefore, the combined consistency condition determines the value of the driving force at the phase boundary

$$f_S = [f] = \theta_0^2[\alpha(3\lambda+2\mu)]\left\langle\frac{\alpha(3\lambda+2\mu)}{2(\lambda+\mu)}\right\rangle. \tag{25}$$

The right hand side of the latter relation can be interpreted as a critical value of the driving force. Therefore, the proposed criterion for the beginning of the stress-induced phase transition is the following one:

$$|f_S| \geq |f_{critical}|, \quad f_{critical} = \theta_0^2[\alpha(3\lambda+2\mu)]\left\langle\frac{\alpha(3\lambda+2\mu)}{2(\lambda+\mu)}\right\rangle. \tag{26}$$

Details of computations of the values of the contact quantities are given in [3, 5], where a finite-volume numerical scheme similar to the wave-propagation algorithm [9] is used, but represented in terms of contact quantities. Other recent examples of the simulation of thermoelastic wave propagation by means of the thermodynamic consistency conditions can be found in [3, 6]. Results of the simulation of one-dimensional phase-transition front propagation are given in [4, 7].

Acknowledgment

Support of the Estonian Science Foundation under contract No.4504 (A.B.) and of the European Network TMR. 98-0229 on "Phase transitions in crystalline substances" (G.A.M.) is gratefully acknowledged.

References

[1] Abeyaratne, R., Knowles, J.K., 1990. On the driving traction acting on a surface of strain discontinuity in a continuum. J. Mech. Phys. Solids 38, 345-360.

[2] Abeyaratne, R., Knowles, J.K., 1993. A continuum model of a thermoelastic solid capable of undergoing phase transitions. J. Mech. Phys. Solids 41, 541-571.

[3] Berezovski, A., Engelbrecht, J., Maugin, G.A., 2000. Thermoelastic wave propagation in inhomogeneous media, Arch. Appl. Mech. 70, 694-706.

[4] Berezovski, A., Engelbrecht, J., Maugin, G.A., 2002. A thermodynamic approach to modeling of stress-induced phase-transition front propagation in solids. In: Sun, Q.P., (ed.), Mechanics of Martensitic Phase Transformation in Solids, Kluwer, Dordrecht, pp. 19-26.

[5] Berezovski, A., Maugin, G.A., 2001. Simulation of thermoelastic wave propagation by means of a composite wave-propagation algorithm. J. Comp. Physics 168, 249-264.

[6] Berezovski, A., Maugin, G.A., 2002a. Thermoelastic wave and front propagation. J. Thermal Stresses 25, 719-743.

[7] Berezovski, A., Maugin, G.A., 2002b. Thermodynamics of discrete systems and martensitic phase transition simulation. Technische Mechanik 22, 118-131.

[8] Berezovski, A., Maugin, G.A., 2003. On the thermodynamic conditions at moving phase-transition fronts in thermoelastic solids. J. Non-Equilib. Thermodyn. 28, 299-313.

[9] LeVeque, R.J., 1997. Wave propagation algorithms for multidimensional hyperbolic systems. J. Comp. Physics 131, 327-353.

[10] Maugin, G.A., 1993. Material Inhomogeneities in Elasticity, Chapman and Hall, London.

[11] Maugin, G.A., 1998. On shock waves and phase-transition fronts in continua. ARI 50, 141-150.

[12] Muschik, W., 1993. Fundamentals of non-equilibrium thermodynamics. In: Muschik, W., (ed.), Non-Equilibrium Thermodynamics with Application to Solids, Springer, Wien, pp. 1-63.

[13] Truskinovsky, L., 1987. Dynamics of nonequilibrium phase boundaries in a heat conducting nonlinear elastic medium. J. Appl. Math. Mech. (PMM) 51, 777-784.

Chapter 30

MODELING OF THE THERMAL TREATMENT OF STEEL WITH PHASE CHANGES

Serguei Dachkovski
University of Bremen
Zentrum für Technomathematik
dsn@math.uni-bremen.de

Michael Böhm
mbohm@math.uni-bremen.de

Abstract We formulate a general frame for modelling of processes with phase changes. Then we discuss some kinetic equations for phase transformations in steels. We compare some of them with experimental data of pearlite transformation in 100Cr6 steel.

Keywords: Phase transformations, kinetic, TRIP, steel, thermomechanics

Introduction

For the thermal treatment of steels (possibly under loading) the following two problems are of special importance for technology. To produce a workpiece of prescribed size one needs to control and to estimate the deformations at the end of the process. To produce it with desirable properties (hardness, forgeability, ductility) one has to control and to estimate the phase composition of the treated ingot (fractions of austenite, pearlite, martensite and so on). One of the simplest examples is quenching. The evolution equation of phases is needed for this purpose, see DB04, e.g., where general frame for modeling of thermoelastic processes with phase changes based on isomorphisms of elastic ranges can be found. We would like to stress that the accuracy of the description of the transformation kinetic is of great importance since it is coupled with constitutive stress-strain equations.

There is the well-known Johnson-Mehl-Avrami equation describing the *isothermal* phase transformation with good accuracy. In the *non-isothermal* case this equation turns to be rather imprecise. There are different generalizations of JMA-equation and iteration models based on it proposed by different authors (see BDHLW03 for an overview). We would like to compare some of them on the base of the experimental data of pearlite transformation in steel 100Cr6.

First we would like to suggest a formulation of a rather general frame for modeling thermo-elasto-plastic processes with phase changes, which is based on the material isomorphisms. Here we generalize a corresponding thermoplasticity approach B03 for the case of phase transformations. This approach does not use the notion of intermediate configurations and hence allows to avoid some objectivity problems.

1. GENERAL FRAMEWORK

Let F be the deformation gradient of a thermoplastic process with phase changes. The following variables are considered as independent: the right Cauchy-Green strain tensor $C = F^T F$, temperature T and $g = \nabla T$. During the process different phases may be produced from a parent phase. Their mass fractions are denoted by Y_i, where i stands for a corresponding phase (austenite, perlite, bainite, e.g.). By definition $Y_i \geq 0$ and $\sum_{i=1}^{n} Y_i = 1$. These variables are treated as intrinsic. Let $\frac{dY}{dt} = \beta\Big(C, T, g, Y, \frac{dC}{dt}, \frac{dT}{dt}\Big)$ be their evolution equation, which is the main subject of the next section.

The dependent variables are: internal energy ψ, entropy η, second Piola-Kirchhoff stress tensor $S = JF^{-1}TF^{-T}$ and material heat flux $q = JF^{-1}q_E$, where $J = det(F)$, T is the Cauchy stress tensor and q_e is the heat flux in Eulerian coordinates. In general there is no one-to-one relation between dependent and independent variables and these quantities are functionals of the independent and intrinsic variables. However such materials as steel at any deformed configuration and phase compound have a thermoelastic range, i.e. a certain domain of variation of independent variables inside of which the one-to-one relations hold:

$$S = S_p(C(t), T(t), g(t), Y(t)), \tag{1}$$
$$q = q_p(C(t), T(t), g(t), Y(t)), \tag{2}$$
$$\psi = \psi_p(C(t), T(t), g(t), Y(t)), \tag{3}$$
$$\eta = \eta_p(C(t), T(t), g(t), Y(t)), \tag{4}$$

where S_p, q_p, ψ_p, η_p are the elastic laws corresponding to the current (generalized) elastic range related to the initial laws S_0, q_0, ψ_0, η_0 by a material isomorphism.

Let P be the material isomorphism between these thermoelastic ranges with phase changes (see DB04 for definitions and details) and H be the set of hardening variables. For their evolution we assume the following flow and hardening rules

$$\frac{dP}{dt} = p\Big(P, C, T, g, Y, H\frac{dC}{dt}, \frac{dT}{dt}\Big), \tag{5}$$

$$\frac{dH}{dt} = h\Big(P, C, T, g, Y, H\frac{dC}{dt}, \frac{dT}{dt}\Big). \tag{6}$$

Note that the history of the process is taken into account by P and H. This completes the set of equations. The above equations should be specified for a given material. For details and discussion we refer to DB04. In the following we confine our consideration to the specification of the kinetic equation $\frac{dY}{dt} = \beta\Big(C, T, g, Y, \frac{dC}{dt}, \frac{dT}{dt}\Big)$ for phase fractions, where we consider the diffusion controlled transformation from one parent to one product phase (e.g. from austenite to pearlite) and use the wollowing notation for the product phase fraction $Y = p$.

2. KINETIC EQUATION FOR PHASE TRANSFORMATION

The classical Johnson-Mehl-Avrami equation JM39

$$p(t) = 1 - \exp\Big(-\Big(\frac{t}{\tau(T)}\Big)^{n(T)}\Big), \tag{7}$$

is in a good agreement with experiment in case of isothermal phase transformation, however it gives only a qualitative description in the non-isothermal case. Here τ and n are two temperature dependent parameters, p is the product phase fraction, t is time and T is temperature. To achieve the high accuracy in simulation of the diffusion phase transformation a lot of attempts have been undertaken in the last decades. The motivation is to have a better control of the additional distortions due to phase transformation during thermal treatment of workpieces. In general, as mentioned above, stresses affect the phase transformation and the elasto-plastic problem is coupled with evolution of phases, we refer to D02 and ABM02. In this paper we consider the simplest case (no temperature gradient, stress free) of austenite-pearlite transformation as a starting point for further investigations of a coupled problem.

In BHSW03 a comparison of five different procedures for kinetic simulation was considered. We used the same experimental data as in BHSW03 and two different kinetic equations. One of them takes into account the history of the temperature evolution, another is the model of Leblond. There are many kinetic equations based on the differential form of the JMA-equation

$$\frac{dp}{dt} = (1-p)\frac{n(T)}{\tau(T)}\Big(-\ln(1-p(t))\Big)^{1-\frac{1}{n(T)}} \tag{8}$$

combined sometimes with the additive Scheil rule, see RHF97, BHSW03, H95, LD84 and BDHLW03 for plenty of examples.

We are going to compare some of these methods for pearlite transformation in 100Cr6 steel. The evaluation of five procedures (JMA-equation, Denis model FDS85,D92, D02, Hougardy model HY86 and two generalizations HLHM99, SYS00 of the JMA-equation with additional factors) has been presented in BHSW03. In that paper the best accuracy was obtained with the Model B, which is an extension of (8) with the correcting factor

$$(1-g(T))\frac{dT}{dt}, \tag{9}$$

where the function $g(T)$ is to be fitted. In the current paper we are going to continue the evaluation and compare two other procedures with the five mentioned above. For this purpose we consider the same steel 100Cr6 and use the same set of continuous cooling experiments with different temperature rates (austenite-pearlite transformation).

The first equation we are going to consider is the equation proposed by Leblond and Devaux in LD84

$$\frac{dp}{dt} = \frac{1-p(t)}{\mu(T)}, \quad \mu(T) > 0, \tag{10}$$

where $T(t)$ is a given temperature variation, $p(t)$ is the pearlite fraction and $\mu(T)$ is a temperature dependent parameter representing the characteristic time of the transformation. The initial condition is $p(0) = 0$. We remark that for a single experiment with strictly monotone $T(t)$ one can always find such a function $\mu(T(t))$ that the simulation performs a prescribed accuracy. However it is desired to find a universal $\mu(T)$ for a range of temperature rates.

The second model is the following. We are going to take the history of the temperature evolution into account. For this purpose we introduce the "averaged" temperature $\theta(t)$ by the following formula

$$\theta(t) = \alpha(1 - e^{-\alpha(t-t_0)})^{-1} \int_{t_0}^{t} T(s)\, e^{-\alpha(t-s)}\, ds, \quad t > t_0, \qquad (11)$$

$\alpha > 0$ is the parameter of the weight function. For $\alpha \to \infty$ follows $\theta(t) \to T(t)$, otherwise $\theta(t)$ depends on the whole path $T(t)$, where the beginning of this path has a smaller contribution as the end, because of the weight function $e^{-\alpha(t-s)}$ in (11). Then we use the classical differential equation (8) with $\theta(t)$ on the place of $T(t)$. The advantage of this approach is that we can use the same material parameters τ and n obtained in isothermal experiments. We remark that for constant temperature we obtain the same JMA differential equation (8).

We consider the exponential cooling curves with different rates starting from 850°C to 100°C. The duration $t_{850/100}$ of cooling is respectively 2000s, 1000s, 500s and 300s. The method of calculation of the pearlite fraction from the dilatometer test is described in BHSW03 and we used the same result from this paper. We confine the consideration on the temperature interval from 800° to 500° as in BHSW03. It corresponds to the durations $t_{800/500}$ of 413s, 206s, 103s and 62s. We use also $\tau(T)$ and $n(T)$ from the same paper:

$$\tau(T) = \tau_0 \exp\left(\frac{Q}{T}\right) \exp\left(\frac{P}{T(T_P - T)^2}\right), \quad n(T) = n_0 + n_1 T, \qquad (12)$$

with $\tau_0 = 0.0018s$, $Q = 7000K$, $P = 1.3 \cdot 10^7 K^3$, $T_P = 760°C$, $n_0 = -16.04$, $n_1 = 0.0324\frac{1}{K}$.

3. SIMULATION RESULTS

The mean square error (or L_2-norm of the deviation) scaled by the length of the correspondent time-interval was used to compare the models (the same as in BHSW03), see the Table 1. Here time intervals correspond to the cooling from 850°C to 500°C. For the Leblond model we assume the following form of $\mu(T)$

$$\mu(T) = a(850 - T)^2, \quad (T \text{ in } °C) \qquad (13)$$

where $a > 0$ has to be fitted from the experimental data. For this simple model we found that the optimal a for these experiments is $a = 8.8$ in the sense of the L_2-norm of the deviation between the simulation and the experiment (see also Fig. 1-2). We see that the result is qualitatively good, but the L_2-norm deviation between simulation curves and experimental curves averaged over four processes was found to be 0.177. That

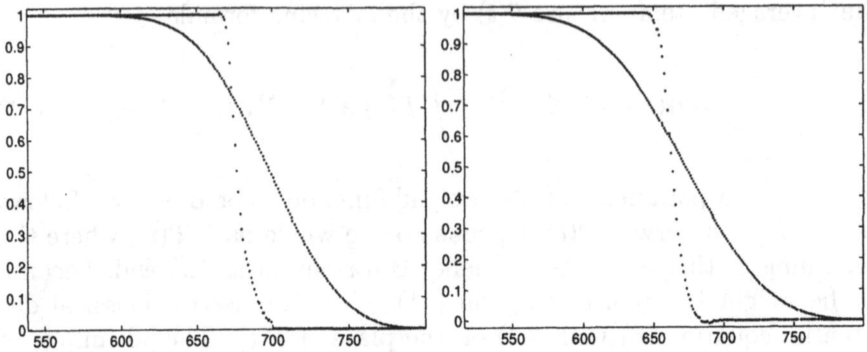

Figure 30.1. (Leblond model) The dotted line represents the experiment, the solid line corresponds to the simulation. Left – $t_{850/500} = 2000s$, right – $t_{850/500} = 1000s$ (Vertical is volume fraction of pearlite, horizontal is temperature in $°C$)

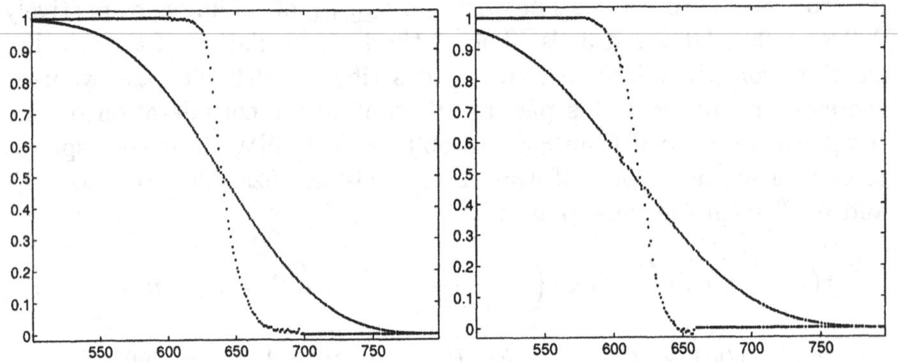

Figure 30.2. (Leblond model) The dotted line represents the experiment, the solid line corresponds to the simulation. Left – $t_{850/500} = 500s$, right – $t_{850/500} = 300s$ (Vertical is volume fraction of pearlite, horizontal is temperature in $°C$.)

is larger than as in the simulations performed in BHSW03. For more accuracy one needs to use better parametrization of the function $\mu(T)$ in the Leblond equation. Now let us proceed to the second method (8), (11). The optimal α in sense of minimizing of L_2-norm of the deviation of the simulation curves from experimental ones, was found as $\alpha = 0.713\frac{1}{s}$ and the corresponding results are presented on the Figures 3-6. We see essentially a better agreement with the experiment in comparison with the Leblond model. The average L_2-error $\left(\frac{1}{k}\sum_{j=1}^{k}\delta_j^2\right)^{\frac{1}{2}}$ (here $k = 4$ is the number of experiments) where the sum is taken over all considered processes and δ_j are the L_2 norms of the difference between the corre-

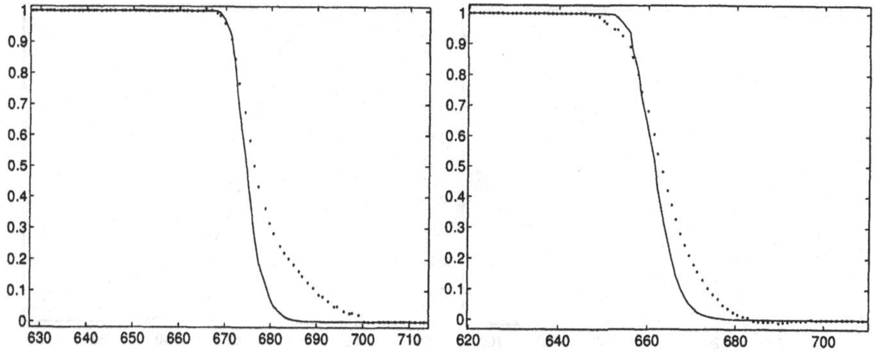

Figure 30.3. Left – $t_{850/500} = 2000s$, right – $t_{850/500} = 500s$. The dotted line represents the experiment, the solid line corresponds to the simulation (Vertical is volume fraction of pearlite, horizontal is the temperature in $°C$).

spondent experimental curve and its simulation curve, for this model is 0.032, that is better in comparison to the JMA-simulation presented in BHSW03. In the first line in the following table we quote results from there. The second and the third line show the L_2-error for each single experiment with the same α (or respectively a) for all processes. In the last line we quote the results for the Model B (see (9) from BHSW03).

Table 1: JMA, JMA-α and Leblond models: Mean square errors.

	2000s	1000s	500s	300s
JMA	0.0375	0.0271	0.0627	0.0877
JMA-α	0.0417	0.0306	0.0283	0.0225
Leblond	0.1781	0.1681	0.1711	0.1934
Model B	0.0313	0.0198	0.0164	0.0078

We would like to remark that equation (8) with initial condition $p(0) = 0$ has two solutions, one of them is trivial. Hence for simulation one have to use $p(0) = \varepsilon$ with a small ε. Then one has a unique solution, but of course it depends on ε. We chose ε small enough so that for smaller values of ε the difference between the simulated solutions is negligible, so we took $\varepsilon = 10^{-7}$.

4. DISCUSSION AND CONCLUSIONS

The advantage of these two methods considered above is the simplicity of the equations, where only one additional parameter has been introduced. The results performed here yield a qualitatively good agreement with the experiment, however the Leblond model with only one parame-

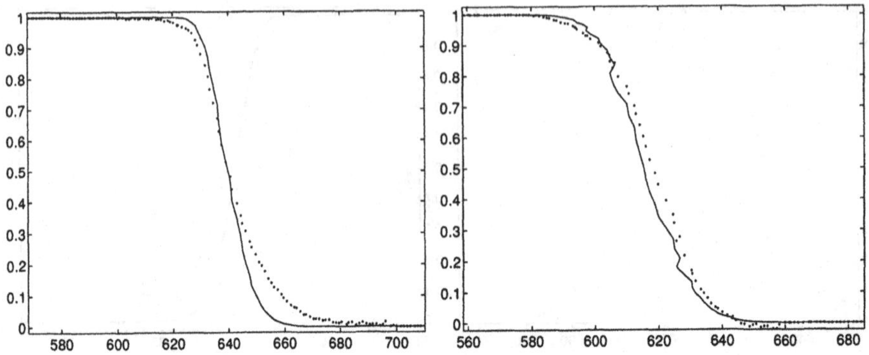

Figure 30.4. Left - $t_{850/500} = 500s$, right - $t_{850/500} = 300s$. The dotted line represents the experiment, the solid line corresponds to the simulation (Vertical is the volume fraction of pearlite, horizontal is the temperature in $°C$).

ter remains to be quantitatively imprecise. One has to find some better formulas for the time-scale parameter $\mu(T)$ to reach a better agreement with experiments.

The second method has better average approximation than the JMA-equation. For higher cooling rates it yields essentially better precision than JMA, but worse for lower ones. It could be probably improved by a more sophisticated functional taking the history of the temperature evolution into account. We conclude that both methods can be used for the kinetic simulation of the pearlite phase transformation, however some improvements of (13) are still needed in case of the Leblond model. We also remark that the Model B has a better accuracy then both models discussed above and others from BHSW03. This models contains the temperature rate on the right side of the evolution equation (we refer to HLHM99 for the motivation) and has essentially higher freedom in fitted parameters (function g from (9) in comparison to one parameter a or α from (10) or (11)) , see Table 1. With reference to BHSW03 we can also conclude that for a given material and a short range of temperature variation one can always find some extension of the classical models to achieve the desired accuracy in simulation. However one has to change the obtained model drastically in case of a different variation of temperature or in case of some changes in composition of material. That is why we believe that some new approaches for the kinetic modeling are still needed.

Acknowledgments

This work has partially been supported by the DFG via the Sonderforschungsbereich "Distortion Engineering"(SFB 570) at the university of Bremen. We would like to thank also the colleagues from the Institut für Werkstofftechnik (IWT) Bremen for the experimental data.

References

U. Ahrens, G. Besserdich, and H. Maier. Sind aufwandige Experimente zur Beschreibung der Phasenumwandlungen von Stählen noch zeitgemass? *HTM 57 p. 99-105.*, 2002.

A. Bertram. Finite thermoplasticity based on isomorphisms *Int. J. Plast. 19 (11)*, 2003.

M. Böhm, S. Dachkovski, M. Hunkel, T. Lübben, and M. Wolff. Phasenumwandlungen im Stahl - Übersicht über einige makroskopische Modelle. *Berichte aus der Technomathematik, FB 3, Universitat Bremen, Report 02.14*, 2003.

M. Böhm, M. Hunkel, A. Schmidt, and M. Wolff. Evaluation of various phase-transition models for 100cr6 for application in commercial FEM programs. *Proc. of ICTMCS 2003, Nancy*, 2003.

S. Dachkovski, M. Böhm. Finite thermoplasticity with phase cchanges based on isomorphisms *Int. J. Plast. 20(2)*, 2004

S. Denis, P. Archambault, E. Gautier, A. Simon, and G. Beck. Precdiction of residual stress and distortion of ferrous and non-ferrous metals: current status and future developments. *J. of Materials Eng. and Performance 11, p. 92-102*, 2002.

S. Denis, D. Farias, and A. Simon. Mathematical model coupling phase transformations and temperature in steels. *ISIJ International 32, p. 316-325*, 1992.

F. Fernandes, S. Denis, and A. Simon. Mathematical model coupling phase transformation and temperature evolution during quenching of steel. *Mat. Sci. Tech. 1*, 1985.

M. Hunkel, T. Lübben, F. Hoffmann, and P. Mayr. Modellierung der bainitischen und perlitischen Umwandlung bei Stahlen. *HTM, 54 (6), p. 365-372*, 1999.

D. Hömberg. A mathematical model for the phase transitions in eutectoid carbon steel. *J. of Appl. Math.*, 54:31–57, 1995.

H. Hougardy and K. Yamazaki. An improved calculation of the transformations in steels. *Steel Research 57 (9), p. 466-471*, 1986.

W. Johnson and R. Mehl. Reaction kinetics in processes of nucleation and growth. *Trans. AIME*, 315:416–441, 1939.

J. Leblond and J. Devaux. A new kinetic model for anisothermal metallurgical transformations in steels including effect of austenite grain size. *Acta metall.*, 32:137–146, 1984.

T. Reti, L. Horvath, and I. Felde. A comparative study of methods used for the prediction of nonisothermal austenite decomposition. *J. of materials engineering and performance*, 6(4):433–441, 1997.

SYSWELD^{TM}. 2000.

XI

PLASTICITY & DAMAGE

Chapter 31

CONFIGURATIONAL STRESS TENSOR IN ANISOTROPIC DUCTILE CONTINUUM DAMAGE MECHANICS

Michael Brünig
Lehrstuhl für Baumechanik-Statik, Universität Dortmund,
D-44221 Dortmund, Germany
michael.bruenig@uni-dortmund.de

Abstract The paper deals with a generalized macroscopic theory of anisotropically damaged elastic-plastic solids. A macroscopic yield condition describes the plastic flow properties and a damage criterion represents isotropic and anisotropic effects. The unbalance of pseudomomentum is established based on the second law of thermodynamics. Evaluation of a strain energy function and assuming the existence of pseudo-potentials of plastic and damage dissipation leads to the definition of the configurational stress tensor, inhomogeneity force and material dissipation force.

Keywords: Anisotropic damage, ductile materials, large strains, configurational stress tensor

1. INTRODUCTION

The theory of configurational forces has been shown to be an effective and valuable basis for the systematic study of different kinds of material defects (see e.g. Kienzler and Herrmann, 2000) and their relation to J-integrals in fracture mechanics has been studied by Maugin (1994), Steinmann (2000) and Müller et al. (2002). It is based on the introduction of the concept of the configurational stress tensor in continuum mechanics of solids. The main advantage of this theory is a unified approach in the investigation of material inhomogeneities, defects or further processes that change the material structure. The concept of energy-momentum tensor has been introduced by Eshelby (1951) to study the defects in elastic continua, and the configurational stress tensor in finite strain elastoplasticity is discussed by Maugin (1994). On the other hand, during the past decades the constitutive modelling of ductile damaged solids in the finite deformation range has received considerable attention and several continuum damage models have been proposed, see Brünig (2001) for an overview. Therefore, the configurational stress tensor and corresponding configurational forces will be presented in the context of ductile damage mechanics which may play a role as the driving force in ductile fracture studies.

2. NEWTONIAN MECHANICS: BALANCE EQUATIONS AND CONSTITUTIVE LAWS

The framework presented by Brünig (2003) is used to describe the inelastic deformations including anisotropic damage due to microdefects. Briefly, the kinematic description employs the consideration of damaged as well as undamaged configurations related via metric transformations which allow for the definition of damage strain tensors $\mathbf{A}^{da}$. The modular structure is accomplished by the kinematic decomposition of strain rates, $\dot{\mathbf{H}}$, into elastic, $\dot{\mathbf{H}}^{el}$, effective plastic, $\dot{\bar{\mathbf{H}}}^{pl}$, and damage parts, $\dot{\mathbf{H}}^{da}$.

To be able to formulate physical equilibrium equations and constitutive laws, the Kirchhoff stress tensor

$$\mathbf{T} = T^{i}_{.j}\, \mathbf{g}_i \otimes \mathbf{g}^j \tag{1}$$

is introduced which satisfies the equilibrium condition

$$\text{div}\mathbf{T} + \rho_o \bar{\mathbf{b}} = \mathbf{0} \tag{2}$$

where $\rho_o \bar{\mathbf{b}} = \rho_o \bar{b}_i \mathbf{g}^i$ represents the physical body forces and div denotes the divergence operator with respect to the current base vectors $\mathbf{g}_i$. Alternatively, the equilibrium condition (2) can be expressed using the Lagrangian representation. This is obtained by introducing the nominal stress tensor $\mathbf{P} = \mathbf{F}^{-T}\mathbf{T} = T^{i}_{.j}\, \overset{o}{\mathbf{g}}_i \otimes \mathbf{g}^j$:

$$\text{Div}\mathbf{P} + \rho_o \bar{\mathbf{b}} = \mathbf{0} \tag{3}$$

where Div means the divergence operator with respect to the initial base vectors $\overset{o}{\mathbf{g}}_i$.

To be able to address equally the two physically distinct mechanisms of irreversible changes, i.e. plastic flow and evolution of damage, respective Helmholtz free energy functions of the fictitious undamaged and of the current damaged configuration are formulated separately. The effective specific free energy density $\bar{\phi}$ of the fictitious undamaged configuration is introduced which depends on effective elastic and effective plastic deformations as well as in an inhomogeneous medium explicitly on the initial position $\overset{o}{\mathbf{x}}$ of any regular material point. It is assumed to be additively decomposed into an effective elastic and an effective plastic part

$$\bar{\phi}(\bar{\mathbf{A}}^{el}, \gamma; \overset{o}{\mathbf{x}}) = \bar{\phi}^{el}(\bar{\mathbf{A}}^{el}; \overset{o}{\mathbf{x}}) + \bar{\phi}^{pl}(\gamma; \overset{o}{\mathbf{x}}) \,. \tag{4}$$

where $\mathbf{A}^{el}$ is the elastic strain tensor and γ denotes an internal plastic variable. Then, the hyperelastic constitutive law leads in the case of isotropic elastic material behavior to the effective stress tensor

$$\bar{\mathbf{T}} = 2G\, \mathbf{A}^{el} + (K - \frac{2}{3}G)\, \mathrm{tr}\mathbf{A}^{el}\, \mathbf{1} \,, \tag{5}$$

where G and K represent the shear and bulk moduli of the undamaged matrix material, respectively. In addition, plastic yielding of the hydrostatic stress-dependent matrix material is assumed to be adequately described by the yield condition

$$f^{pl}(\bar{\mathbf{T}}; \overset{o}{\mathbf{x}}) = \sqrt{\bar{J}_2} - c\,(1 - \frac{a}{c}\bar{I}_1) = 0 \,, \tag{6}$$

where $\bar{I}_1 = \mathrm{tr}\bar{\mathbf{T}}$ and $\bar{J}_2 = \frac{1}{2}\,\mathrm{dev}\bar{\mathbf{T}} \cdot \mathrm{dev}\bar{\mathbf{T}}$ are invariants of the effective stress tensor $\bar{\mathbf{T}}$, $c(\overset{o}{\mathbf{x}})$ denotes the strength coefficient of the matrix material and $a(\overset{o}{\mathbf{x}})$ represents the hydrostatic stress coefficient. In elastic-plastically deformed and damaged metals irreversible volumetric strains are mainly caused by damage and, in comparison, volumetric plastic strains are negligible (Spitzig et al. (1975)). Thus, the plastic potential function $g^{pl} = \sqrt{\bar{J}_2}$ depends only on the second invariant of the effective stress deviator which leads to the isochoric effective plastic strain rate

$$\dot{\bar{\mathbf{H}}}^{pl} = \dot{\lambda}\, \frac{\partial g^{pl}}{\partial \bar{\mathbf{T}}} = \dot{\lambda}\, \frac{1}{2\sqrt{\bar{J}_2}}\, \mathrm{dev}\bar{\mathbf{T}} = \dot{\gamma}\, \frac{1}{\sqrt{2\,\bar{J}_2}}\, \mathrm{dev}\bar{\mathbf{T}} \,. \tag{7}$$

Furthermore, considering the damaged configurations the Helmholtz free energy of the damaged material sample is assumed to consist of three parts:

$$\phi(\mathbf{A}^{el}, \mathbf{A}^{da}, \gamma, \mu; \overset{o}{\mathbf{x}}) = \phi^{el}(\mathbf{A}^{el}, \mathbf{A}^{da}; \overset{o}{\mathbf{x}}) + \phi^{pl}(\gamma; \overset{o}{\mathbf{x}}) + \phi^{da}(\mu; \overset{o}{\mathbf{x}}) \,. \tag{8}$$

The elastic part of the free energy of the damaged material ϕ^{el} is expressed in terms of the elastic and damage strain tensors, $\mathbf{A}^{el}$ and $\mathbf{A}^{da}$, whereas the plastic part, ϕ^{pl} due to plastic hardening, and the damaged part, ϕ^{da} due to damage strengthening, only take into account the respective internal effective plastic and damage state variables, γ and μ. All three parts of the Helmholtz free energy function (8), additionally, depend explicitly on the initial position $\overset{o}{\mathbf{x}}$ of any regular material point to be able to describe corresponding inhomogeneous characteristics. In addition, following Maugin (1994) the existence of a pseudo-potential of plastic dissipation $D^{pl}(\dot{\overline{\mathbf{H}}}^{pl}, \dot{\gamma}; \overset{o}{\mathbf{x}})$ as well as a pseudo-potential of damage dissipation $D^{da}(\dot{\mathbf{H}}^{da}, \dot{\mu}; \overset{o}{\mathbf{x}})$ is assumed which are positive and convex functions and are homogeneous of degree one in their respective first two arguments. This leads to the definition of the work-conjugate plastic stress tensor

$$\mathbf{T}^p = \frac{\partial D^{pl}}{\partial \dot{\overline{\mathbf{H}}}^{pl}} \tag{9}$$

and the plastic strength coefficient of the damaged material

$$\tilde{c} = \frac{\partial D^{pl}}{\partial \dot{\gamma}} \tag{10}$$

as well as to the work-conjugate damage stress tensor

$$\mathbf{T}^d = \frac{\partial D^{da}}{\partial \dot{\mathbf{H}}^{da}} \tag{11}$$

and the damage strength coefficient

$$\sigma = \frac{\partial D^{da}}{\partial \dot{\mu}} \tag{12}$$

which may be determined using respective experimental data. The explicit dependence of D^{pl} and D^{da} on the initial material point position $\overset{o}{\mathbf{x}}$ indicates possible plastic and damage inhomogeneities, respectively, i.e. material point dependence of plastic and damage thresholds or macroscopic hardening and softening characteristics.

The hyperelastic constitutive law then yields the Kirchhoff stress tensor

$$\begin{aligned}\mathbf{T} = \rho_o \frac{\partial \Phi}{\partial \mathbf{A}^{el}} &= 2(G + \eta_2 \,\mathrm{tr}\mathbf{A}^{da})\, \mathbf{A}^{el} + [(K - \frac{2}{3}G + 2\eta_1 \,\mathrm{tr}\mathbf{A}^{da})\mathrm{tr}\mathbf{A}^{el} \\ &+ \eta_3\, (\mathbf{A}^{da} \cdot \mathbf{A}^{el})]\, \mathbf{1} \,+ \eta_3 \,\mathrm{tr}\mathbf{A}^{el}\mathbf{A}^{da} + \eta_4\, (\mathbf{A}^{el}\mathbf{A}^{da} + \mathbf{A}^{da}\mathbf{A}^{el})\end{aligned} \tag{13}$$

which is linear in $\mathbf{A}^{el}$ and $\mathbf{A}^{da}$, and $\eta_1 ... \eta_4$ are newly introduced material constants taking into account the deterioration of the elastic properties due to damage. Furthermore, in analogy to the yield surface and flow rule concepts employed in plasticity theory, evolution of damage is assumed to be adequately described by the damage criterion $f^{da}(\mathbf{T}^d, \sigma; \overset{o}{\mathbf{x}})$ and introduction of a damage potential function $g^{da}(\mathbf{T}^d)$ leads to the corresponding damage rule. Several numerical studies and experimental observations have shown that in structural metals the volume changing contribution to void growth is found to remarkably overwhelm the shape changing part when the mean remote normal stress is large. Therefore, isotropic damage behavior is assumed to give a reasonable approximation for void volume fractions up to the critical porosity $f = f_c$. Based on kinematic considerations (see Brünig (2001)), the damage strain rate tensor is then given by

$$\dot{\mathbf{H}}^{da} = \frac{1}{3}(1-f)^{-1}\dot{f}\mathbf{1} \tag{14}$$

and the isotropic damage behavior is assumed to be governed by the damage condition

$$f^{da} = I_1 - \sigma = 0 \quad \text{for } f < f_c \, . \tag{15}$$

When the current void volume fraction exceeds the critical value f_c the anisotropy effects caused by the changes in shape of the initially spherical voids are assumed to become important. Further void growth and coalescence as well as the activation of shear instabilities in the matrix material between the voids lead to anisotropic damage which is assumed to be adequately described by the damage criterion

$$f^{da} = I_1 + \tilde{\beta}\sqrt{J_2} - \sigma = 0 \quad \text{for } f \geq f_c \tag{16}$$

where $\tilde{\beta}$ describes the influence of the deviatoric stress state on the damage condition. Furthermore, the damage potential function $g^{da}(\mathbf{T}^d) = \alpha I_1 + \beta\sqrt{J_2}$ with kinematically based damage parameters α and β leads to the damage rule

$$\dot{\mathbf{H}}^{da} = \frac{1}{3}(1-f)^{-1}\dot{f}\mathbf{1} + \dot{f}\beta\frac{1}{2\sqrt{J_2}}\text{dev}\mathbf{T}^d \, , \tag{17}$$

where the first term represents inelastic volumetric deformations caused by isotropic growth of microvoids whereas the second term takes into account the dependence of the evolution of shape and orientation of microdefects on the direction of stress.

3. ESHELBIAN MECHANICS: FUNDAMENTAL EQUATIONS AND BALANCE LAWS

The equilibrium conditions (2) and (3) are formulated in the actual configuration. To be able to rewrite the balance laws of continuum mechanics using variables of the reference configuration and to obtain a complete projection of tensorial equations on the material manifold a complete pull-back has to be performed (Maugin (1995)):

$$(\text{div}\mathbf{T})\mathbf{F} + \rho_o\bar{\mathbf{b}}\mathbf{F} = \mathbf{0} \tag{18}$$

or

$$(\text{Div}\mathbf{P})\mathbf{F} + \rho_o\bar{\mathbf{b}}\mathbf{F} = \mathbf{0} \tag{19}$$

To be able to derive the balance law for the configurational forces the gradients of the Helmholtz free energy density $\phi(\mathbf{A}^{el}, \mathbf{A}^{da}, \gamma, \mu; \overset{o}{\mathbf{x}})$ and the pseudo-potentials of plastic and damage dissipation, $D^{pl}(\dot{\bar{\mathbf{H}}}^{pl}, \dot{\gamma}; \overset{o}{\mathbf{x}})$ and $D^{da}(\dot{\mathbf{H}}^{da}, \dot{\mu}; \overset{o}{\mathbf{x}})$, are used as a starting point. In particular, the gradient of the free energy density with respect to the initial position of a material point is given by

$$\begin{aligned} \text{Grad}\phi = & \frac{\partial \phi}{\partial \mathbf{A}^{el}} \cdot \text{Grad}\mathbf{A}^{el} + \frac{\partial \phi}{\partial \mathbf{A}^{da}} \cdot \text{Grad}\mathbf{A}^{da} \\ & + \frac{\partial \phi}{\partial \gamma}\text{Grad}\gamma + \frac{\partial \phi}{\partial \mu}\text{Grad}\mu + \text{Grad}_e\phi \,, \end{aligned} \tag{20}$$

where Grad_e denotes the explicit material gradient. Taking into account the definition of the energy density

$$\psi = \rho_o\phi + \int (D^{pl} + D^{da})dt \tag{21}$$

leads after some algebraic manipulations to the gradient:

$$\begin{aligned} \text{Grad}\psi - \mathbf{T} \cdot \text{Grad}\mathbf{H} = & -\int (\dot{\mathbf{T}}^p \cdot \text{Grad}\bar{\mathbf{H}}^{pl} + \dot{\mathbf{T}}^d \cdot \text{Grad}\mathbf{H}^{da})dt \\ & - \int (\dot{\bar{c}}\text{Grad}\gamma + \dot{\sigma}\text{Grad}\mu)dt + \text{Grad}_e\psi \end{aligned} \tag{22}$$

which clearly shows that there are now three types of material inhomogeneities, namely caused by the nonhomogeneous free energy function as well as due to the nonhomogeneous plastic and damage pseudo-potential functions. These can be exhibited by constructing the corresponding configurational force of true inhomogeneity

$$\mathbf{f}^{inh} = \text{Grad}_e\psi = \text{Grad}_e[\rho_o\phi + \int (D^{pl} + D^{da})dt] \,. \tag{23}$$

In addition, the plastic and damage quasi-inhomogeneity force

$$\begin{aligned} \mathbf{f}^d = & -\int (\dot{\bar{c}}\ \text{Grad}\gamma + \dot{\sigma}\ \text{Grad}\mu)dt \\ & - \int (\dot{\mathbf{T}}^p \cdot \text{Grad}\bar{\mathbf{H}}^{pl} + \dot{\mathbf{T}}^d \cdot \text{Grad}\mathbf{H}^{da})dt \end{aligned} \tag{24}$$

can be defined which takes into account dissipation effects due to plastic and damage deformations.

Furthermore, the product $\mathbf{T} \cdot \mathrm{Grad}\mathbf{H}$ can alternatively be formulated in terms of the nominal stress tensor $\mathbf{P} = \mathbf{F}^{-T}\mathbf{T} = T^i_{.j} \overset{o}{\mathbf{g}}_i \otimes \mathbf{g}^j$ and the mixed strain tensor $\mathbf{L} = \mathbf{H}\mathbf{F} = H^i_{.j}\mathbf{g}_i \otimes \overset{oj}{\mathbf{g}}$:

$$\mathbf{T} \cdot \mathrm{Grad}\mathbf{H} = \mathrm{Div}(\mathbf{PL}) - \mathrm{Div}\mathbf{P}\,\mathbf{L} + \mathbf{P}\,\mathrm{Dif}\mathbf{L} \tag{25}$$

where the difference of the strain gradients

$$\mathrm{Dif}\mathbf{L} = (H_{i\,.\,,k}^{\;j} - H^j_{\,.\,k,i})\, \overset{oi}{\mathbf{g}} \otimes \mathbf{g}_j \otimes \overset{ok}{\mathbf{g}} \tag{26}$$

has been used. Taking into account the physical equilibrium condition (3), Eq. (25) can be rewritten in the form

$$\mathbf{T} \cdot \mathrm{Grad}\mathbf{H} = \mathrm{Div}(\mathbf{PL}) - \rho_o \bar{\mathbf{b}}\,\mathbf{L} + \mathbf{P}\,\mathrm{Dif}\mathbf{L}\,. \tag{27}$$

This leads to the definition of the configurational stress tensor

$$\mathbf{C} = \psi \mathbf{1} - \mathbf{P}\,\mathbf{L} \tag{28}$$

which is formulated in terms of the base vectors of the reference configuration:

$$\mathbf{C} = C^i_{\;.j}\, \overset{o}{\mathbf{g}}_i \otimes \overset{oj}{\mathbf{g}} \tag{29}$$

as well as of the configurational body force vector

$$\bar{\mathbf{f}} = -\rho_o \bar{\mathbf{b}}\,\mathbf{L} - \mathbf{P}\,\mathrm{Dif}\mathbf{L} - \mathbf{f}^d - \mathbf{f}^{inh}\,. \tag{30}$$

Please note that the configurational stress tensor (28) in the elastic-plastic-damage context involves the full stress $\mathbf{P}$ and the total strain $\mathbf{L}$ as well as energy-based contributions and the time integral of dissipation in the energy density ψ.

Finally, taking into account Eqs. (21)-(23), (28) and (30), Eq. (27) results in a configurational force balance

$$\mathrm{Div}\mathbf{C} + \bar{\mathbf{f}} = \mathbf{0} \tag{31}$$

or

$$C^i_{\;.j,i}\, \overset{oj}{\mathbf{g}} + \bar{f}_j\, \overset{oj}{\mathbf{g}} = \mathbf{0} \tag{32}$$

which clearly shows that the configurational force is derived from the divergence of the configurational stress tensor. Please note that the configurational force system satisfies an equilibrium condition analogous to Eqs. (2) or (3) for the physical force system. This local balance of pseudo-momentum (31), however, is a fully material balance law in which the flux is the configurational stress tensor referred to the reference configuration and the source term is the configurational force. The formulation with respect to the reference configuration allows the interpretation of the configurational force as a driving force on a defect in the material. In addition, the configurational force of explicit inhomogeneity $\mathbf{f}^{inh}$ and plastic and damage quasi-inhomogeneity $\mathbf{f}^{d}$ may play a role in ductile fracture studies.

4. CONCLUSIONS

An efficient framework for anisotropically damaged elastic-plastic solids and the definition of the configurational stress tensor in anisotropic continuum damage mechanics have been presented. Based on the introduction of strain energy functions and pseudo-potentials of plastic and damage dissipation the inhomogeneity force and material dissipation force have been formulated which may be seen as driving forces in ductile fracture analyses.

References

Brünig, M., A framework for large strain elastic-plastic damage mechanics based on metric transformations, Int. J. Eng. Sci. 39 (2001), 1033-1056.

Brünig, M., An anisotropic ductile damage model based on irreversible thermodynamics, Int. J. Plasticity 19 (2003), 1679-1713.

Eshelby, J.D., The force on an elastic singularity, Phil. Trans. Roy. Soc. London A 244 (1951), 87-112.

Kienzler, R., Herrmann, G., Mechanics in material space, Springer: New York, Berlin, Heidelberg (2000).

Maugin, G.A., Eshelby stress in elastoplasticity and ductile fracture, Int. J. Plasticity 10 (1994), 393-408.

Maugin, G.A., Material forces: Concepts and applications, ASME Appl. Mech. Rev. 48 (1995), 213-245.

Mueller, R., Kolling, S., Gross, D., On configurational forces in the context of the finite element method, Int. J. Numer. Meth. Engng. 53 (2002), 1557-1574.

Spitzig, W.A., Sober, R.J., Richmond, O., Pressure dependence of yielding and associated volume expansion in tempered martensite, Acta Metall. 23 (1975) 885-893.

Steinmann, P., Application of material forces to hyperelastostatic fracture mechanics. I. Continuum mechanical setting, Int. J. Solids Structures 37 (2000) 7371-7391.

Chapter 32

SOME CLASS OF SG CONTINUUM MODELS TO CONNECT VARIOUS LENGTH SCALES IN PLASTIC DEFORMATION

Lalaonirina Rakotomanana
IRMAR - Unité Mixte de Recherche 6625
CNRS - Université de Rennes 1, 35042 Rennes (France)
lalaonirina.rakotomanana@univ-rennes1.fr

Abstract The present work proposes a continuum theory of strain gradient plasticity with additional plastic spin rate and plastic strain rate tensors defined by the Cartan coefficients of structure.

1. INTRODUCTION

Experimental testings have shown the existence of a material length scale for microcracking and plasticity of materials. Continuum plasticity is used at macroscopic level ($> 100\mu m$) whereas microscopic slip models constitute the basic tool for crystallin plasticity ($\sim 10^{-4}\mu m$). In between, gradient continua have been proposed to model mesoscopic plasticity. Development of gradient plasticity includes at least three basic steps : use of displacement second gradient as additional variable e.g. [6], decomposition of deformation gradient into plastic and elastic parts e.g. [1], [17], and definition of strain gradient plastic variables e.g. [2]. However, some questions still remain such as the consistency problem of using strain gradient as variables e.g. [3], [9], [10], the definition of plastic deformation at the intermediate level e.g. [2], [4], [13], and the plastic spin rate e.g. [2], [11]. The present work is an attempt to give some answer to these questions by using a geometric approach [15] and then proposes additional variables which may enable to connect the various length scales in gradient plasticity. It conforms to previous works based on non Riemaniann geometry to model continuum dislocations e.g. [1], [12] and plastic deformation e.g. [7], [8], [17].

2. STRAIN GRADIENT CONTINUUM

We fix once and for all on an Euclidean reference (Σ) endowed with a metric tensor $\mathbf{g}$ and a volume-form ω_0, uniform and constant. Let O be the origin of (Σ). We define a global (curvilinear) coordinate system (x^1, x^2, x^3) on (Σ) associated to a local basis $(\mathbf{u}_{01}, \mathbf{u}_{02}, \mathbf{u}_{03})$.

2.1. DEFORMATION OF A CONTINUUM

Geometry. A continuum $\mathcal{B}$ is a set of material points M modeled by a manifold e.g. [15]. The tangent vector space of $\mathcal{B}$ at M is denoted $T_M\mathcal{B}$. The deformation of $\mathcal{B}$ is a map φ from an initial configuration $\mathcal{B}_0$ to the deformed configuration $\mathcal{B}$. The location of M is defined by $\mathbf{OM} = \varphi(\mathbf{OM}_0)$. The differential map $d\varphi$ transforms a tangent vector $\mathbf{u}_0$ at M_0 according to $\mathbf{u} = d\varphi(\mathbf{u}_0)$. $\mathbf{F} \equiv d\varphi$ is the deformation gradient. We shall always consider field of local basis $(\mathbf{u}_{10}, \mathbf{u}_{20}, \mathbf{u}_{30})$ at $\mathcal{B}_0$ which deforms to $(\mathbf{u}_1, \mathbf{u}_2, \mathbf{u}_3)$ with $\mathbf{u}_a = d\varphi(\mathbf{u}_{a0})$ at $\mathcal{B}$.

Classical continuum. Deformation of $\mathcal{B}$ is defined by change of the metric components $g_{ab} \equiv \mathbf{g}(\mathbf{u}_a, \mathbf{u}_b)$ e.g. [3], the volume-form component $\omega_0(\mathbf{u}_1, \mathbf{u}_2, \mathbf{u}_3) = \sqrt{det g_{ab}}$, and the symbols of Christoffel e.g. [14]:

$$\gamma^c_{ab} \equiv \frac{1}{2}\, g^{cd} \left(\frac{\partial g_{ad}}{\partial x^b} + \frac{\partial g_{db}}{\partial x^a} - \frac{\partial g_{ab}}{\partial x^d} \right) \tag{1}$$

These symbols $\gamma^c_{ab}(\mathbf{x})$ are the coefficients of a *metric connection* γ which is necessary to calculate the gradient of tensor fields on $\mathcal{B}$. Metric components g_{ab} are equal to components of Cauchy-Green tensor e.g. [16].

Mesoscopic description. Typically, large plastic strain produced in crystallin solids requires regenerative multiplication of dislocations essentially due to Frank-Read sources or multiple cross glide. This later is known to initiate and to increase shear slip banding in plasticity. Physically, each slip results in a discontinuity of φ either between atoms for monocrystals or between grains for polycrystals. An attempt to bridge length scale levels should account for this discontinuity e.g. [4]. Mathematically, given a vector $\mathbf{u}$ and scalar fields ψ on $\mathcal{B}$, the scalar $\mathbf{u}(\psi)$ is the direction derivative of ψ along $\mathbf{u}$. The *Lie-Jacobi* bracket of two vector fields $\mathbf{u}$ and $\mathbf{v}$ on $\mathcal{B}$ is a vector field $[\mathbf{u}, \mathbf{v}]$ on $\mathcal{B}$ defined by:

$$[\mathbf{u}, \mathbf{v}](\psi) \equiv \mathbf{uv}(\psi) - \mathbf{vu}(\psi) \tag{2}$$

Physically, the Lie-Jacobi bracket $[\mathbf{u}, \mathbf{v}]$ measures the failure of closure in the deformed configuration of any initial closed parallelogram e.g. [16]. This happens when the deformation φ includes nucleation of dislocations or grain subdivions within $\mathcal{B}$. Cartan has defined the coefficients of structure $\aleph^c_{0ab}$ associated to $(\mathbf{u}_1, \mathbf{u}_2, \mathbf{u}_3)$ by the formula:

$$[\mathbf{u}_a, \mathbf{u}_b] \equiv \aleph^c_{0ab}\mathbf{u}_c \tag{3}$$

Starting with an initial vector base $[\mathbf{u}_{a0}, \mathbf{u}_{b0}] \equiv 0$, we can thus define: (a) *holonomic* deformations such that $[\mathbf{u}_a, \mathbf{u}_b] = 0$, φ is a $C^k, k > 2$ immersion e.g. [3]; (b) *non holonomic* deformations for which $[\mathbf{u}_a, \mathbf{u}_b] = \aleph^c_{0ab}\mathbf{u}_c \neq 0$ e.g. [16]. In this second case, the formula (1) does no more hold and should be revisited. To extend the Christoffel's symbols, we now introduce a more general definition of an affine connection which is the necessary tool for any gradient calculus on the continuum $\mathcal{B}$.

Definition 2.1 *(Affine connexion) An affine connection ∇ on $\mathcal{B}$ is a map $\nabla : T_M\mathcal{B} \times T_M\mathcal{B} \longrightarrow T_M\mathcal{B}$ which associates any couple of vector fields $\mathbf{u}$ and $\mathbf{v}$ on $\mathcal{B}$ to a vector field $\nabla_\mathbf{u}\mathbf{v}$ such that, λ and μ being any two real numbers and ψ any scalar field on $\mathcal{B}$, e.g. [14]:*

$$\nabla_{\lambda\mathbf{u}_1+\mu\mathbf{u}_2}\mathbf{v} = \lambda\nabla_{\mathbf{u}_1}\mathbf{v} + \mu\nabla_{\mathbf{u}_2}\mathbf{v} \qquad \nabla_{\psi\mathbf{u}}\mathbf{v} = \psi\nabla_\mathbf{u}\mathbf{v}$$
$$\nabla_\mathbf{u}(\lambda\mathbf{v}_1 + \mu\mathbf{u}_2) = \lambda\nabla_u\mathbf{v}_1 + \mu\nabla_\mathbf{u}\mathbf{v}_2 \qquad \nabla_\mathbf{u}(\psi\mathbf{v}) = \psi\nabla_\mathbf{u}\mathbf{v} + \mathbf{u}(\psi)\,\mathbf{v}$$

Scalars Γ^c_{ab} defined by $\nabla_{\mathbf{u}_a}\mathbf{u}_b \equiv \Gamma^c_{ab}\mathbf{u}_c$ generalize the Christoffel symbols. The gradient of vector $\mathbf{v}$ and of linear form ω are deducted:

$$\nabla\mathbf{v}(\mathbf{u}) \equiv \nabla_\mathbf{u}\mathbf{v} \qquad (\nabla_\mathbf{u}\omega)(\mathbf{v}) = \mathbf{u}[\omega(\mathbf{v})] - \omega(\nabla_\mathbf{u}\mathbf{v}) \tag{4}$$

Conversely to γ, an affine connection ∇ on $\mathcal{B}$ cannot be defined solely by the metric $\mathbf{g}$. The torsion and curvature tensors are also required.

Definition 2.2 *Consider on $\mathcal{B}$ an affine connection ∇. Let ω be a 1-form field and $\mathbf{u}_1$, $\mathbf{u}_2$ and $\mathbf{w}$ vector fields on $\mathcal{B}$. The torsion and curvature fields associated to ∇ are defined respectively by:*

$$\aleph(\mathbf{u}_1, \mathbf{u}_2, \omega) \equiv (\nabla_{\mathbf{u}_1}\mathbf{u}_2 - \nabla_{\mathbf{u}_2}\mathbf{u}_1 - [\mathbf{u}_1, \mathbf{u}_2])(\omega) \tag{5}$$
$$\Re(\mathbf{w}, \mathbf{u}_1, \mathbf{u}_2, \omega) \equiv \{\nabla_{\mathbf{u}_1}\nabla_{\mathbf{u}_2}\mathbf{w} - \nabla_{\mathbf{u}_2}\nabla_{\mathbf{u}_1}\mathbf{w} - \nabla_{[\mathbf{u}_1,\mathbf{u}_2]}\mathbf{w}\}(\omega) \tag{6}$$

Components of torsion are $\aleph^c_{ab} = (\Gamma^c_{ab} - \Gamma^c_{ba}) - \aleph^c_{0ab}$ and components of curvature $\Re^b_{acd} = \mathbf{u}_c\left(\Gamma^b_{da}\right) - \mathbf{u}_d\left(\Gamma^b_{ca}\right) + \Gamma^b_{ce}\Gamma^e_{da} - \Gamma^b_{de}\Gamma^e_{ca} - \aleph^e_{0cd}\Gamma^b_{ea}$.

Theorem 1 *Let θ be a scalar field on $\mathcal{B}$. If the variation of θ from one point to another point depends on the path then the torsion field in $\mathcal{B}$ is not null and it characterizes the field of discontinuity of θ on $\mathcal{B}$. Let $\mathbf{w}$ be a vector field on $\mathcal{B}$. If the variation of $\mathbf{w}$ from one point to another point depends on the path then $\aleph$ or $\Re$ do not vanish on $\mathcal{B}$ and they characterize the field of discontinuity of $\mathbf{w}$ on $\mathcal{B}$.*

Proof See e.g. [15] □. The continuum theory presented here is based on the Cartan's circuit when the $\mathcal{B}$ does not have any lattice structure [7]. A continuum $\mathcal{B}$ is affinely equivalent to the ambient space (Σ) if and only if the torsion and curvature tensor fields are identically null at every point of $\mathcal{B}$ e.g. [3], [12].

2.2. STRAIN GRADIENT CONTINUUM

Strain gradient continuum. Classical models of strain gradient continuum introduce $\nabla \mathbf{g}$ or alternatively the second gradient $\nabla^2 \mathbf{u}$ as additional variables e.g. [6]. In this work, we adopt a geometric approach based on Cartan geometry to account for mesoscopic mechanisms [15].

Definition 2.3 *A strain gradient continuum $\mathcal{B}$ is a continuum which allows continuous distribution of scalar and vector discontinuity.*

From the previous results, the local geometry of $\mathcal{B}$ is thus defined by the metric $g_{ab}(\mathbf{x})$, torsion $\aleph^c_{ab}(\mathbf{x})$ and curvature $\Re^c_{abd}(\mathbf{x})$. The question is now to determine if a connection ∇, reconstructed from its torsion $\aleph$ and curvature $\Re$, and the metric $\mathbf{g}$ are compatible e.g. [12]. We recall that ∇ is compatible with $\mathbf{g}$ if and only if $\nabla_{\mathbf{v}}\mathbf{g} \equiv 0$, $\forall \mathbf{v} \in T_M\mathcal{B}$ e.g. [14].

Theorem 2 *(Levi-Civita connection) Let $\mathcal{B}$ be endowed with* $\mathbf{g}$*. Then there is a unique torsion-free connection $\overline{\nabla}$ on $\mathcal{B}$ compatible with* $\mathbf{g}$.

Proof: See proof in e.g. [14]□. The components of the Levi-Civita connection are uniquely determined as:

$$\overline{\Gamma}^c_{ab} = \gamma^c_{ab} + \kappa^c_{ab} \qquad \kappa^c_{ab} \equiv \frac{1}{2}\left(\aleph^c_{0ab} + g^{cd}g_{ae}\aleph^e_{0db} + g^{cd}g_{eb}\aleph^e_{0da}\right) \tag{7}$$

They include not only Christoffel symbols γ^c_{ab} but also additional non holonomic terms κ^c_{ab}, which can be considered as plastic variables. The continuum macroscopic model based on the multiplicative elastoplasticity e.g. [17] does not account for this intermediate plastic variables. Hereafter, $\mathcal{B}$ always denotes a strain gradient continuum for which $\overline{\nabla}$ is compatible with but may be independent of $\mathbf{g}$. Components of torsion and curvature of $\overline{\nabla}$ become:

$$\begin{aligned}
\aleph^c_{ab\overline{\nabla}} &= \aleph^c_{ab\gamma} + \aleph^c_{ab\kappa} = (\kappa^c_{ab} - \kappa^c_{ba}) - \aleph^c_{0ab} \\
\Re^b_{acd\overline{\nabla}} &= \Re^b_{acd\gamma} + \Re^b_{acd\kappa} + \gamma^b_{ce}\kappa^e_{da} - \gamma^b_{de}\kappa^e_{ca} + \kappa^b_{ce}\gamma^e_{da} - \kappa^b_{de}\gamma^e_{ca}
\end{aligned}$$

Remark 2.1 *These relations allow to calculate the discontinuity of scalar and vector fields within $\mathcal{B}$ by means of $\overline{\nabla}$. By the way, we observe that the translation dislocations are additive whereas the rotation dislocations (disclinations) are not e.g. [15]. It should be stressed that κ^c_{ab} is not symmetric with respect to lower indices even if κ is torsion free.*

Objective time derivatives. The motion of $\mathcal{B}$ is defined by the map $\mathbf{OM}(t) = \varphi(M,t)$ and its velocity field $\mathbf{v} \equiv \frac{d}{dt}[\mathbf{OM}]$. For the kinematics of $\mathcal{B}$, we define a frame indifferent time derivative such that any tensor embedded in $\mathcal{B}$ must have a vanishing time derivative. To start with, let $\mathbf{u}$ be a vector field on $\mathcal{B}$ with velocity field $\mathbf{v}$. Its time derivative with respect to $\mathcal{B}$ is defined as:

$$\frac{d^B}{dt}\mathbf{u} \equiv \frac{d}{dt}\mathbf{u} - \nabla \mathbf{v}(\mathbf{u}) \tag{8}$$

$\mathbf{u}$ is embedded in $\mathcal{B}$ if its $\mathcal{B}$-derivative vanishes. For any ω^a element of the dual base we have $\omega^a(\mathbf{u}_b) = \delta^a_b$. If $\mathbf{u}_b$ is an embedded vector, then:

$$\frac{d\mathbf{u}_b}{dt} = \nabla \mathbf{v}\,(\mathbf{u}_b) \qquad \frac{d\omega^a}{dt} = -\nabla \mathbf{v}^T\,(\omega^a)$$

Definition 2.4 *($\mathcal{B}$- derivative) Let $\mathbf{A}$ be a mixed tensor of the type (p,q) on $\mathcal{B}$. The time derivative of $\mathbf{A}$ with respect to $\mathcal{B}$ is a tensor of the same type as $\mathbf{A}$, which stisfies for any p-uplet of vectors $(\mathbf{u}_1,...,\mathbf{u}_p)$ and for any q-uplet of 1-forms $(\omega_1,...,\omega_q)$, embedded in $\mathcal{B}$, the condition:*

$$\left(\frac{d^B}{dt}\mathbf{A}\right)(\mathbf{u}_1,...,\mathbf{u}_p,\omega^1,...,\omega^q) \equiv \frac{d}{dt}[\mathbf{A}(\mathbf{u}_1,...,\mathbf{u}_p,\omega^1,...,\omega^q)] \tag{9}$$

For kinematics of a strain gradient continuum, the $\mathcal{B}$-derivatives of the geometric variables are calculated according to the relationships:

$$\zeta_{\omega_0} \equiv \frac{d^B}{dt}\omega_0 \qquad \zeta_g \equiv \frac{1}{2}\frac{d^B}{dt}\mathbf{g} \qquad \zeta_{\aleph} \equiv \frac{d^B}{dt}\aleph \qquad \zeta_{\Re} \equiv \frac{d^B}{dt}\Re \tag{10}$$

Non holonomic deformation rate. The expression of the Levi-Civita connection (7) highlights the influence of non holonomic deformation on the velocity gradient $\overline{\nabla}\mathbf{v}$. First terms γ^c_{ab} correspond to classical holonomic deformation. We now give a mechanical interpretation of additional terms. For any vector field $\mathbf{v}$ on $\mathcal{B}$, we first define the 1-form field $\mathbf{v}^*$ on $\mathcal{B}$, $\mathbf{g}$-isomorphic of $\mathbf{v}$, by $\mathbf{v}^*(\mathbf{u}) \equiv \mathbf{g}(\mathbf{u},\mathbf{v})$ Consider the decomposition $\overline{\nabla}\mathbf{v}^* = \mathbf{D} + \Omega$. The components of the strain rate are denoted $\zeta_g(\mathbf{u}_a,\mathbf{u}_b) = \mathbf{D}(\mathbf{u}_a,\mathbf{u}_b) = \frac{1}{2}\left[(\overline{\nabla}_{\mathbf{u}_a}\mathbf{v}^*)(\mathbf{u}_b) + (\overline{\nabla}_{\mathbf{u}_b}\mathbf{v}^*)(\mathbf{u}_a)\right]$. In addition to classical terms (calculated with the metric connection γ), we obtain the non holonomic strain rate projected onto $(\mathbf{u}_1,\mathbf{u}_2,\mathbf{u}_3)$:

$$\mathbf{D}^{sg}(\mathbf{u}_a,\mathbf{u}_b) = \frac{1}{2}\left(g^{cd}g_{ae}\aleph^e_{0db} + g^{cd}g_{eb}\aleph^e_{0da}\right)\,v_c \tag{11}$$

From the spin rate $\Omega(\mathbf{u}_a,\mathbf{u}_b) = \frac{1}{2}\left[(\overline{\nabla}_{\mathbf{u}_a}\mathbf{v}^*)(\mathbf{u}_b) - (\overline{\nabla}_{\mathbf{u}_b}\mathbf{v}^*)(\mathbf{u}_a)\right]$, there is also an additional non holonomic spin rate whose components are given by the relationship:

$$\Omega^{sg}(\mathbf{u}_a, \mathbf{u}_b) = \frac{1}{2}\aleph^c_{0ab}\, v_c \tag{12}$$

Remark 2.2 *We remark that the non holonomic strain and spin rates* $\mathbf{D}^{sg}$ *and* Ω^{sg} *express a rate of local topology change of the continuum* $\mathcal{B}$. *They are not due to evolution of the metric components* $g_{ab}(\mathbf{x}, t)$, *due to the shape change, but are due to some microstructure rearrangements and shear band at the mesoscopic level.*

3. STRAIN GRADIENT PLASTIC MODELS

Strain gradient continuum. The local motion of $\mathcal{B}$ is described by $g_{ab}(\mathbf{x}, t)$, $\aleph^c_{ab}(\mathbf{x}, t)$ and $\Re^c_{abd}(\mathbf{x}, t)$. For completness, the thermomechanic process within $\mathcal{B}$ is described by the stress σ, Helmholtz free energy ϕ, body force $\rho\mathbf{b}$, entropy s, and volumic heating per unit mass r.

Definition 3.1 *A strain gradient continuum of the rate type* $\mathcal{B}$ *is defined by the constitutive tensor functions,* $\Im = \{\sigma, \phi, s\}$,*:*

$$\Im = \tilde{\Im}(\omega_0, \mathbf{g}, \aleph, \Re, \zeta_g, \zeta_\aleph, \zeta_\Re) \tag{13}$$

Theorem 3 *Let* $\mathcal{B}$ *be a strain gradient continuum of the rate type. Then the free energy* ϕ *defined by (13) takes necessarily the form* $\phi = \tilde{\phi}(\omega_0, \mathbf{g}, \aleph, \Re)$ *and the entropy inequality is written as:*

$$\mathbf{J}_g : \zeta_g + \mathbf{J}_\aleph : \zeta_\aleph + \mathbf{J}_\Re : \zeta_\Re \geq 0 \tag{14}$$

$$\mathbf{J}_g \equiv \sigma - \rho\frac{\partial\phi}{\partial\omega_0} : \omega_0\mathbf{i} - 2\rho\frac{\partial\phi}{\partial\mathbf{g}} \qquad \mathbf{J}_\aleph \equiv -\rho\frac{\partial\phi}{\partial\aleph} \qquad \mathbf{J}_\Re \equiv -\rho\frac{\partial\phi}{\partial\Re} \tag{15}$$

Proof : It is shown by applying the method of Coleman and Noll [5] □.

Discussion on strain gradient dependence. From thermodynamics point of view, the Coleman and Noll's method can be applied because $\mathbf{g}$ and ∇ are independent variables for the present model. A problem of thermomechanic consistency arises if these constitutive arguments are not independent e.g. [9]. Indeed, the requirement of a positive dissipation at any point of $\mathcal{B}$ would induce too restrictive conditions for evolution laws and $\mathbf{J}_\aleph$ and $\mathbf{J}_\Re$ if they are dependent. From geometry background, consider smooth deformation of $\mathcal{B}$ such that $g_{ab} \equiv \mathbf{g}(\mathbf{u}_a, \mathbf{u}_b)$ are C^2. The boundary value problem may be recast into the minimization of a scalar valued function. In the seek of invariant formulations of physical field theories e.g. [10], if we consider a simply connected $\mathcal{B} \in R^3$ and any scalar function $\phi(\mathbf{g}, \nabla\mathbf{g}, \nabla^2\mathbf{g})$ such as strain energy density, assumed to be C^4, the invariance of $\int_{\mathcal{B}} \phi(\mathbf{g}, \nabla\mathbf{g}, \nabla^2\mathbf{g})\,\omega_0$ under proper coordinate transformations $\overline{\mathbf{x}}(\mathbf{x})$ (covariance) implies that there does not exist a scalar density $\phi(\mathbf{g}, \nabla\mathbf{g})$ which depends solely on the metric components g_{ab} and its first derivatives $\frac{\partial g_{ab}}{\partial x^c}$ [10]. This result is related to the smoothness of g_{ab} and on the connectedness of $\mathcal{B}$ e.g. [3].

Continuum vs. discrete crystal plasticity. For microscopic crystal plasticity, $\mathbf{F} \equiv d\varphi$ is classically decomposed into elastic and plastic parts $\mathbf{F} = \mathbf{F}^e\mathbf{F}^p$ e.g. [17]. It implies $\mathbf{L} \equiv \nabla\mathbf{v}^* = \mathbf{L}^e + \mathbf{L}^p$. Assume that plastic deformation takes place only by dislocation glide mechanism along slip directions $\mathbf{s}_0$ and slip normal planes $\mathbf{m}_0$. The Taylor equation holds $\mathbf{L}^p = \sum_\alpha \dot{\gamma}^{(\alpha)}\ \mathbf{s}^{(\alpha)} \otimes \mathbf{m}^{(\alpha)}$ in which $\mathbf{s} = \mathbf{F}^e(\mathbf{s}_0)$ and $\mathbf{m} = \mathbf{F}^e(\mathbf{m}_0)$. $\dot{\gamma}^{(\alpha)}$ denotes the slip rate on the system α. Discrete micro-plastic spin and strain rates projected onto $(\mathbf{u}_1, \mathbf{u}_2, \mathbf{u}_3)$ hold:

$$\Omega^p = \sum_\alpha \frac{1}{2}\dot{\gamma}^{(\alpha)} \left(s_a^{(\alpha)} m_b^{(\alpha)} - m_a^{(\alpha)} s_b^{(\alpha)}\right) \mathbf{u}^a \otimes \mathbf{u}^b \quad (16)$$

$$\mathbf{D}^p = \sum_\alpha \frac{1}{2}\dot{\gamma}^{(\alpha)} \left(s_a^{(\alpha)} m_b^{(\alpha)} + m_a^{(\alpha)} s_b^{(\alpha)}\right) \mathbf{u}^a \otimes \mathbf{u}^b \quad (17)$$

These microscopic relations are often used to define plastic deformation rate. However, it is worth noting that a micro-volume of metal contains more than 10^6 atoms. Therefore, mixing microscopic and macroscopic levels implicitly lies on a strong homogenization assumption. In this work, between microscopic and macroscopic decriptions, there are strain gradient plastic rates Ω^{sg} and $\mathbf{D}^{sg}$ at the mesoscopic level. Levi-Civita connection properties ensure their existence and uniqueness.

4. CONCLUDING REMARKS

The present work proposes a geometric approach for strain gradient plasticity by accounting for microscopic or mesoscopic deformations. Results show the existence of intermediate plastic spin and plastic strain tensor rates which can not be deduced from macroscopic model assuming the multiplicative decomposition of $d\varphi$. We wonder if they could be related to the kinematics of grain subdividion in polycrystals elastoplasticity when using multiscale decomposition e.g. [4].

References

[1] Bilby BA, Gardner LRT, Stroh AN. Continuous distributions of dislocations and the theory of plasticity, In *Actes du IXème du Congrès de Mécanique Appliquée*, 1957, pp 35-44.

[2] Cermelli P, Gurtin ME. On the characterization of geometrically necessary dislocations in finite plasticity, *J Mech Phys Solids* 49, 2001, pp 1539-1568.

[3] Ciarlet PG, Laurent F. Continuity of a deformation as a function if its Cauchy-Green tensor, *Arch Rat Mech Analysis* 167, 2003, pp 255-269.

[4] Clayton JD, McDowell DL. A multiscale multiplicative decomposition for elastoplasticity of polycrystals, *Int J Plasticity* 19, 2003, pp 1401-1444.

[5] Coleman BD, Noll W. The thermodynamics of elastic materials with heat conduction and viscosity, *Arch Rat Mech Analysis* 13, 1963, pp 167-170.

[6] Fleck NA, Hutchinson JW. Strain gradient plasticity. In : Hutchinson JW, Wu TY editors, *Adv in Appl Mech* 33, Academic Press, 1997, pp 295-361.

[7] Kroener E. Dislocation: a new concept in the continuum theory of plasticity, *J Math Phys* 42, 1963, pp 27-37.

[8] Le KC, Stumpf H. A model of elastoplastic bodies with continously distributed dislocations, *Int J Plasticity* 12, 5, 1996, pp 611-627.

[9] Lorentz E, Andrieux S. A variational formulation for nonlocal damage models, *Int J Plasticity* 15, 1999, pp 119-138.

[10] Lovelock D, Rund H. *Tensors, differential forms and variational principles*, Wiley, 1975.

[11] Lubarda VA. On the partition of rate of deformation in crystal plasticity, *Int J Plasticity* 15, 1999, pp 721-736.

[12] Maugin G *Material Inhomogeneities in Elasticity*. Chapman and Hall, 1993.

[13] Menzel A, Steinmann P. On the continuum formulation of higher gradient plasticity for single and polycrystals, *J Mech Phys Solids* 48, 2000, pp 1777-1796.

[14] Nakahara M. *Geometry, Topology and Physics*, in *Graduate student series in Physics* IOP Publishing, 1996.

[15] Rakotomanana RL. Contribution à la modélisation géométrique et thermodynamique d'une classe de milieux faiblement continus, *Arch Rat Mech Analysis* 141, 1997, pp 199-236.

[16] Rakotomanana RL. *A geometric approach to thermomechanics of dissipating continua*, Birkhauser, Boston, 2003.

[17] Steinmann P. Views on multiplicative elastoplasticity and the continuum theory of dislocations, *Int J Engng Sci* 34, 1996, pp 1717-1735.

Chapter 33

WEAKLY NONLOCAL THEORIES OF DAMAGE AND PLASTICITY BASED ON BALANCE OF DISSIPATIVE MATERIAL FORCES

Helmut Stumpf, Jerzy Makowski, Klaus Hackl
Lehrstuhl für Allgemeine Mechanik, Ruhr-Universität Bochum, D-44780 Bochum, Germany

Jaroslaw Gorski
Lehrstuhl für Grundbau und Bodenmechanik, Ruhr-Universität Bochum, D-44780 Bochum, Germany

Abstract It is shown that the concept of material forces together with associated balance laws, besides the classical laws of linear and angular momentum in the physical space, provide a firm theoretical framework within which weakly nonlocal (gradient type) models of damage and plasticity can be formulated in a clear and rigorous manner. The appropriate set of balance laws for physical and material forces as well as the first and second law of thermodynamics are formulated in integral form. The corresponding local laws are next derived and the general structure of thermodynamically consistent constitutive equations is formulated.

Keywords: Physical and material forces, microstructure, defects, weakly nonlocal damage and plasticity models

1. INTRODUCTION

In local theories of damage and elastoplasticity published in the literature appropriate kinematical variables are considered as internal variables, for which evolution laws have to be formulated. Numerical investigations have shown that FE solutions based on local theories exhibit a strong mesh-dependency, if localization in the material space occurs, and that they are not able to simulate appropriately size-dependent effects. To overcome these difficulties weakly nonlocal (gradient type) and nonlocal models were proposed in the literature based either on variational approach (e.g. Frémond and Nedjar, 1996; Lorentz and Andrieux, 1999; Nedjar, 2001; Saczuk et al., 2003) or on the postulate of additional balance laws for material forces (e.g. Makowski and Stumpf, 2001; Stumpf and Hackl, 2003). In the literature the non-classical material forces are also denoted configurational forces (Maugin, 1993; Gurtin, 2000, Steinmann 2000). The weakly nonlocal model for isotropic brittle damage proposed by Frémond and Nedjar (1996) is based on a virtual work principle leading to the classical balance law of physical forces and an additional balance law for a scalar-valued material force associated with isotropic damage. In Nedjar (2001) this isotropic gradient-damage model is combined with the classical local theory of small strain plasticity. The variational formulation of Lorentz and Andrieux (1999) is an internal variable approach without postulating balance of material forces. As point of departure a global form of the free energy as functional of strain, internal variable and its gradient and a global form of the dissipation potential as functional of the rates of the internal variable and its gradient are introduced. The second law of thermodynamics is verified for the whole structure circumventing in this way the limitations of the Coleman and Noll (1963) procedure.

While the above-described models can be considered as weakly nonlocal (Euclidian gradient type) the nonlocal theory of inelasticity and damage presented in Saczuk et al. (2003) is based on a Lagrange minimization technique, where as underlying kinematics a two-level manifold structure is applied, where the deformation gradient is introduced as linear connection with macro- and micro-covariant derivatives, what leads to an Euclidian space gradient (weakly nonlocal) formulation only under simplifying assumptions.

In Makowski and Stumpf (2001) the microstructure of material bodies is modeled by a finite set of directors, where balance of the associated material forces is postulated. In Stumpf and Hackl (2003) a weakly nonlocal, thermodynamically consistent theory for the analysis of anisotropic damage evolution in thermo-viscoelastic and quasi-brittle materials is presented, where balance of non-dissipative and dissipative material forces associated with anisotropic damage is postulated.

The aim of the present paper is to derive a thermodynamically consistent weakly nonlocal theory for damage and elastoplasticity at finite strain including both the gradient of anisotropic damage and the plastic strain gradient, where the anisotropic damage is described by a symmetric damage tensor $\mathbf{D}$, referred to the undeformed reference configuration, and the plastic deformation by a locally defined plastic deformation tensor $\mathbf{F}^p$.

2. BALANCE LAWS OF PHYSICAL AND MATERIAL FORCES

For the purpose of this paper, the material body may be identified with a region B in the physical space, which the body occupies in a fixed reference configuration. The motion of the body is then described by a mapping $\mathbf{x} = \mathbf{x}(\mathbf{X}, t)$, which carries each material particle whose reference place is $\mathbf{X}$ into its place $\mathbf{x}$ in the spatial configuration of the body at actual time t. Once such a mapping is specified, all variables associated with the macroscopic kinematics of the body such as velocity, deformation gradient, strain tensors, etc., are defined in the standard manner. The basic laws of Newtonian mechanics representing the balance of forces and couples in the physical space take the form:

$$\int_P \mathbf{b}\,dv + \int_{\partial P} \mathbf{T}\mathbf{n}\,da = \frac{d}{dt}\int_P \mathbf{p}\,dv, \tag{2.1}$$

$$\int_P \mathbf{x}\times\mathbf{b}\,dv + \int_{\partial P} \mathbf{x}\times\mathbf{T}\mathbf{n}\,da = \frac{d}{dt}\int_P \mathbf{x}\times\mathbf{p}\,dv. \tag{2.2}$$

The notations used in this paper are standard. In particular, the field variables appearing in (2.1) and (2.2) are:

$\mathbf{T}(\mathbf{X}, t)$: physical first Piola-Kirchhoff stress tensor,
$\mathbf{b}(\mathbf{X}, t)$: external physical body force,
$\mathbf{p}(\mathbf{X}, t)$: physical momentum density.

Moreover, $\mathbf{n}(\mathbf{X})$ is the outward unit normal vector to the boundary ∂P of a subdomain $P \subset B$ occupied by any part of the body in the reference configuration.

It is well-known that the classical laws of Newtonian mechanics alone are insufficient to describe appropriately the microstructural changes in the material during an irreversible deformation with damage and plasticity. Moreover, the physical mechanisms underlying damage and plasticity of the material are essentially different. Accordingly, in order to account for these two different mechanisms it is necessary to introduce into the theory two additional independent kinematical field variables. In this paper it will be assumed that microdefects as microcracks, microvoids responsible for the degradation of the material properties are characterized by a second order "damage tensor" $\mathbf{D}(\mathbf{X},t)$ and the plastic deformation, based on dislocation motion in the case of poly-crystalline materials, by a second order plastic deformation tensor $\mathbf{F}^{\mathrm{p}}(\mathbf{X},t)$. It will be also assumed that the evolution of $\mathbf{D}(\mathbf{X},t)$ and $\mathbf{F}^{\mathrm{p}}(\mathbf{X},t)$ during the deformation of the material is caused by corresponding material forces and stresses, respectively, satisfying their own balance laws: one associated with damage indicated by an upper index d,

$$\int_P(-\mathbf{K}^{\mathrm{d}}+\mathbf{G}^{\mathrm{d}})dv+\int_{\partial P}\mathbb{H}^{\mathrm{d}}\mathbf{n}da=\frac{d}{dt}\int_P\mathbf{P}^{\mathrm{d}}dv \tag{2.3}$$

and one associated with plastic deformation indicated by an upper index p,

$$\int_P(-\mathbf{K}^{\mathrm{p}}+\mathbf{G}^{\mathrm{p}})dv+\int_{\partial P}\mathbb{H}^{\mathrm{p}}\mathbf{n}da=\frac{d}{dt}\int_P\mathbf{P}^{\mathrm{p}}dv. \tag{2.4}$$

The integration of the terms in (2.3) and (2.4) has to be performed over the same subdomain P and the boundary surface ∂P as in (2.1) and (2.2). The field variables appearing in (2.3) and (2.4) are:

$\mathbf{K}^{\mathrm{d}}(\mathbf{X},t)$, $\mathbf{K}^{\mathrm{p}}(\mathbf{X},t)$: second order material stress tensors,

$\mathbf{G}^{\mathrm{d}}(\mathbf{X},t)$, $\mathbf{G}^{\mathrm{p}}(\mathbf{X},t)$: external influence tensors representing e.g. chemical reactions breaking internal material bonds associated with defect evolution and plastic deformation,

$\mathbb{H}^{\mathrm{d}}(\mathbf{X},t)$, $\mathbb{H}^{\mathrm{p}}(\mathbf{X},t)$: third-order material stress tensors,

$\mathbf{P}^{\mathrm{d}}(\mathbf{X},t)$, $\mathbf{P}^{\mathrm{p}}(\mathbf{X},t)$: material momenta associated with evolution of damage $\mathbf{D}$ and plastic deformation $\mathbf{F}^{\mathrm{p}}$.

Evolving microdefects have no mass but they have inertia. This effect is included in the present theory through the material momenta $\mathbf{P}^{\mathrm{d}}$ and $\mathbf{P}^{\mathrm{p}}$. The physical interpretation of the material stress tensors $\mathbf{K}^{\mathrm{d}}$, $\mathbf{K}^{\mathrm{p}}$ and $\mathbb{H}^{\mathrm{d}}$, $\mathbb{H}^{\mathrm{p}}$ becomes obvious in the next section, especially in eqn (3.3).

Passing in the balance laws (2.1) and (2.2) to the limit of a material point, the classical field equations are obtained representing in the local form the balance laws of physical forces and couples:

$$\mathrm{Div}\,\mathbf{T}+\mathbf{b}=\dot{\mathbf{p}}, \qquad \mathbf{T}\mathbf{F}^{\mathrm{T}}-\mathbf{F}\mathbf{T}^{\mathrm{T}}=\mathbf{0}, \tag{2.5}$$

where $\mathbf{F}(\mathbf{X},t)$ denotes the physical deformation gradient, $\mathbf{F}=\nabla\mathbf{x}$, and the superimposed dot stands for the time derivative. The same procedure of localization applied to the balance of material forces (2.3) and (2.4) yields the following local field equations

$$\mathrm{Div}\,\mathbb{H}^{\mathrm{d}}-\mathbf{K}^{\mathrm{d}}+\mathbf{G}^{\mathrm{d}}=\dot{\mathbf{P}}^{\mathrm{d}}, \qquad \mathrm{Div}\,\mathbb{H}^{\mathrm{p}}-\mathbf{K}^{\mathrm{p}}+\mathbf{G}^{\mathrm{p}}=\dot{\mathbf{P}}^{\mathrm{p}}. \tag{2.6}$$

The inertia terms on the right side of the material balance laws (2.6) are due to the assumption that dynamically moving defects and dislocations have inertia.

We have to point out that the local laws of physical forces and couples (2.5) and material forces (2.6) are valid at all points in the reference configuration B of the body provided that the relevant fields are smooth. However, the integral form of the corresponding balance laws admits also singular surfaces, at which physical and/or material stresses fail to be continuous. In this case the corresponding jump conditions at the singular surfaces have to be added to the local equations (2.5) and (2.6). This issue will not be considered in the present paper.

3. WEAK FORM OF THE MOMENTUM BALANCE LAWS

As direct implication of the local equations $(2.5)_1$ and (2.6) we have

$$\begin{aligned}&\int_P\Big((\mathrm{Div}\,\mathbf{T}+\mathbf{b}-\dot{\mathbf{p}})\bullet\dot{\mathbf{x}}+(\mathrm{Div}\,\mathbb{H}^{\mathrm{d}}-\mathbf{K}^{\mathrm{d}}+\mathbf{G}^{\mathrm{d}}-\dot{\mathbf{P}}^{\mathrm{d}})\bullet\dot{\mathbf{D}}\\&+\ (\mathrm{Div}\,\mathbb{H}^{\mathrm{p}}-\mathbf{K}^{\mathrm{p}}+\mathbf{G}^{\mathrm{p}}-\dot{\mathbf{P}}^{\mathrm{p}})\bullet\dot{\mathbf{F}}^{\mathrm{p}}\Big)dv=0\end{aligned} \tag{3.1}$$

for every part P of the body. Applying the divergence theorem, equation (3.1) can be transformed to

$$\begin{aligned}&\int_P(\sigma+\sigma_K)dv=\int_P(\mathbf{b}\bullet\dot{\mathbf{x}}+\mathbf{G}^{\mathrm{d}}\bullet\dot{\mathbf{D}}+\mathbf{G}^{\mathrm{p}}\bullet\dot{\mathbf{F}}^{\mathrm{p}})dv\\&+\int_{\partial P}(\mathbf{T}\mathbf{n}\bullet\dot{\mathbf{x}}+\mathbb{H}^{\mathrm{d}}\mathbf{n}\bullet\dot{\mathbf{D}}+\mathbb{H}^{\mathrm{p}}\mathbf{n}\bullet\dot{\mathbf{F}}^{\mathrm{p}})da\end{aligned} \tag{3.2}$$

with the stress power $\sigma(\mathbf{X},t)$ defined by

$$\sigma = \mathbf{T} \bullet \dot{\mathbf{F}} + \mathbf{K}^{\mathrm{d}} \bullet \dot{\mathbf{D}} + \mathbb{H}^{\mathrm{d}} \bullet \nabla\dot{\mathbf{D}} + \mathbf{K}^{\mathrm{p}} \bullet \dot{\mathbf{F}}^{\mathrm{p}} + \mathbb{H}^{\mathrm{p}} \bullet \nabla\dot{\mathbf{F}}^{\mathrm{p}}, \tag{3.3}$$

and the stress power $\sigma_\kappa(\mathbf{X},t)$ given by

$$\sigma_\kappa = \dot{\mathbf{p}} \bullet \dot{\mathbf{x}} + \dot{\mathbf{P}}^{\mathrm{d}} \bullet \dot{\mathbf{D}} + \dot{\mathbf{P}}^{\mathrm{p}} \bullet \dot{\mathbf{F}}^{\mathrm{p}}. \tag{3.4}$$

It is seen that (3.2) with (3.3) and (3.4) represents the global balance laws (2.1), (2.3) and (2.4) in the weak form, what is known as the principle of stress power.

The scalar function $\sigma_\kappa(\mathbf{X},t)$ according to (3.4) represents the total power of inertia forces including the inertia forces of moving material points in the physical space and evolving defects and dislocations in the material space. With the assumption that the referential mass density $\rho_0(\mathbf{X})$ is time-independent so that the balance law of mass is satisfied identically, the total power of inertia forces can be derived from the rate of the kinetic energy density $\kappa = \kappa(\dot{\mathbf{x}}, \dot{\mathbf{D}}, \dot{\mathbf{F}}^{\mathrm{p}})$ as function of the rates of the independent kinematical field variables,

$$\sigma_\kappa = \dot{\kappa}(\dot{\mathbf{x}}, \dot{\mathbf{D}}, \dot{\mathbf{F}}^{\mathrm{p}}). \tag{3.5}$$

With the classical assumption that the macromomentum in the physical space is given by $\mathbf{p} = \rho_0\dot{\mathbf{x}}$ and the additional assumptions that the material momenta $\mathbf{P}^{\mathrm{d}}$ and $\mathbf{P}^{\mathrm{p}}$ are proportional to $\dot{\mathbf{D}}$ and $\dot{\mathbf{F}}^{\mathrm{p}}$, respectively, the total kinetic energy takes the form

$$\kappa = \frac{1}{2}\rho_0\dot{\mathbf{x}} \bullet \dot{\mathbf{x}} + \frac{1}{2}\rho_0\mathbb{J}^{\mathrm{d}}\dot{\mathbf{D}} \bullet \dot{\mathbf{D}} + \frac{1}{2}\rho_0\mathbb{J}^{\mathrm{p}}\dot{\mathbf{F}}^{\mathrm{p}} \bullet \dot{\mathbf{F}}^{\mathrm{p}}. \tag{3.6}$$

Here $\mathbb{J}^{\mathrm{d}}(\mathbf{X},t)$ and $\mathbb{J}^{\mathrm{p}}(\mathbf{X},t)$ are fourth order inertia tensors associated with dynamically moving defects (e.g. microcracks and microvoids) and dynamically moving dislocations, respectively. These two tensors have to satisfy certain symmetry conditions in order to ensure that the kinetic energy (3.6) serves as potential for the material momenta $\mathbf{P}^{\mathrm{d}}$ and $\mathbf{P}^{\mathrm{p}}$. Such conditions are satisfied identically, if the inertia tensors $\mathbb{J}^{\mathrm{d}}$ and $\mathbb{J}^{\mathrm{p}}$ are symmetric and time-independent.

4. BALANCE OF ENERGY AND DISSIPATION INEQUALITY

Besides the balance laws for physical and material forces and couples (2.5) and (2.6) it is necessary to formulate the first and second law of thermodynamics taking into account the kinetic energy and the mechanical power due to the dynamics of the body in the physical space as well as due to the evolution of defects, dislocations and heat production and transport in the material space. Accordingly, the appropriate form of the first law of thermodynamics representing the balance of energy follows from (3.2) as

$$\begin{aligned}\frac{d}{dt}\left\{\int_P(\varepsilon+\kappa)dv\right\} &= \int_P(\mathbf{b}\bullet\dot{\mathbf{x}}+\mathbf{G}^{d}\bullet\dot{\mathbf{D}}+\mathbf{G}^{p}\bullet\dot{\mathbf{F}}^{p})dv \\ &+\int_{\partial P}(\mathbf{T}\mathbf{n}\bullet\dot{\mathbf{x}}+\mathbb{H}^{d}\mathbf{n}\bullet\dot{\mathbf{D}}+\mathbb{H}^{p}\mathbf{n}\bullet\dot{\mathbf{F}}^{p})da+\int_P r\,dv-\int_{\partial P}\mathbf{q}\bullet\mathbf{n}\,da,\end{aligned}\tag{4.1}$$

while the second law of thermodynamics representing the principle of entropy growth in the form of the Clausius-Duhem inequality reads

$$\frac{d}{dt}\int_P\eta\,dv\geq\int_P\theta^{-1}r\,dv-\int_{\partial P}\theta^{-1}\mathbf{q}\bullet\mathbf{n}\,da.\tag{4.2}$$

The various field variables appearing in (4.1) and (4.2), which are not present in the physical and material balance laws (2.1)-(2.4), are:

$\varepsilon(\mathbf{X},t)$: specific internal energy,
$r(\mathbf{X},t)$: external body heating,
$\mathbf{q}(\mathbf{X},t)$: referential heat flux vector,
$\eta(\mathbf{X},t)$: entropy,
$\theta(\mathbf{X},t)$: absolute temperature (by additional assumption, $\theta(\mathbf{X},t)>0$).

The localization of the balance law of energy (4.1) yields

$$\dot{\varepsilon}=\sigma+r-\mathrm{Div}\,\mathbf{q},\tag{4.4}$$

provided that the local laws (2.5) and (2.6) hold. Here, the total stress power $\sigma(\mathbf{X},t)$ is given by (3.3). In the case of existing singular surfaces within the part P of the material body corresponding jump conditions have to be added.

Finally, the localization of the entropy inequality (4.2) yields the local dissipation inequality

$$\dot{\eta}-\left(\theta^{-1}r-\mathrm{Div}(\theta^{-1}\mathbf{q})\right)\geq 0.\tag{4.5}$$

It should be noted that in the form (4.5) the local dissipation inequality is independent of the laws of mechanics and the law of energy balance. Introducing the free energy $\psi(\mathbf{X},t)$ (measured per unit volume of the reference configuration) defined by $\psi \equiv \varepsilon - \theta\eta$, the dissipation inequality (4.5) can be rewritten in an equivalent form as

$$D \equiv \sigma - \dot{\psi} - \eta\dot{\theta} - \theta^{-1}\mathbf{q} \bullet \nabla\theta \geq 0, \tag{4.6}$$

referred to as the reduced dissipation inequality. It represents the second law of thermodynamics under the assumption that the balance laws of physical and material forces and the balance of energy hold.

5. GENERAL CONSTITUTIVE EQUATIONS

The independent thermo-kinematical variables of the theory presented in this paper are $(\mathbf{F}, \mathbf{D}, \mathbf{F}^{\mathrm{p}}, \theta)$. Let us denote with $\boldsymbol{\varepsilon}$ the following set of kinematical variables and with $\dot{\boldsymbol{\varepsilon}}$ their rates:

$$\boldsymbol{\varepsilon} \equiv (\mathbf{F}, \mathbf{D}, \nabla\mathbf{D}, \mathbf{F}^{\mathrm{p}}, \nabla\mathbf{F}^{\mathrm{p}}), \qquad \dot{\boldsymbol{\varepsilon}} \equiv (\dot{\mathbf{F}}, \dot{\mathbf{D}}, \nabla\dot{\mathbf{D}}, \dot{\mathbf{F}}^{\mathrm{p}}, \nabla\dot{\mathbf{F}}^{\mathrm{p}}). \tag{5.1}$$

Then the set of physical and material stress tensors power-conjugate to (5.1) is

$$\boldsymbol{\sigma} = (\mathbf{T}, \mathbf{K}^{\mathrm{d}}, \mathbb{H}^{\mathrm{d}}, \mathbf{K}^{\mathrm{p}}, \mathbb{H}^{\mathrm{p}}), \tag{5.2}$$

what enables us to rewrite the stress power (3.3) in compact form as $\sigma = \boldsymbol{\sigma} \bullet \dot{\boldsymbol{\varepsilon}}$.

To formulate the constitutive equations, we have to express the free energy ψ and the stress tensors (5.2) together with the entropy η and the heat flux vector $\mathbf{q}$ as functions of $(\boldsymbol{\varepsilon}, \theta, \dot{\boldsymbol{\varepsilon}}, \nabla\theta)$,

$$\begin{aligned} \psi &= \hat{\psi}(\boldsymbol{\varepsilon}, \theta, \dot{\boldsymbol{\varepsilon}}, \nabla\theta), & \boldsymbol{\sigma} &= \hat{\boldsymbol{\sigma}}(\boldsymbol{\varepsilon}, \theta, \dot{\boldsymbol{\varepsilon}}, \nabla\theta), \\ \eta &= \hat{\eta}(\boldsymbol{\varepsilon}, \theta, \dot{\boldsymbol{\varepsilon}}, \nabla\theta), & \mathbf{q} &= \hat{\mathbf{q}}(\boldsymbol{\varepsilon}, \theta, \dot{\boldsymbol{\varepsilon}}, \nabla\theta). \end{aligned} \tag{5.3}$$

The general constitutive equations (5.3) have to satisfy the restrictions of thermodynamical admissibility and objectivity. The thermodynamical admissibility requires that the response functions in the constitutive equations (5.3) satisfy the dissipation inequality (4.6). From this requirement the following constitutive restrictions are obtained, $\partial_{\dot{\boldsymbol{\varepsilon}}}\hat{\psi} = 0$ and $\partial_{\nabla\theta}\hat{\psi} = 0$.

It means the free energy function can not depend on $\dot{\varepsilon}$ and $\nabla\theta$, and the constitutive equation for the entropy is determined by the free energy

$$\psi = \hat{\psi}(\varepsilon, \theta), \qquad \eta = \hat{\eta}(\varepsilon, \theta) = -\partial_\theta \hat{\psi}(\varepsilon, \theta). \tag{5.4}$$

In view of (5.4) the dissipation inequality reduces to the form

$$D = \left(\hat{\sigma}(\varepsilon, \theta, \dot{\varepsilon}, \nabla\theta) - \partial_\varepsilon \hat{\psi}(\varepsilon, \theta)\right) \bullet \dot{\varepsilon} - \theta^{-1} \hat{\mathbf{q}}(\varepsilon, \theta, \dot{\varepsilon}, \nabla\theta) \bullet \nabla\theta \geq 0. \tag{5.5}$$

The inspection of this inequality leads to the result that the physical and material stresses (5.2) consist each of two parts, a non-dissipative part, which can be derived from the free energy potential, $\partial_\varepsilon \hat{\psi}(\varepsilon, \theta)$, and a dissipative part indicated here by a lower asterisk,

$$\sigma = \hat{\sigma}(\varepsilon, \theta, \dot{\varepsilon}, \nabla\theta) = \partial_\varepsilon \hat{\psi}(\varepsilon, \theta) + \hat{\sigma}_*(\varepsilon, \theta, \dot{\varepsilon}, \nabla\theta), \tag{5.6}$$

where the dissipative parts of the physical and material stresses, $\sigma_* = (\mathbf{T}_*, \mathbf{K}^d_*, \mathbb{H}^d_*, \mathbf{K}^p_*, \mathbb{H}^p_*)$, have to satisfy the dissipation inequality

$$D = \hat{\sigma}_*(\varepsilon, \theta, \dot{\varepsilon}, \nabla\theta) \bullet \dot{\varepsilon} - \theta^{-1} \hat{\mathbf{q}}(\varepsilon, \theta, \dot{\varepsilon}, \nabla\theta) \bullet \nabla\theta \geq 0. \tag{5.7}$$

The form of the entropy production inequality (5.7) suggests to assume the existence of a dissipation pseudo-potential ϕ of the form

$$\phi = \phi(\varepsilon, \theta, \dot{\varepsilon}, \nabla\theta), \tag{5.8}$$

where ε and θ can be considered as parameters. If such a dissipation pseudo-potential exists, the dissipative driving stresses σ_* and the heat flux vector $\mathbf{q}$ can be derived from ϕ as

$$\sigma_* = \partial_{\dot{\varepsilon}} \phi(\varepsilon, \theta, \dot{\varepsilon}, \nabla\theta), \qquad \theta^{-1}\mathbf{q} = -\partial_{\nabla\theta} \phi(\varepsilon, \theta, \dot{\varepsilon}, \nabla\theta). \tag{5.9}$$

Introducing $(5.9)_1$ into (5.6) leads to the constitutive equations expressed by two potentials, the free energy ψ and the dissipation pseudo-potential ϕ,

$$\sigma = \partial_\varepsilon \hat{\psi}(\varepsilon, \theta) + \partial_{\dot{\varepsilon}} \phi(\varepsilon, \theta, \dot{\varepsilon}, \nabla\theta). \tag{5.10}$$

Furthermore, with (5.9) the dissipation inequality (5.7) takes the form

$$D = \partial_{\dot{\varepsilon}} \phi(\varepsilon, \theta, \dot{\varepsilon}, \nabla\theta) \bullet \dot{\varepsilon} + \partial_{\nabla\theta} \phi(\varepsilon, \theta, \dot{\varepsilon}, \nabla\theta) \bullet \nabla\theta \geq 0, \tag{5.11}$$

where the first term is the entropy production due to the dissipative driving forces in physical and material space and the second term is the entropy production due to the heat flux in the material space.

6. DYNAMIC GOVERNING EQUATIONS OF DEFORMATION AND DEFECT AND PLASTIC EVOLUTION

Within the considered theory, the governing equations to determine the deformation of the structure with a damage and plastic evolution inside the material body are obtained by introducing the thermodynamically admissible constitutive equations of the previous section into the balance laws of physical and material forces, obtained in Sect. 2, together with the equation of energy balance derived in Sect. 4. The resulting field equations can be simplified for isothermal processes leading to

$$\operatorname{Div}\mathbf{T}(\boldsymbol{\varepsilon},\dot{\boldsymbol{\varepsilon}})+\mathbf{b}=\rho_0\ddot{\mathbf{x}}, \tag{6.1}$$

$$\begin{aligned}&\operatorname{Div}\mathbb{H}^{d}(\boldsymbol{\varepsilon},\dot{\boldsymbol{\varepsilon}})-\mathbf{K}^{d}(\boldsymbol{\varepsilon},\dot{\boldsymbol{\varepsilon}})+\mathbf{G}^{d}=\rho_0(\mathbb{J}^{d}\dot{\mathbf{D}})^{\boldsymbol{\cdot}},\\&\operatorname{Div}\mathbb{H}^{p}(\boldsymbol{\varepsilon},\dot{\boldsymbol{\varepsilon}})-\mathbf{K}^{p}(\boldsymbol{\varepsilon},\dot{\boldsymbol{\varepsilon}})+\mathbf{G}^{p}=\rho_0(\mathbb{J}^{p}\dot{\mathbf{F}}^{p})^{\boldsymbol{\cdot}}.\end{aligned} \tag{6.2}$$

Thus, the fields $\mathbf{x}(\mathbf{X},t)$, $\mathbf{D}(\mathbf{X},t)$ and $\mathbf{F}^{p}(\mathbf{X},t)$ can be determined as solution of the system of equations (6.1) and (6.2) taking into account appropriate boundary and initial conditions. The boundary conditions consistent with this set of governing equations can be derived in the manner shown in Stumpf and Hackl (2003).

7. CONCLUSION

The presented weakly nonlocal (gradient type) theory of damage and plasticity can be considered as a framework with various gradient and local models as special cases. If e.g. it is assumed that for brittle material the plastic deformation plays no role the gradient model of anisotropic damage of Stumpf and Hackl (2003) is obtained. As it is outlined in that paper a further restriction to isotropic damage, isotropic elastic material behavior, small strains, quasi-static and isothermal process the elastic damage model of Frémond and Nedjar (1996) is recovered. In that paper also corresponding material parameters and numerical applications are presented.

From the general theory of the present paper a simplest gradient model of ductile damage is obtained, if for quasi-static and isothermal process at small elastic strains, isotropic gradient damage is assumed, while the plastic material behavior is described by the classical local theory of plasticity for small plastic strains neglecting gradients of the plastic deformation. This leads to the model of Nedjar (2001).

If both the damage gradient and the plastic strain gradient are assumed to be small and can be neglected the dynamical balance laws of material forces presented in this paper yield dynamical evolution laws for the corresponding field variables. In the case of quasi-static elastoplasticity without damage the results of Cermelli et al. (2001) are obtained.

ACKNOWLEDGEMENT: The financial support, provided by Deutsche Forschungsgemeinschaft (DFG) under Grant SFB 398-A7/8, is gratefully acknowledged.

8. REFERENCES

Cermelli, P., Fried, E., Sellers, S. (2001), Configurational stress, yield and flow in rate-independent plasticity. *Proceedings of the Royal Society of London*, A457, 1447-1467.

Coleman, B.D., Noll, W. (1963), The thermodynamics of elastic materials with heat conduction and viscosity. *Archive for Rational Mechanics and Analysis*, 13, 167-178.

Frémond, M., Nedjar, B. (1996), Damage, gradient of damage and principle of virtual power. *International Journal of Solids and Structures* 33, 1083-1103.

Gurtin, M.E. (2000), *Configurational Forces as Basic Concepts of Continuum Physics.* Springer-Verlag, Berlin.

Lorentz, E., Andrieux, S. (1999), A variational formulation for nonlocal damage models. *International Journal of Plasticity* 15, 119-138.

Makowski, J., Stumpf, H. (2001), Thermodynamically based concept for the modelling of continua with microstructure and evolving defects. *International Journal of Solids and Structures* 38, 1943-1961.

Maugin, G.A. (1993), *Material Inhomogeneities in Elasticity.* Chapman & Hall.

Nedjar, B. (2001), Elastoplastic-damage modelling including the gradient of damage: formulation and computational aspects. *International Journal of Solids and Structures* 38, 5421-5451.

Saczuk, J., Hackl, K., Stumpf, H. (2003), Rate theory of nonlocal gradient damage - gradient viscoinelasticity. *International Journal of Plasticity* 19, 675-706.

Steinmann, P. (2000), Application of material forces to hyperelastostatic fracture mechanics. Part I: Continuum mechanical setting. *International Journal of Solids and Structures* 37, 7371-7391.

Stumpf, H., Hackl, K. (2003), Micromechanical concept for the analysis of damage evolution in thermo-viscoelastic and quasi-brittle materials. *International Journal of Solids and Structures* 40, 1567-1584.